The Universe
from Your Backyard

Chris Floyd

The Universe from Your Backyard

A Guide to Deep-Sky Objects
from ASTRONOMY Magazine

by David J. Eicher

Assistant Editor, ASTRONOMY
Editor-in-Chief, *Deep Sky*

AstroMedia
Milwaukee

The right of the University of Cambridge to print and sell all manner of books was granted by Henry VIII in 1534. The University has printed and published continuously since 1584.

Cambridge University Press
Cambridge
New York New Rochelle Melbourne Sydney

FOR RICHARD BERRY, whose knowledge and helpful advice have made it a pleasure to write about the stars.

Published by AstroMedia, a Division of Kalmbach Publishing Co.,
1027 North 7th Street, Milwaukee, WI 53233, USA
and by
The Press Syndicate of the University of Cambridge
The Pitt Building, Trumpington Street, Cambridge CB2 1RP
32 East 57th Street, New York, NY 10022, USA
10 Stamford Road, Oakleigh, Melbourne 3166, Australia

 First published 1988.
Printed in Singapore

Editorial Staff: Richard Berry, Katherine M. Bond, Robert Burnham, Stephen Cole, David J. Eicher, Jeff Kanipe, Kristine R. Majdacic, Francis Reddy, Richard Talcott.

Art and Production Staff: Todd Graveline (art coordinator), Larry Luser (production coordinator), Jennifer J. Devlin, Thomas L. Hunt, Patti L. Keipe, Patricia LaBrecque, Jane B. Lucius, Mary MacAdam, Susan R. Myers, Keith Ward.

Front endpaper photo: Tom Dey (Rosette Nebula)
Back endpaper photo: Mike Sisk (The Andromeda Galaxy)

Library of Congress Cataloging-in-Publication Data

Eicher, David J., 1961-
The universe from your backyard.

Bibliography: p.
Includes index.
1. Astronomy — Observers' manuals. I. Title.
QB64.E53 1988 523 88-8920
ISBN 0-913135-05-4
ISBN 0-521-36299-7 (Cambridge)

Contents

Foreword

When I first became interested in astronomy, the space age was just around the corner. I learned my constellations at scout camp, observing with the naked eye from a sand pit, the only available (dry!) place not hemmed in with trees. The sky was so inky black that tracing the constellations was hard — the constellations were cluttered with zillions of stars the maps didn't show.

That experience was unforgettable. But the skies at home were those of a large city, so naturally it was easy to slip into thinking that the universe began and ended with the solar system. As I and countless others learned the hard way, city lights from the typical urban or suburban backyard cast such an obscuring pall over the sky that only the Moon, planets, and a few of the brightest stars are visible.

But over the years and with more chances to get away from the cities I've lived in, I have rediscovered the worlds outside the Sun's realm. These contain some of the finest sights in the sky and anyone's astronomy experience is much poorer for having missed them. Spiral galaxies, glowing nebulae studded with dust, double stars with pale tints that challenge you to describe them — all these and more lie within the grasp of even a small telescope.

The book in your hands is the key that opens this realm to you. Taking the sky in small manageable sections, it introduces you to the sights of deep space and makes them familiar.

The author, David Eicher, is a highly experienced backyard astronomer. For years he has studied the skies with large telescopes and small. Whatever size instrument you have, he has probably used it and knows just what you can see with it. Dave also knows about the difficulties of finding objects. He'll explain time-tested techniques like averted vision and star hopping. He knows how to find a dark-sky viewing site, how to cope with nights when the atmosphere refuses to settle down, and so on.

Many books on deep-sky observing tend to beat you over the head with astrophysics. It's true that the science behind what you are seeing is often quite interesting, but this book is intended for ordinary astronomy buffs like you. It talks mainly about what *you can see*, not what an astronomer with a computer-imaging system on the Hale 200-inch can see. It's a realistic and visually oriented book, written at your level, that distills a wealth of experience and eyepiece time and puts it at your service.

It works at a nice level of detail. One of my favorite constellations is Sagitta, a small dim group squeezed in next to Aquila in the summer Milky Way. Sagitta is largely ignored because it's so small and because it lacks the whizz-bang sights of Cygnus to the north and Sagittarius to the south. But turn to page 85 in this book. There you'll find a thorough description of Sagitta and its neighbor, the equally dim Delphinus. Dave's even included that loose open cluster next to M-71 I like so much!

Or, to make a more significant point, take a favorite like the Orion Nebula. When observers write about this object, they sometimes seem as if they're competing to see who can say the floweriest things about it. Of course, after such airy flights the reality in the eyepiece is a letdown. Not here — Dave's description on page 137 is clear and straightforward. I think you'll find it closely matches what you actually see.

When I started exploring the realm outside the solar system, very little was available in the way of observing manuals or guidebooks. T. W. Webb's *Celestial Objects* was hard to find and concentrated heavily on double stars. Other guidebooks were unobtainable or just didn't exist.

I wish I could somehow reach into the past and give my earlier self this book. I needed it badly and had it been available back then, skywatching would have been a much richer and more enjoyable experience.

Robert Burnham
Author of *The Star Book*

Preface

In January 1984 ASTRONOMY Magazine published the first article in a regular monthly series, "The Backyard Astronomer," which featured the most interesting deep-sky objects in a constellation. Reader response was great: the series hadn't been running six months before we started to receive letters asking us whether we planned to reprint the articles as a book. The idea had already crossed our minds, and the book you now hold in your hands is the result. The forty-six articles cover an average of fifteen deep-sky objects per constellation, 690 in all.

My hope is that you will use this book on two levels. First, take it outside under the stars and use it at the telescope. The star maps in each section will help you find the objects with relative ease. As you tour the galaxies of Ursa Major, the unusual stars in Corona Borealis, or the magnificent nebulae in Sagittarius, you'll discover how much of the universe you can see from your backyard.

Second, as your knowledge of the sky grows, use the book inside as a reference guide. You can compare the galaxies of Virgo with those in Sculptor, read carefully about why globular clusters are so concentrated in one section of sky, or compile a list of the faintest planetary nebulae visible in your telescope. Fundamental data for every object appears in a table in each chapter.

As experts on deep-sky objects are quick to point out, the 1980s are a fast-paced time for our developing knowledge of galaxies, clusters, and nebulae. The quality of fundamental data on the brightnesses, sizes, and distances of deep-sky objects is improving by leaps and bounds as more professional astronomers study these objects. Whenever possible, I have included the most recent available information from professional astronomical journals. For guidance I am indebted to my friends Brian Skiff of Lowell Observatory and David Levy of the University of Arizona, who keenly understand who's doing what in professional astronomy.

If this book conveys any single message, it's that you can see a great deal of the universe armed with only a small telescope. You don't need fancy equipment to witness — "live" — the stuff that fills the cosmos. Of course larger telescopes reveal the universe in greater detail, but you're perfectly capable of seeing hundreds of objects with a 3-inch or 4-inch telescope.

Most important, when you're out under the stars, relax and enjoy what you're seeing. Astronomy is a potentially powerful hobby: it allows you to see nature from a perspective that only a few people will ever experience. Along with the information and maps in this book, your telescope and eyes will show you places in our Galaxy and beyond that you may not have imagined in your wildest dreams. And as you observe more and more, you'll form a picture of the universe in your mind's eye — one that will enlarge your perspective as a tiny being on our little blue planet Earth.

David J. Eicher
March 1988

Far left: Planetary nebula NGC-7662, shown here as an overexposed ''star,'' appears as a double-ringed shell of nebulosity in large backyard scopes. Photo by Martin C. Germano. Left: Galaxy NGC-925 in Triangulum has knotty, dusty arms. Photo by K.A. Brownlee. Below left: Although M-33 has a reputation for being elusive, it is easily visible under dark skies. Photo by Ron Royer. Below right: NGC-891, one of the sky's best edge-on spirals. Photo by Bill Iburg. Bottom: M-31, the Andromeda Galaxy, is our ''sister spiral'' in the Local Group — it reveals a wealth of detail under dark skies. Photo by Martin C. Germano.

Andromeda
And
Andromedae

Triangulum
Tri
Trianguli

Keith Ward

The constellations **Andromeda** and **Triangulum** do not contain many bright deep-sky objects, as they lie away from the plane of the Milky Way; they mainly feature faint galaxies. Two spirals in this region of sky, however, are the brightest and largest members of the Local Group of galaxies — the home cluster **M-31** in Andromeda and M-33 in Triangulum.

With a total magnitude of 3.5, M-31 (NGC-224), known as the Andromeda Galaxy, is the brightest spiral galaxy in the sky, visible without optical aid as a fuzzy patch spanning 3°. Easy to find about 1° west of Nu (ν) Andromedae, M-31 is our "sister spiral" in the Local Group. It contains 300 billion stars whose combined mass might well total over one trillion suns. The galaxy lies some 760 kiloparsecs away and measures 55 kiloparsecs across; we could fit a dozen M-31s end-to-end between the great spiral and our own Milky Way.

M-31's light is unevenly distributed across its surface, the brightest part being the galaxy's yellowish core — easily visible in binoculars and finder telescopes as an elongated streak of light. This nuclear bulge represents the inner few kiloparsecs of M-31; most of the galaxy has a much lower surface brightness and is more difficult to see. A 6-inch telescope shows a large, extended outer halo of greyish nebulosity surrounding this bright core. The light comes from millions of relatively young stars held in M-31's spiral arms. An 8-inch scope on a dark night reveals two parallel dust lanes alongside the southern edge of the galaxy's core, and a knotty bright patch of nebulosity near M-31's southwestern tip is NGC-206 — a huge starcloud in one of the galaxy's spiral arms. Used at high power, 10-inch telescopes show the core of M-31 to be star-like.

If you look at M-31 with a low-power eyepiece, you'll see two small fuzzy patches within the same field of view as the nucleus. These are **M-32** (NGC-221) and **NGC-205**, two elliptical satellite galaxies located roughly five and seven kiloparsecs from M-31, respectively. M-32 shines at magnitude 8.2 and measures some 8′ x 6′ across, giving it an average surface brightness of 12.3 magnitudes per square arc-minute. NGC-205 has a higher total magnitude but measures 17′ x 10′, yielding a much fainter surface brightness of 13.6. In comparison, the tenuous spiral arms of M-31 have a surface brightness of 13.4. M-32 appears as a small, glowing, condensed ball of light with a notably brighter center in most telescopes. In a 6-inch scope, NGC-205 looks like a large oval nebulosity, slightly brighter in the center and very diffuse and unconcentrated near its edges. M-31 contains two other elliptical satellites: NGC-147 and NGC-185. They are located about 90 kiloparsecs from M-31 and lie some 7° north in Cassiopeia.

After the M-31 group, Andromeda's next easiest galaxy to find is **NGC-404**, a small roundish lenticular located 6′ northwest of the second-magnitude star Beta (β) Andromedae. NGC-404 measures 1.3′ x 1.3′ wide and glows at magnitude 11.9. Although its surface brightness is high, proximity to Beta makes it difficult to locate on less-than-dark nights. Try a 6-inch at low-power; after you find the galaxy, switch to high power and move Beta slightly out of the field. This will help you see the feeble light from the galaxy.

Another galaxy to look for is the faint, edge-on spiral **NGC-891**. This lies halfway between the bright double star Gamma (γ) Andromedae and the glowing open cluster M-34 in Perseus. It shines at magnitude 10.0 and covers a sky area of 12′ x 1′; its low surface brightness makes it difficult to spot in 3-inch and 4-inch telescopes. A good 6-inch scope shows a thin spindle of nebulosity spanning about 10′, a bright central bulge, and a narrow dust lane bisecting NGC-891's long axis. NGC-891 belongs to the NGC-1023 group of galaxies, a cluster some 13 megaparsecs distant. The group contains NGC-891, NGC-1023, and the dusty Sc-type spiral NGC-925 in Triangulum.

NGC-7640 is a dim Sb-type spiral — a challenging object for 4-inch scopes under dark skies. It is a good example of a nearly edge-on galaxy and distributes its 12.5 magnitude glow over a 9.0′ x 1.0′ area. An 8-inch telescope shows NGC-7640 as a slender, even nebulosity with a bright bulge centered on the galaxy's nucleus. A 17.5-inch telescope reveals that the pancake-shaped nucleus is mottled, suggesting patches of dark matter like the two dust bands encircling M-31.

BEST VISIBLE DURING
AUTUMN

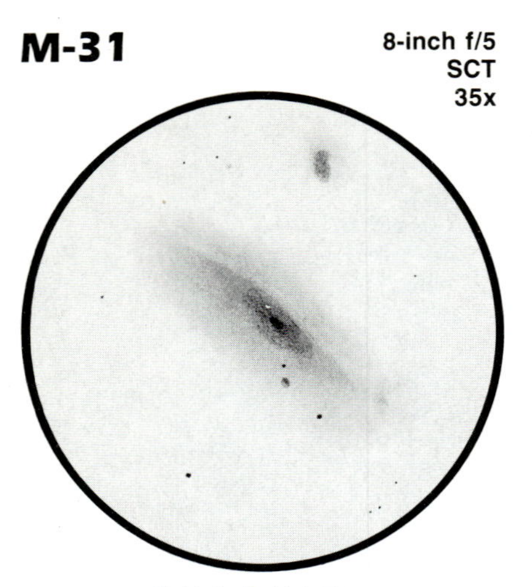

Sketch by David J. Eicher

One of the finest large open clusters in the northern sky is **NGC-752**. This group spans 45′ — 1½ Moon diameters — and contains over 70 stars, the brightest of which shine at eighth magnitude. With binoculars or a finder telescope, look about 5° south and slightly west of Gamma. You'll see a loose, scattered grouping of the several dozen brightest stars in the cluster. Rich-field telescopes between 4- and 8-inches in aperture provide the nicest views, showing over 70 stars.

An unusual planetary nebula is **NGC-7662**. It lies ½° southwest of the fifth-magnitude star 13 Andromedae. NGC-7662 is one of the double-ringed planetaries, which display a bright, well-defined ring of gas encapsulated by a much larger, dimmer, and hazier envelope. The problem in observing this nebula is its small size of 32″ x 28″. It shines at eighth magnitude, but you'll need great magnification and clear skies to see more than just a fuzzy "star." In 3- and 4-inch scopes, NGC-7662 appears as a bluish-green point source; it is similar to its stellar neighbors except for the contrasting color. Eight-inch telescopes at 300x, seeing and transparency permitting, show NGC-7662 as an elliptical ring of light with a dark center. An 8-incher doesn't show the outer nebulosity or the dim 13th magnitude central star, but larger scopes may.

Gamma (γ) Andromedae is a favorite double star for owners of small telescopes. Also known as Almach, this star's colorful components are second only to those of Albireo in Cygnus. The magnitude 2.2 primary is golden-yellow or slightly orange, while the 5.1 magnitude secondary is a striking greenish-blue. The fainter star lies about 10″ away from the primary, in position angle 63°. Virtually any telescope used at medium powers will show both of these stars, with an inky blackness of space between. Keep in mind that this small dark gap represents over 74 billion miles! The primary is itself a double, but the separation of its two stars is only 0.3″. The brightest member of Gamma is a spectroscopic double, making this a system with four stars.

Typical of the long period variables, **R Andromedae** has a period of 409 days during which it fluctuates between magnitudes 5.3 and 15.1. R Andromedae is easily found near a bright triangle formed by the stars Theta (θ), Rho (ρ), and Sigma (σ) Andromedae. If you have keen eyesight and a dark sky, try to record when R Andromedae appears and disappears without optical aid. You can also conveniently follow its brightness changes with a small telescope, which should show the star during most of its cycle.

There are three great spirals in the Local Group of galaxies: the Milky Way, M-31, and the large face-on galaxy **M-33** (NGC-598) in Triangulum. At a distance of 1.1 megaparsecs, M-33 is the closest face-on galaxy but has a reputation for being very elusive. Its total magnitude is 5.7, yet on poor viewing nights, small telescopes will barely show its core. This is because its dimensions of 62′ x 39′ and face-on orientation give the galaxy's spiral arms a surface brightness of only 14.0 magnitudes per square arc-minute. Dark skies coupled with large apertures and low f/ratios, however, show detail in M-33's arms: knotty patches of dark nebulosity, a tiny stellar nucleus, a faintly glowing spiral pattern, knots of bright material tucked in the arms, and the bright starcloud NGC-206. This feature lies 10′ northeast of the starlike core, and contains enough material to make a small galaxy of its own!

NGC-672 is a bright barred spiral in Triangulum. It measures 4.5′ x 1.7′ and glows at 11.6 magnitude, making it an easy mark for a 4-inch telescope, even at low power. A 10-inch scope at high power shows an oval halo of faint nebulosity with a bright, distinctly mottled inner core. Larger telescopes under dark skies may show details in this galaxy, which is often overlooked because of its large, bright neighbor M-33.

Far more challenging than either M-33 or NGC-672 is a small double galaxy named **NGC-750/1**. This object consists of two faint ellipticals, NGC-750 and NGC-751. They are separated by only 24″, so spotting them is a difficult task. NGC-750 shines at magnitude 13.0 and measures 24″ x 18″, while NGC-751 is a 13.8 magnitude galaxy covering 12″ of sky. With small telescopes, these galaxies appear as two fuzzy "stars" that almost touch each other. A 17.5-inch reflector provides enough light-gathering power to see the galaxies as side-by-side nebular balls of pale grey light.

A final galaxy in Triangulum is **NGC-925** — a large, faint Sb-type spiral. This galaxy measures 9.4′ x 4.0′ and is steeply inclined to our line-of-sight, which keeps its surface brightness fairly high. A 4-inch scope shows it as an elliptical, ghostly patch of milky light. A 10-incher shows some mottling about the nucleus, but doesn't reveal much more detail. Reflectors in the 16-inch to 20-inch range show some dusty structure surrounding the galaxy's hub, and a large outer envelope of even, faint light representing millions of unresolved stars.

Object	M#	Type	R.A. (2000) Dec.		Mag.	Size/Sep./Per.	H
R And		LPV	0h 24.0m	+38°34′	5.3↔15.1	409d	
NGC-205		()	0h 40.3m	+41°41′	8.0	17.4′x9.8′	E5pec
NGC-221	M-32	()	0h 42.7m	+40°52′	8.2	7.6′x5.8′	E2
NGC-224	M-31	§	0h 42.7m	+41°16′	3.5	177.8′x63.1′	Sb
NGC-404		()L	1h 09.4m	+35°43′	11.9	1.3′x1.3′	SO3
NGC-598	M-33	§	1h 33.9m	+30°39′	5.7	61.7′x38.9′	Sc(s)
NGC-672		§B	1h 47.8m	+27°26′	11.6	4.5′x1.7′	SBc
NGC-750/1		()	1h 57.5m	+33°13′	13.0,13.8	24″x18″,12″	EO
NGC-752		⊙	1h 57.8m	+37°41′	6.5	45′	
Gamma (γ) And		★²	2h 03.8m	+42°20′	2.1,5.1	10″	
NGC-891		§	2h 22.4m	+42°21′	12.2	12.0′x1.0′	Sb
NGC-925		§B	2h 27.3m	+33°35′	12.0	9.4′x4.0′	SBc(s)
NGC-7640		§B	23h 22.1m	+40°51′	12.5	9.0′x1.0′	SBc(s)
NGC-7662		■	23h 25.8m	+42°28′	8.6	17″x14″	

H = Hubble classification type for galaxies

★²	*Double Star*
LPV	*Long Period Variable*
⊙	*Open Cluster*
■	*Planetary Nebula*
§	*Spiral Galaxy*
§B	*Barred Spiral Galaxy*
()	*Elliptical Galaxy*
()L	*Lenticular Galaxy*

3h
2h
1h
0h
Galactic Equator
CASSIOPEIA
Double Cluster
PERSEUS
+50°
φ
NGC-891
Almach
γ
ANDROMEDA
NGC-205
M-31
M-32
υ
τ
+40°
θ
μ
ϱ
σ
NGC-752
NGC-404
β
NGC-925
β
γ
π
NGC-750/1
+30°
δ
TRIANGULUM
M-33
ε
α
NGC-672
ξ
η
ARIES
PISCES
+20°
2h
1h

Above: The Helical Nebula is the largest and brightest planetary nebula in the sky, but its low surface brightness makes it difficult to see on nights of poor transparency. Photo by Jack B. Marling.

Far left: M-2 is a magnitude 6.5 globular cluster whose edges resolve into stars in a 6-inch telescope. Photo by Kim Zussman.

Left: The Saturn Nebula, NGC-7009, is known for its faint ''ears'' extending on either side of the nebula. These nebular extensions photograph easily but are very difficult to observe. Photo by David Healy.

Aquarius

Aqr
Aquarii

Keith Ward

Lying between the Great Square of Pegasus and the barren starfields of Capricornus and Piscis Austrinus, **Aquarius** the Water Bearer is sufficiently far away from the plane of the Milky Way that it contains relatively few galactic objects. The only notable members of the galaxy in the constellation are two bright planetary nebulae and several unusual stars and clusters. Aquarius is loaded, however, with faint galaxies, a few of which are impressive when viewed with backyard telescopes.

The most atypical deep-sky object in Aquarius is the sprawling **Helical Nebula** (NGC-7293). Also nicknamed the Helix, NGC-7293 is the largest and brightest planetary nebula in the sky, measuring some 769″ (nearly 13′) across and shining at photographic magnitude 6.5. The Helical Nebula is large and bright because it is relatively close, only about 150 parsecs away. (This is an average distance based on many widely discrepant figures.)

You might assume that the Helix, so named because its ring shape resembles a helical coil, would be an easy target even under slightly imperfect skies. This isn't the case, however. Its light is so spread out that individual parts of it appear dim and have a low contrast against the sky background.

To best observe the Helical Nebula, choose a night and time when the Moon is absent from the sky, the transparency is very good, and Aquarius is near its highest nightly elevation. Acquaint yourself with Aquarius and find the south-central part of the constellation; you'll see two 5th-magnitude stars catalogued as Upsilon (υ) and 57 Aquarii. The Helix lies between these stars, about a third of the way from Upsilon toward 57 Aqr.

When searching for the Helix use a low-power, wide-field eyepiece. A good nebular filter may help if your observing site is light polluted. If you have a 6-inch or larger scope you may see the nebula as a roundish patch of greenish gray light some 10′ across without any distinct features. If you can see it without difficulty, increase the magnification to something like 15x per inch of telescope aperture. A 10-incher at 150x shows the dark "hole" inside the nebula and subtle brightness variations along its edges, as well as several faint stars involved in the nebulosity. One of these stars, a 13th-magnitude bluish star at the dark hole's center, is the central star and is visible in 8-inch and larger scopes.

Another Aquarian planetary lies far to the north and is much smaller. With a photographic magnitude of 8.3 and a diameter of only 25″, the **Saturn Nebula** (NGC-7009) — so nicknamed because of ansae (projecting arms of nebulosity extending out on either side of the disk) — is small enough that its surface brightness is very high. It is relatively easy to find and observe with virtually any telescope.

The Saturn Nebula lies in the western end of Aquarius, about 2° west of the 4th-magnitude star Nu (ν) Aquarii. Sweep for this object using a magnification of about 10x per inch of aperture; the nebula's small angular size makes it inconspicuous at very low powers. When you find the Saturn Nebula you'll see a bright, bluish green oval of light surrounding a 12th-magnitude bluish central star. On a night of exceptional seeing, a good 10- or 12-inch telescope may show the ansae as faint projections of nebulosity spanning 44″ and ending in a bright condensation. If you see these delicate features, you'll be in a relatively select group of backyard observers who have seen such detail in a planetary nebula.

Some 2° southwest of the Saturn Nebula is a curious group of four stars that appear as a fuzzy smear of light in small finder scopes. This little asterism is Messier object **M-73** (NGC-6994). Charles Messier described it as "three or four small stars which look like a nebula at first sight; it contains a little nebulosity." Measuring 2.8′ across and glowing at photographic magnitude 8.9, it appeared to Messier to contain nebulosity when he observed the group in October 1780. Modern photographs show no nebulosity in the area, suggesting that Messier was simply mistaken. In fact the group is probably not a cluster but a chance alignment of stars lying at different distances. Although unspectacular in the eyepiece, M-73 is a curiosity of deep-sky cataloguing worth viewing at least once.

Only 1.5° west and slightly north of M-73 is the fine globular cluster **M-72** (NGC-6981). Shining at magnitude 9.4 and measuring nearly 6′ across, finder scopes show this object as a fuzzy, enlarged "star"; small telescopes at low power reveal a 4′-diameter disk without any resolution. But a 6- or 8-inch scope

BEST VISIBLE DURING
AUTUMN

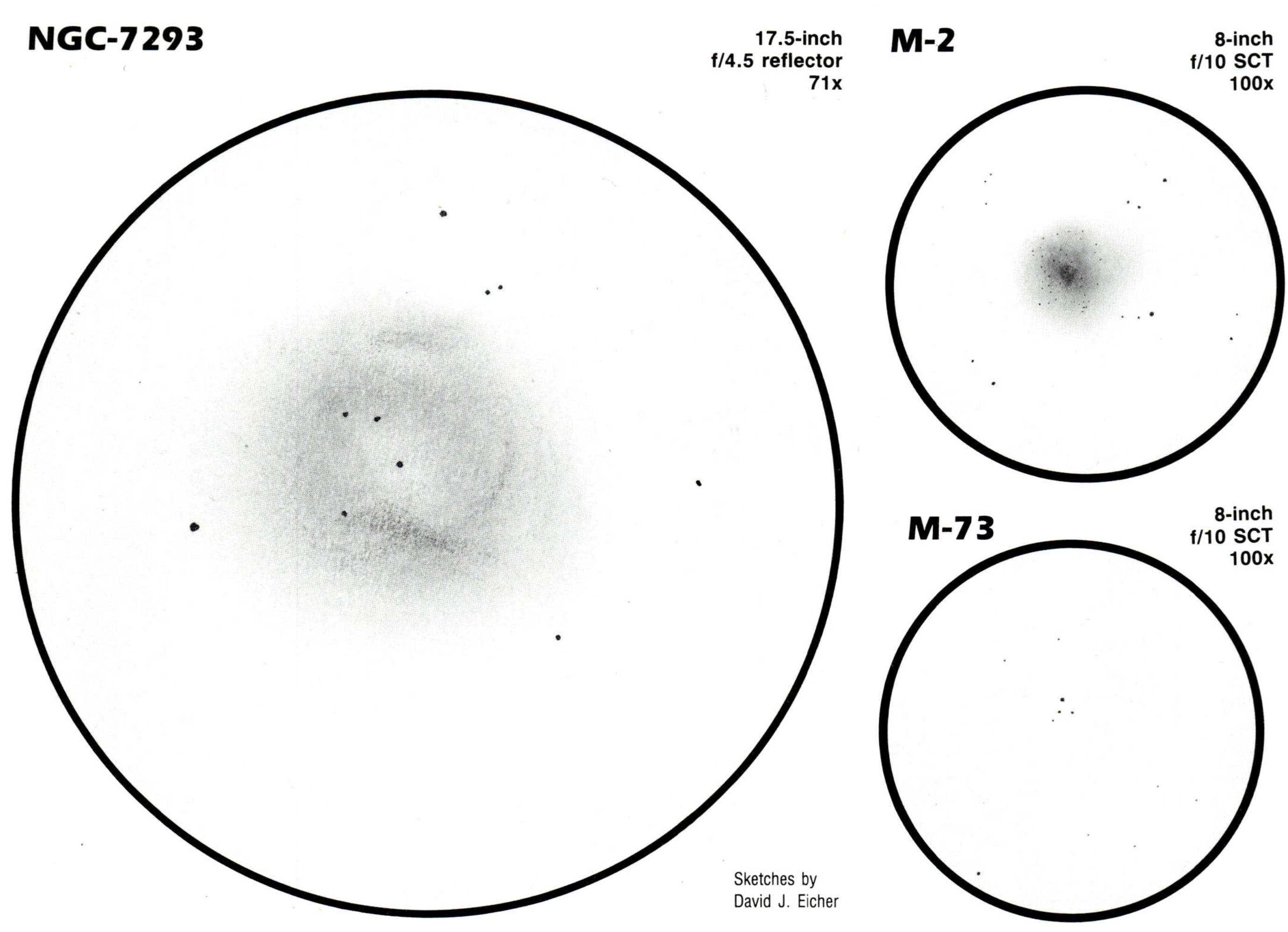

Sketches by
David J. Eicher

at high power in steady seeing will resolve stars along M-72's edges. Telescopes of 16-inches or more aperture show M-72 looking rather like M-13 as seen with a small scope.

Far more impressive is **M-2** (NGC-7089), a globular located 5° north of the bright double star Beta (β) Aquarii. This cluster is twice as large as M-72 and far brighter (magnitude 6.5), offering a satisfying view even in small scopes at low powers. M-2 is an obvious nonstellar blob of light in finder scopes and appears as an 8′-diameter disk in a 6-inch reflector at high power. On a dark night a 17.5-inch scope at 71x nicely resolves the cluster into myriad stars.

A third globular in Aquarius is **NGC-7492**, a large (6.2′ diameter) cluster composed of faint stars summing up to an integrated magnitude of 11.5. NGC-7492 lies 3.5° due east of the bright star Delta (δ) Aquarii and appears as a 4′-diameter pale gray disk in small scopes. Resolution of this cluster is difficult since the brightest giants in the cluster are quite faint. The largest backyard telescopes may resolve a few stars along the edges of this faint cluster on the very best of nights.

Aquarius contains a slew of galaxies but only a few of them are bright enough to show detail in backyard scopes. Lying a degree apart in the northeastern part of the constellation are **NGC-7723** and **NGC-7727**, two 11th-magnitude spirals. NGC-7723 is an Sb-type spiral measuring 3.6′ by 2.6′ across, and NGC-7727 is a barred spiral spanning 4.2′ by 3.4′. **NGC-7606** is another 11th-magnitude spiral; its spiral arms measure 5.8′ by 2.6′, and it contains a bright nucleus. **NGC-7184** is a large Sb-type spiral glowing at blue magnitude 12; with dimensions of 5.8′ by 1.8′ it appears as a silvery needle of light in small telescopes.

Object	M#	Type	R.A. (2000)	Dec.	Mag.	Size/Sep./Per.	H
NGC-6981	M-72	●	20h 53.5m	−12°32′	9.4	5.9′	
NGC-6994	M-73	∷	20h 59.0m	−12°38′	8.9_P	2.8′	
NGC-7009		■	21h 04.2m	−11°22′	8.3_P	25″	
NGC-7089	M-2	●	21h 33.5m	− 0°49′	6.5	12.9′	
NGC-7184		§	22h 02.7m	−20°49′	12.0_B	5.8′x1.8′	Sb+
Zeta (ζ)		★²	22h 28.8m	− 0°07′	4.3, 4.5	1.5″	
NGC-7293		■	22h 29.6m	−20°49′	6.5_P	769″	
NGC-7492		●	23h 08.4m	−15°37′	11.5	6.2′	
NGC-7606		§	23h 19.1m	− 8°29′	10.8	5.8′ x 2.6′	Sb+
NGC-7723		§	23h 38.9m	−12°58′	11.1	3.6′ x 2.6′	Sb
NGC-7727		§	23h 39.9m	−12°18′	10.7	4.2′ x 3.4′	S(B)a pec
R		LPV	23h 43.8m	−15°17′	5.8↔12.4	387d	

★²	*Double Star*
LPV	*Long Period Variable*
∷	*Asterism*
●	*Globular Star Cluster*
■	*Planetary Nebula*
§	*Spiral Galaxy*

H = Hubble type for galaxies

Subscript "P" denotes photographic magnitude; subscript "B" denotes blue magnitude.

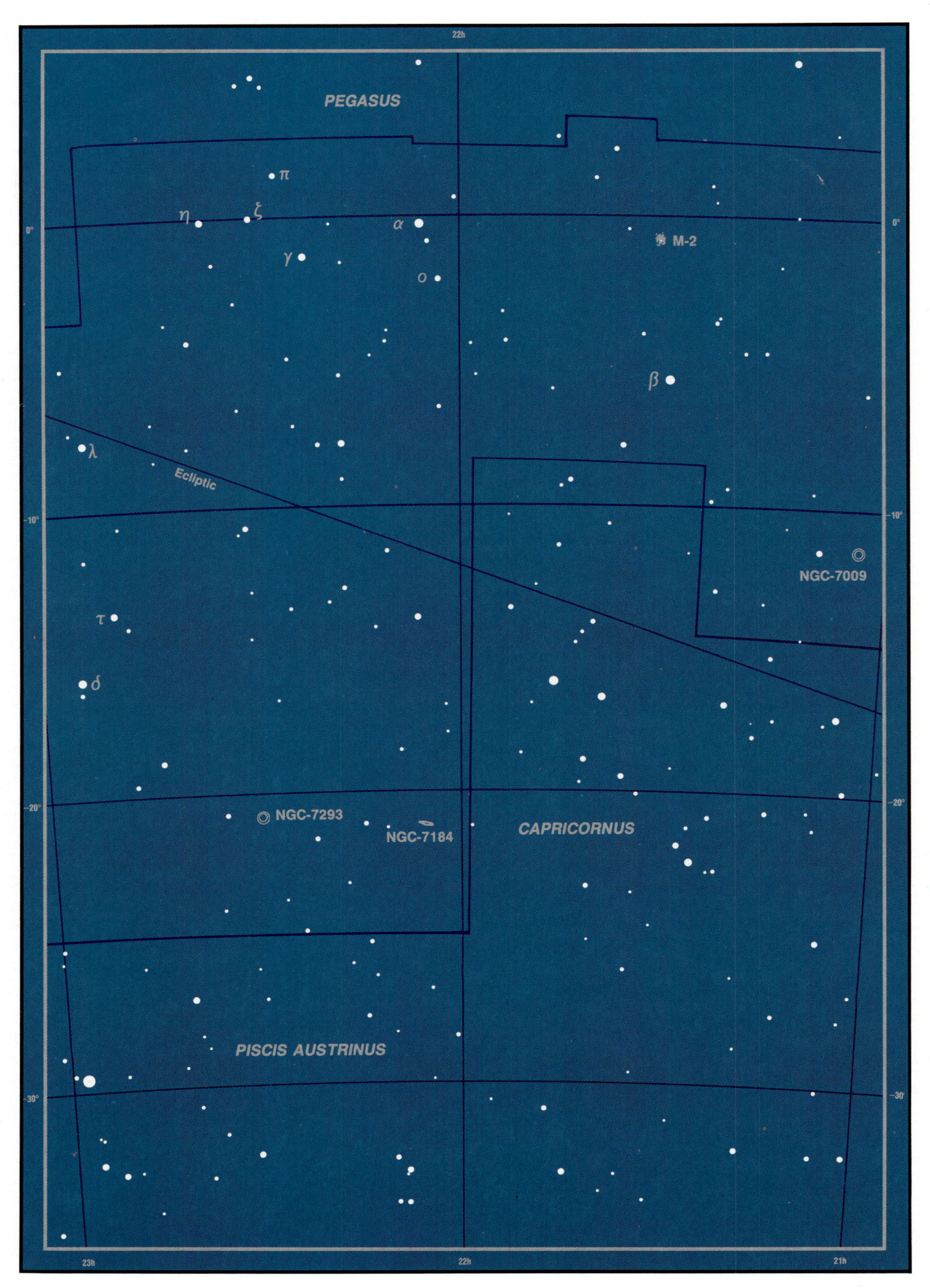

22h
PEGASUS
π
η
ξ
α
0°
M-2
γ
o
β
λ
Ecliptic
-10°
10°
NGC-7009
τ
δ
-20°
-20°
NGC-7293
NGC-7184
CAPRICORNUS
PISCIS AUSTRINUS
-30°
-30
23h
22h
21h

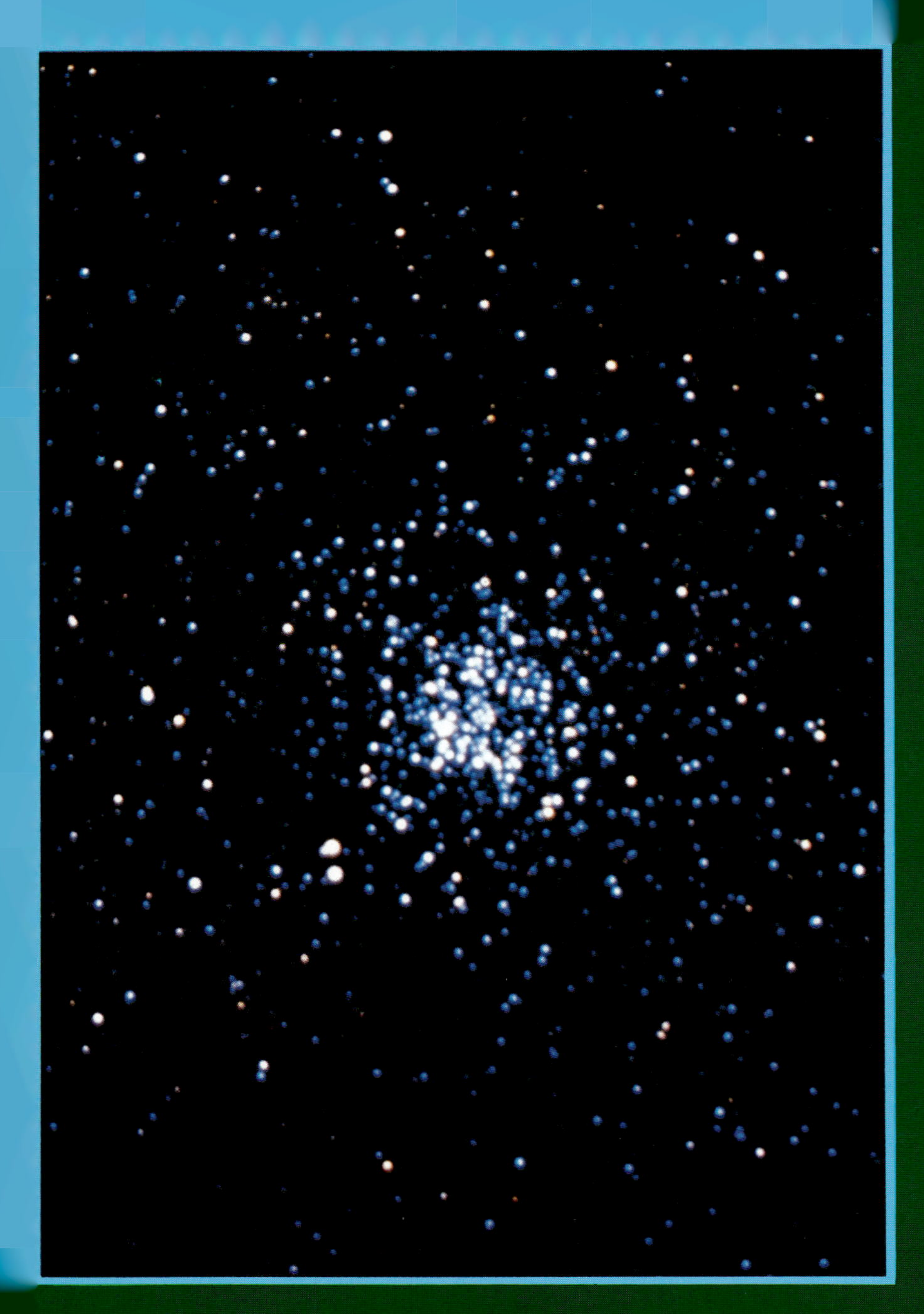

Above left: M-11, the Wild Duck Cluster is one of the richest assemblages of stars in the summer sky and is resolvable in any telescope. Photo by Scott Rosen.
Top: NGC-6781 is a large, low-surface brightness planetary nebula visible as a ghostly arc of light. Photo by Jack B Marling.
Above: NGC-6804, another planetary o similar brightness, measures about hal the size of NGC-6781. Photo by Jack B Marling.
Left: The dark nebulae Barnard 142 anc 143 form a distinctive "E" in the rich Aquila Milky Way. Photo by David Healy

Aquila
Aql
Aquilae
Scutum
Sct
Scuti

Keith Ward

Tucked between the barely noticeable constellations Delphinus and Sagitta to the north and the rich and impressive star clouds of Sagittarius to the south, **Aquila** the Eagle and **Scutum** the Shield are two of the summer Milky Way's finest offerings. Each contains several unique examples of bright deep-sky objects: Aquila offers abundant planetary nebulae and Scutum possesses bright open clusters.

The finest sight in Scutum is the brilliant open cluster **Messier 11** (NGC-6705), which lies on the northern edge of the Scutum starcloud. Discovered in 1681 by Gottfried Kirch at the Berlin Observatory, M-11 was then described as being a "small, obscure spot with a star shining through and rendering it more luminous." It is actually a large, rich, relatively distant open cluster composed of at least 870 stars brighter than magnitude 17.

At first glance, observers with small instruments may mistake M-11 for a globular cluster. In binoculars and finderscopes the group appears as a misty clump of nebulosity with only a hint of resolution — a grainy smattering of barely perceivable points of light. A 3-inch refractor at high power shows M-11 as a patch of tiny, faint stars some 10′ across, with one bright standout near the group's center. Larger scopes clearly show M-11's triangular shape and numerous glittering stars, an impressive sight that the great English observer Admiral Smyth likened to a group of wild ducks in flight.

M-11's total magnitude is 5.8 (visible to the naked eye under a very dark sky), and its brightest star shines at magnitude 8.0. Since the cluster lies some 1.7 kiloparsecs distant, it and the other brightest members must be very luminous giants to shine as brightly as they do. If the Sun were placed at M-11's distance it would shine at barely 15th magnitude.

Since the cluster is one of the richest of its type, the density of stars in M-11 has been the subject of discussion for decades. It appears that there are some 83 stars per cubic parsec in M-11. An observer inside the group would see several hundred 1st-magnitude stars in his night sky.

About 30′ northwest of M-11 are two 6th-magnitude double stars. Another 30′ northwest of these two doubles and 1° south of Beta (*β*) Scuti is the peculiar semi-regular variable star **R Scuti**. This ruddy star's variability was discovered by the English observer E. Pigott in 1795, and it has since become one of the most widely observed variables in the sky. It is classed as an RV Tauri-type variable, one that shows a variety of complex behavior for several reasons.

Displaying at least two superimposed periods, R Scuti has a primary period of about 144 days, but its amplitude is somewhat variable. Occasionally it reaches magnitude 4.8 and drops to magnitude 6.0. Sometimes the star fades to fainter than magnitude 8 at minimum. R Scuti also shows a long wave-like cycle of some 1,300 days.

Try observing R Scuti every week or so this summer. Using binoculars or a small telescope, estimate its brightness relative to the surrounding stars in Scutum. Is it brighter or fainter than the two double stars to the southeast? By how much? Keeping a chart of R Scuti's magnitude will, over time, permit you to compile a "life history" of its behavior.

Three degrees south of the M-11/R Scuti area lie two more impressive objects. **M-26** (NGC-6694) is an open cluster that is often passed by because of its brighter neighbor to the north. But M-26 itself is a worthwhile object, as it contains thirty stars in an area spanning 15′ across for a combined magnitude of 8.0. The brightest star in M-26 glows at magnitude 10.3 — much brighter than the Sun would be at such a distance but many times fainter than the brightest stars in M-11.

Small telescopes show M-26 as a tightly-packed group of stars in a very rich field, with the brightest stars lying on the cluster's southwestern edge. Larger telescopes show progressively fainter stars strewn amongst the primary cluster members, but they don't greatly enhance the view of this little group of suns.

Two degrees east and a little north of M-26 is the fine globular star cluster **NGC-6712**. Measuring 7.2′ in diameter and shining at magnitude 8.2, this object is a fine sight in any telescope; it even appears as a moderately large disk of gray light in finderscopes. A 6-inch scope at high power resolves the edges of NGC-6712, which lies some 7.6 kiloparsecs away and is rather heavily obscured by galactic absorption. Large scopes resolve some stars across the face of the cluster, making them appear suspended on a background of misty nebulosity. Within the same low-power field is the faint planetary nebula **IC 1295**, which measures some 90″ across but

BEST VISIBLE DURING
SUMMER

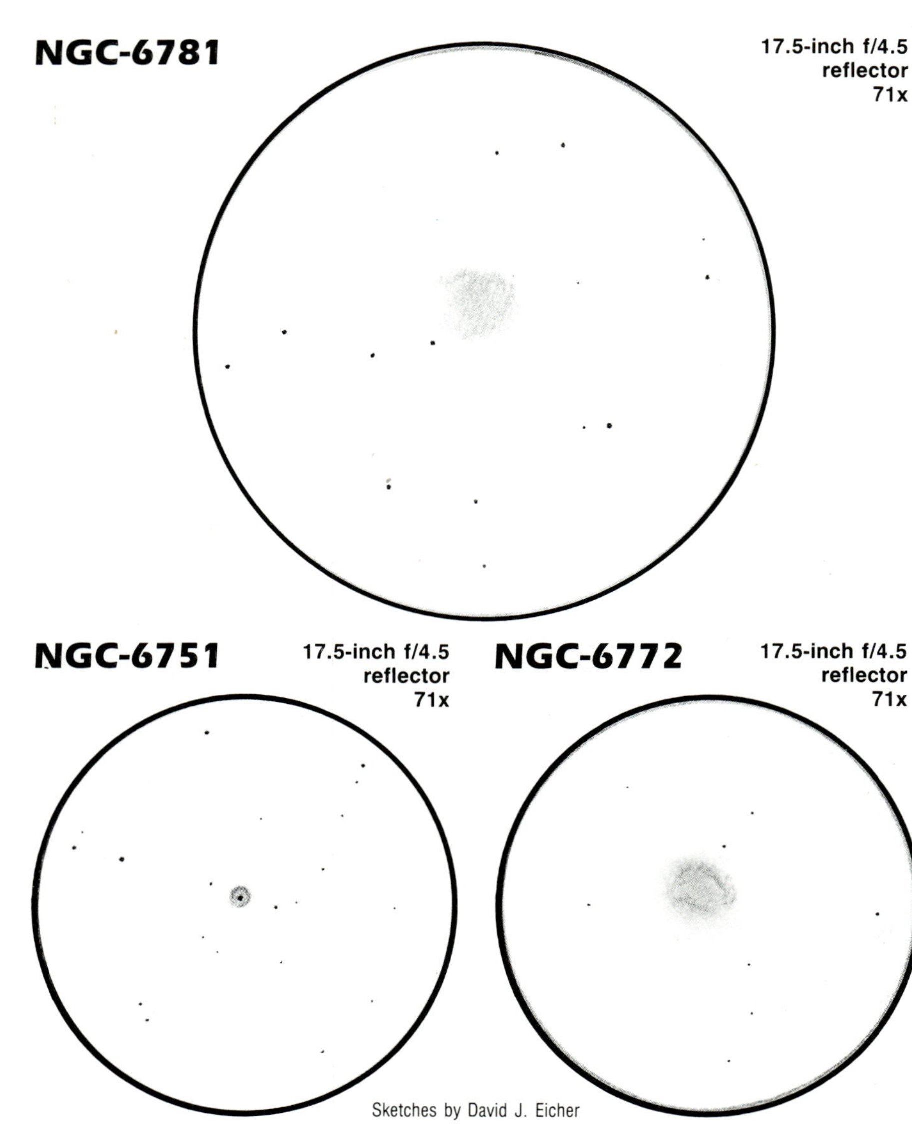

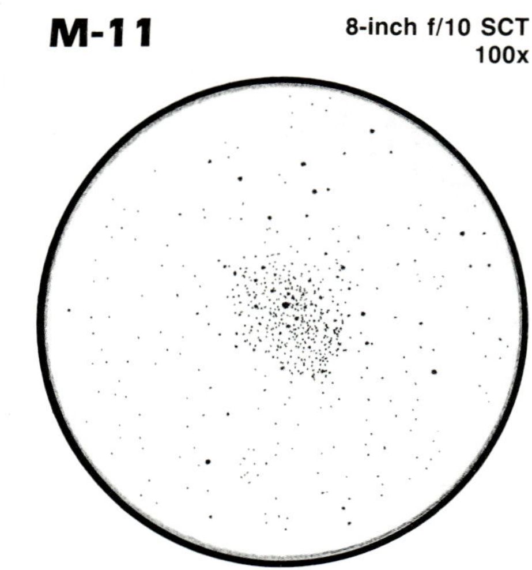

Sketches by David J. Eicher

glows dimly at 15th magnitude.

Near the center of the constellation Scutum lies Alpha (α) Scuti, the star group's 4th-magnitude luminary. In the same low-power field, just to the east, is the bright, scattered open cluster **NGC-6664**, which contains fifty stars in an area spanning 16′. At a distance of 1.3 kiloparsecs, this cluster is closer than M-26, and its brightest star shines at magnitude 10.2. A small telescope at low power is all that's required to view NGC-6664 comfortably; larger scopes show a few additional members in the group but generally decrease the field of view.

Unlike Scutum, Aquila contains numerous planetary nebulae observable in backyard telescopes. The best is **NGC-6781**, a giant, ring-shaped nebulosity glowing at photographic magnitude 11.8. NGC-6781 measures some 109″ — nearly two arcminutes — across, so its surface brightness is pretty low. Observe it using medium powers on a very dark, transparent night. Under those conditions you should pick it up in a 4-inch or 6-inch telescope with little trouble.

Three-and-a half degrees northeast of NGC-6781 is the slightly fainter nebula **NGC-6804**, lying just southwest of a 7th-magnitude double star. NGC-6804 glows at photographic magnitude 12.2 and measures a mere 31″ across, giving it a much higher surface brightness than NGC-6781. In the southern part of Aquila lies **NGC-6772**, a one-arcminute diameter planetary faintly glowing at photographic magnitude 14.2. About 4° southwest of NGC-6772 is the bright, diminutive planetary **NGC-6751**, a photographic magnitude 12.5 object measuring only 20″ across.

About 3° northwest of Altair, Aquila's brightest star, is the stark and easily observed dark nebulae **Barnard 142** and **B143**. These two entries from E.E. Barnard's catalogue of dark nebulae form a giant "E" shape some 80′ x 50′ in extent, through which very few stars shine. Try observing it with a low-power eyepiece — you'll be delightfully surprised.

Object	M#	Type	R.A. (2000)	Dec.	Mag.	Size/Sep./Per.	N★
NGC-6664		⊙	18h 36.7m	−8°13′	7.8	16′	50
NGC-6694	M-26	⊙	18h 45.2m	−9°24′	8.0	15′	30
R Sct		SRV	18h 47.5m	−5°42′	4.5↔8.2	144d	
NGC-6705	M-11	⊙	18h 51.1m	−6°16′	5.8	14′	
NGC-6712		●	18h 53.1m	−8°42′	8.2	7.2′	
IC 1295		■	18h 54.6m	−8°50′	15.0_P	86″	
NGC-6751		■	19h 05.9m	−6°00′	12.5_P	20″	
NGC-6755		⊙	19h 07.8m	+4°14′	7.5	15′	100
NGC-6772		■	19h 14.6m	−2°42′	14.2_P	62″	
NGC-6781		■	19h 18.4m	+6°33′	11.8_P	109″	
NGC-6804		■	19h 31.6m	+9°13′	12.2_P	31″	
B142/B143		□D	19h 40.7m	+10°57′	—	80′x50′	
NGC-6814		§	19h 42.7m	−10°19′	11.2	3.2′x3.0′	

N★ = Number of stars in the cluster
Subscript "P" denotes photographic magnitude; subscript "B" denotes blue magnitude.

SRV	*Semiregular Variable Star*
⊙	*Open Star Cluster*
●	*Globular Star Cluster*
■	*Planetary Nebula*
□D	*Dark Nebula*
§	*Spiral Galaxy*

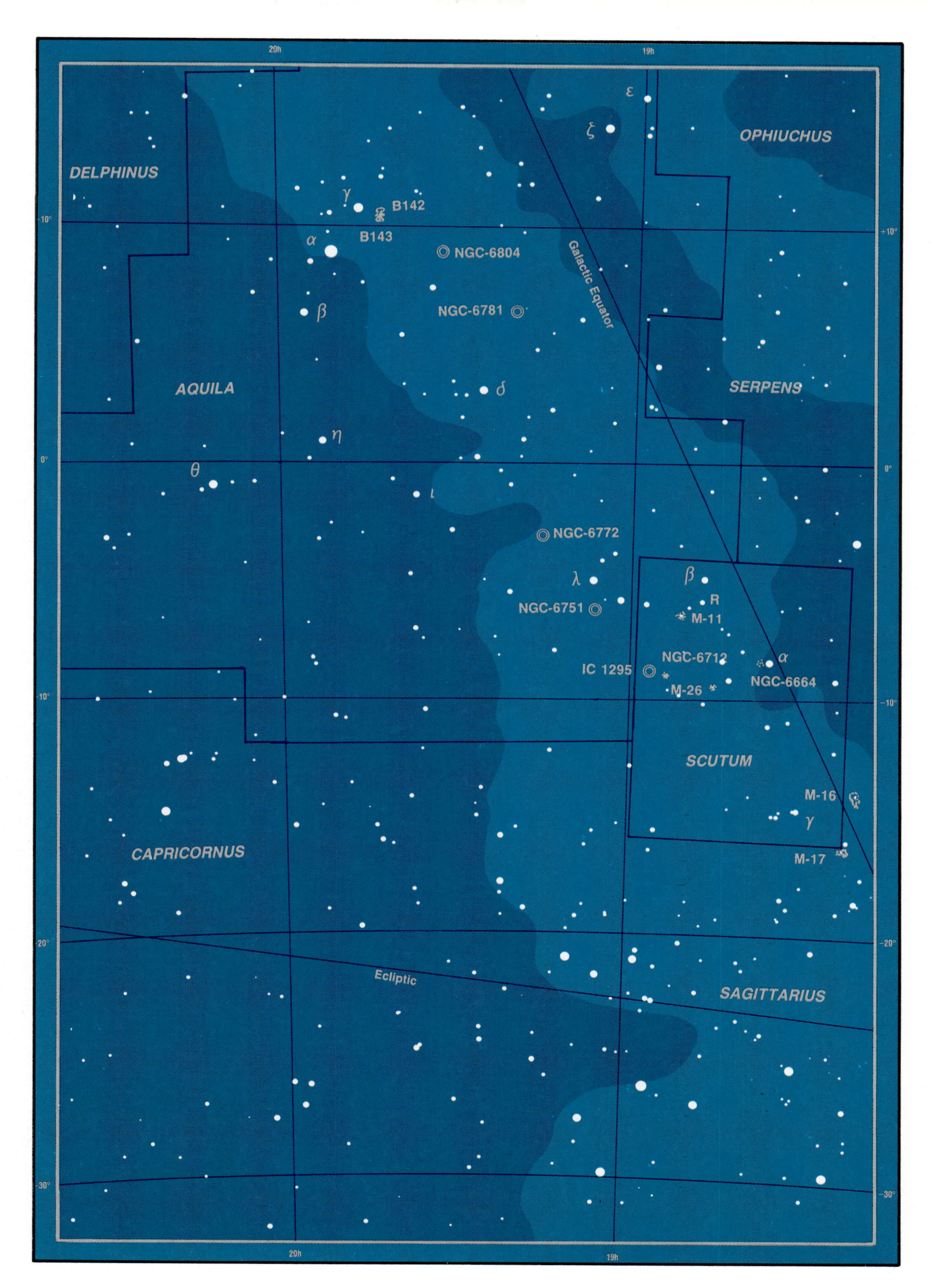
20h
19h
DELPHINUS
OPHIUCHUS
B142
B143
NGC-6804
NGC-6781
Galactic Equator
AQUILA
SERPENS
NGC-6772
NGC-6751
R
M-11
NGC-6712
IC 1295
M-26
NGC-6664
SCUTUM
M-16
M-17
CAPRICORNUS
Ecliptic
SAGITTARIUS
+10°
0°
-10°
-20°
-30°

NGC-772 is a magnitude 10.3 spiral in Aries, visible in backyard scopes as a bright core surrounded by a diffuse halo of nebulosity. The little galaxy to its southwest is NGC-770, a 14th-magnitude elliptical system. U.S. Naval Observatory photograph.

Aries

Ari
Arietis

Keith Ward

Lying in a rather barren part of the sky between the constellations Triangulum and Cetus, **Aries** the Ram is an inconspicuous group of stars. Aries is composed of a single 2nd-magnitude star, one 3rd-magnitude star, and a handful of 4th- and 5th-magnitude stars. The constellation also contains several deep-sky objects. Those it does hold include a dozen widely scattered galaxies and two bright double stars.

The largest and brightest galaxy in Aries is **NGC-772**, a magnitude 10.3 spiral measuring 7.1′ by 4.5′. This galaxy is relatively easy to find because it lies slightly less than 2° east-southeast of the 4th-magnitude star Gamma (γ) Arietis. (The galaxy forms a right triangle with Gamma and the 5th-magnitude star Iota (ι) Arietis, some 2° south of Gamma.)

Small telescopes show NGC-772 as a gray, oval haze some 2′ or 3′ across. A 4-inch scope at high power reveals the galaxy's bright, condensed, ball-like nucleus centered in the nebulous envelope representing the spiral arms. An 8- or 10-inch telescope shows a star just off the galaxy's western side and under good conditions may faintly show the galaxy's spiral arms or at least the brighter northwestern arm. Large scopes in the 16-inch and larger class show the galaxy as a large oval smudge of light superimposed over a bright, almost starlike nucleus and a ghostly spiral arm pattern.

If you observe NGC-772 with a large scope, you may also spot **NGC-770** in the same field of view, 5′ south of the spiral. NGC-770 is an elliptical galaxy some 1.3′ by 1.0′ across with a blue magnitude of 14.1. This object appears as a tiny, fuzzy, out-of-focus "star" at low power. At high power during good seeing it shows as a pale patch of nebulosity with a slightly brighter middle.

After you observe NGC-772, move your telescope back to the bright star **Gamma Arietis**. With a low-power eyepiece you'll see that Gamma is a wide double star with equally bright components (each shines at magnitude 4.8). Separated by 7.8″, Gamma's stars are resolvable with any telescope and generally considered one of the prettiest equal pairs, each star a sparkling blue-white. Gamma Arietis was one of the first known double stars, found accidentally by the English scientist Robert Hooke in 1664 while he was searching for a comet. Gamma Arietis is a relatively close pair of stars lying some 50 parsecs away. The combined luminosity of the two stars equals about fifty Suns.

South of the Gamma Arietis area lies another galaxy. **NGC-877** is relatively easy to find by star hopping: point your telescope toward Gamma; move it southeast to Iota; move it an equal distance east to a wide, flattened triangle of 6th-magnitude stars; and then follow the arc of the triangle south to 19 Arietis. Now simply move the finder scope 1.5° southeast, and you should see two 8th-magnitude stars positioned east-west. With a low-power eyepiece, center the telescope's field of view on the easternmost of the two stars, and you should see the galaxy and the star. With an 8-inch telescope NGC-877 appears as a small, round patch of grayish nebulosity some 1.5′ across the center and slightly brighter than the surrounding light. A star lies on the galaxy's southeastern side and a slightly fainter star lies on the northwestern side. Larger telescopes don't dramatically improve the view, though they do show a larger area of nebulosity and hint at mottling across the galaxy's face.

Some 4° southwest of NGC-877 near the southern border of Aries is the bright elliptical galaxy **NGC-821**. To locate this galaxy, aim your finder scope toward Xi (ξ) Arietis, the brightest star in a semicircle of six stars on the Aries/Cetus border. Next, with a low-power eyepiece move the telescope 4° west and 30′ north, and you should see an oval smudge of light in your eyepiece. A 4-inch scope shows this magnitude 10.8 galaxy as a nebulous smear spanning 3′ by 2′ with a 10th-magnitude foreground star on the galaxy's northwest side. An 8-inch scope reveals that NGC-821's surface is not uniformly illuminated but slightly mottled.

Midway between NGC-821 and NGC-772 lies the small Sb-type spiral **NGC-803**. This galaxy softly glows at magnitude 12.4 and covers 3.3′ by 1.5′. A 4-inch scope shows NGC-803 as a thin, dim spike of gray-green nebulosity, and a 6-inch scope shows it as an elongated patch of nebulosity with a bright center. An 11th-magnitude star lies just to the galaxy's west side.

The northwestern corner of Aries holds two more galaxies visible in small telescopes. **NGC-697** is a highly inclined barred spiral with a photographic magnitude of 12.7. To locate this galaxy, place

BEST VISIBLE DURING
AUTUMN

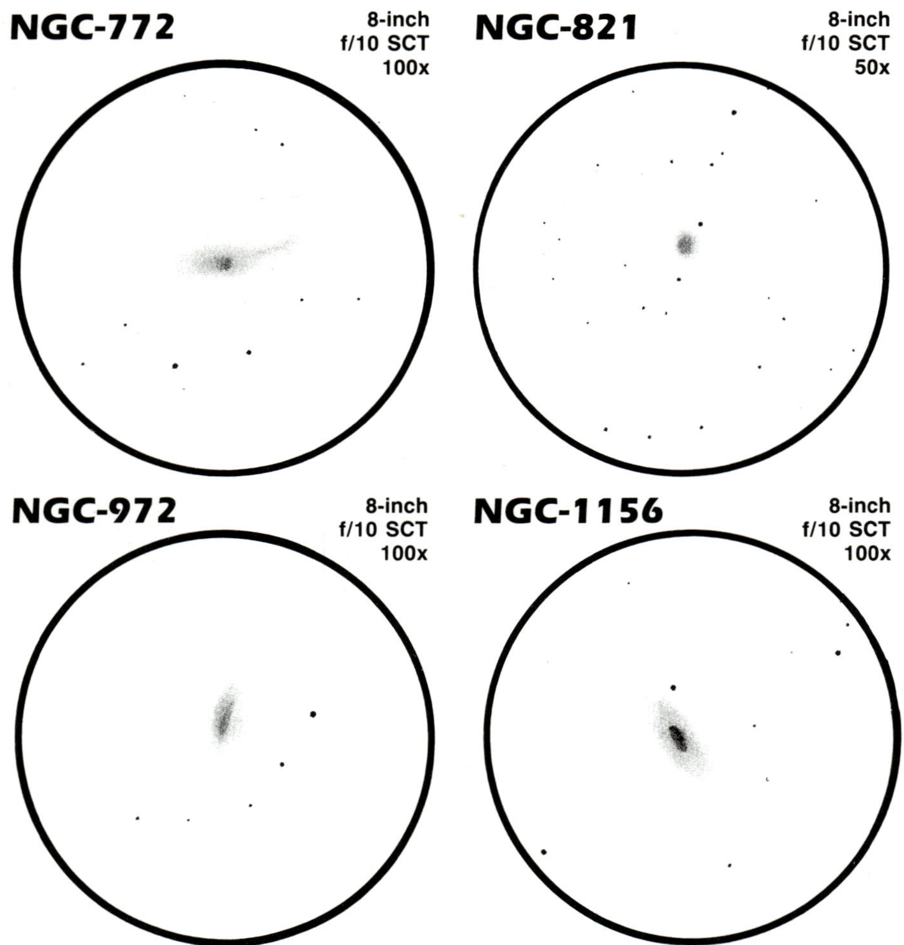

Sketches by David J. Eicher

your finder scope on the 3rd-magnitude star Beta (β) Arietis and move 2° to the northwest. This will bring you to a 6th-magnitude double star cataloged as 1 Arietis. NGC-697 lies to the east, just outside of a low-power field centered on this wide double. Six-inch telescopes show NGC-697 as a relatively bright, spindle-shaped patch of light elongated east-west with a 12th-magnitude star off the galaxy's eastern arm. Now move back to 1 Arietis and position your scope on a 10th-magnitude star 45′ south of the double. This brings you to **NGC-691**, a spiral glowing at blue magnitude 12.4 and measuring 3.5′ by 2.7′. This object appears as a low surface brightness haze with a bright, condensed center just southwest of the 10th-magnitude star.

In north-central Aries some 2° southwest of the 5th-magnitude star Nu (ν) Arietis lies the small spiral galaxy **NGC-976**. This galaxy glows softly at magnitude 12.4 and measures 1.7′ by 1.5′. It has a relatively low surface brightness, so it appears in a 6-inch telescope like a small, nebulous patch without detail.

Seven degrees northeast of NGC-672 is the irregular galaxy **NGC-1156**. This galaxy is observationally interesting because it shows mottled detail in small telescopes and lies in a rich starfield. To locate NGC-1156, start by centering your finder on the 4th-magnitude double star 41 Arietis in the northern part of Aries. Move the telescope 3° east-southeast to a bright pair of stars called 49 and 51 Arietis. Note the bright double star southeast of this pair. This is 52 Arietis. The galaxy forms an equilateral triangle with the 49/51 pair and 52, and represents the triangle's southwestern corner.

When you zero in on NGC-1156, you'll see it lies inside a much smaller equilateral triangle made up of 9th- and 10th-magnitude stars. The galaxy lies between one of these stars and a row of three 12th-magnitude stars; a faint star lies on the northwestern edge of the galaxy. Small scopes show NGC-1156 as an oval blob of light with a bright middle, surrounded by an irregularly shaped halo of faint light. On nights of good transparency 8-inch scopes show patchy mottling in this halo, and 12-inch and larger scopes regularly show a grainy, uneven texture in the nebulosity.

Also in the northern part of Aries is **NGC-972**, an 11th-magnitude highly inclined spiral measuring 3.6′ by 2.0′. To find this object, start at the bright double 41 Arietis and move 3° northwest through 35 Arietis, but not as far as the pair of 6th-magnitude stars 12 and 13 Trianguli. You'll see a close pair of 7th-magnitude stars oriented nearly north-south; the galaxy lies 30′ west of this pair. Visually NGC-972 appears as an elongated smudge of bright nebulosity just north of a faint double star. NGC-972 has a bright middle and a large, faint envelope of surrounding nebulosity.

Object	M#	Type	R.A. (2000)	Dec.	Mag.	Size/Sep./Per.	H
NGC-691		§	1h 50.7m	+21°46′	12.4_B	3.5′x 2.7′	Sb+
NGC-697		§B	1h 51.3m	+22°21′	12.7_P	4.7′x 1.8′	S(B)b+:
Gamma (γ)		★²	1h 53.5m	+19°18′	4.8,4.8	7.8″	
NGC-770		0	1h 59.2m	+18°57′	14.1_B	1.3′x 1.0′	E3:
NGC-772		§	1h 59.3m	+19°01′	10.3	7.1′x 4.5′	Sb
UGC-1551		§B	2h 03.6m	+24°04′	13.3_B	3.0′x 1.5′	SBdm
NGC-803		§	2h 03.8m	+16°02′	12.4	3.3′x 1.5′	Sb
NGC-821		0	2h 08.4m	+11°00′	10.8	3.5′x 2.2′	E2
NGC-877		§	2h 18.0m	+14°33′	11.8	2.3′x 1.8′	Sc
NGC-976		§	2h 34.0m	+20°59′	12.4	1.7′x 1.5′	Sb
NGC-972		§	2h 34.2m	+29°19′	11.3	3.6′x 2.0′	Sc
NGC-1024		§	2h 39.2m	+10°51′	13.8_P	4.7′x 1.9′	Sb
Σ326		★²	2h 55.6m	+26°52′	7.6,9.8	5.9″	
NGC-1156		#	2h 59.7m	+25°14′	11.7	3.1′x 2.3′	Ir+

★² *Double Star*
§ *Spiral Galaxy*
§B *Barred Spiral Galaxy*
0 *Elliptical Galaxy*
Irregular or Peculiar Galaxy

H = Hubble type for galaxies

Subscript "P" denotes photographic magnitude; subscript "B" denotes blue magnitude.

3h
2h
+40°
+30°
+20°
+10°
0°
ANDROMEDA
PERSEUS
M-33
TRIANGULUM
NGC-972
Σ326
NGC-1156
UGC-1551
α
NGC-697
NGC-691
τ
ζ
ε
ν
NGC-976
β
δ
γ
NGC-772
NGC-770
TAURUS
π
PISCES
NGC-803
σ
NGC-877
NGC-1024
NGC-821
Ecliptic
CETUS

Above left: M-37 is one of the brightest, richest open clusters in the sky. Photo by David Healy. Below left: The faint emission nebula IC 405, called the Flaming Star Nebula, is visible in large backyard telescopes as a ghostly grey smear of light. Photo by Jack Newton. Top: The diminutive nebula NGC-1931 appears as a round ball of fuzzy light, much like an elliptical galaxy. Photo by Jack Newton. Above: M-36, lying inside the pentagon of Auriga, is an impressive collection of 60 stars. Photo by Jack Newton.

Auriga

Aur
Aurigae

Keith Ward

Known primarily for its brilliant star clusters — the finest being M-36, M-37, and M-38 — the pentagon-shaped constellation **Auriga** the Charioteer also contains several unusual nebulae owing to its position along a rich section of the winter Milky Way. By chance alignment, Auriga contains some unusual stars that lie relatively near us compared with the scattering of distant star clusters found in this part of the sky.

Lying about 5° west of Theta (θ) Aurigae is the open cluster **Messier 36** (NGC-1960), a large, bright grouping of at least 60 stars between magnitudes 9 and 14. This is one of the two Messier clusters lying within the pentagon of Auriga; M-38, detailed below, lies some 2.3° northwest. In a very low-power telescope both clusters are visible together. M-36's central core of bright stars measures only 10′ across, but it's bright enough to be spotted with small finder telescopes even under suburban skies.

With small telescopes, M-36 appears as a bright knot of two dozen stars in a circle spanning 15′. Most of the stars appear white or bluish-white, although a few hint at yellowish hues. Larger telescopes show progressively more stars in the cluster; a 16-inch reveals over 50 stars, most of them fainter than 11th magnitude, in a diameter of 15′. M-36 is fairly distant as bright clusters go. A study made at Lowell Observatory places it some 1.3 kiloparsecs away.

M-38 (NGC-1912) is larger and brighter than M-36 — bright enough to see in virtually any telescope or finder — but it contains fewer stars. Measuring 20′ across, some two-thirds the size of the Full Moon, M-38's stars are arranged in an irregular shape resembling an upside-down Greek letter π. The cluster contains many bright stars arranged in pairs. The brightest star in the cluster is an 8th-magnitude type G0-star, one much like our Sun but some 900 times brighter. This star is truly luminous; the Sun, if placed the same distance from Earth (1.3 kiloparsecs) would appear as a lowly magnitude 15.3 star.

BEST VISIBLE DURING
WINTER

The third of Messier's star clusters in Auriga is **M-37** (NGC-2099), lying just outside the pentagon several degrees southeast of the M-36/M-38 area. M-37, by far the richest of the three clusters, contains at least 150 stars brighter than 12th magnitude. Telescopes smaller than 2 inches show M-37 as a nebulous glow, but 3-inch or larger scopes resolve the outer parts of the cluster, revealing an enormous number of sparkling stars. Large backyard telescopes show nearly 200 stars, most of which are bluish-white or slightly yellow, in a field measuring 20′ across.

If you're observing with a 3-inch or larger telescope on a clear, dark night you should see a tiny smear of light just southwest of M-38. This is the diminutive cluster **NGC-1907**, a 5′ diameter collection of some 40 stars of magnitude 10 and fainter. Although unspectacular in backyard telescopes, NGC-1907 is a curious group in contrast to nearby M-38.

Midway between M-36 and the bright star Psi (ψ) Aurigae is a peanut-shaped emission nebula, **NGC-1931**. This object is one of the many relatively dim, small emission objects scattered along the galactic plane, a difficult object for observers equipped with small telescopes. An 8-inch scope under good skies shows the nebula as a double-lobed patch of greenish light some 3′ across with a few bright stars involved. A 16-inch scope more clearly reveals the nebula's outline, which resembles M-76 in Perseus.

Due east of both M-36 and NGC-1931, past the star Psi (ψ) Aur, is the unusual star **AE Aurigae**. A bright O-type star, AE Aur normally shines at 6th magnitude but is capable of irregular and bizarre flares and drops in its light output. Its distance is about 550 parsecs, giving it an absolute luminosity of 900 Suns.

AE Aur is the illuminating source for **IC 405**, a large emission nebula called the Flaming Star Nebula. This faint, wispy cloud measures some 18′ across and contains structures similar to the Veil Nebula in Cygnus. Measurements of the nebula and star indicate that AE Aur is a rapidly-moving "runaway star" that originated in Orion along with the fast-paced stars Mu (μ) Columbae and 53 Arietis, both of which can be traced back to a common point of origin with AE Aur. The nebula IC 405 is a dust and gas cloud that happens, by sheer chance, to be illuminated by AE Aur at present. As time passes, AE will move through this cloud, leaving it dark once again.

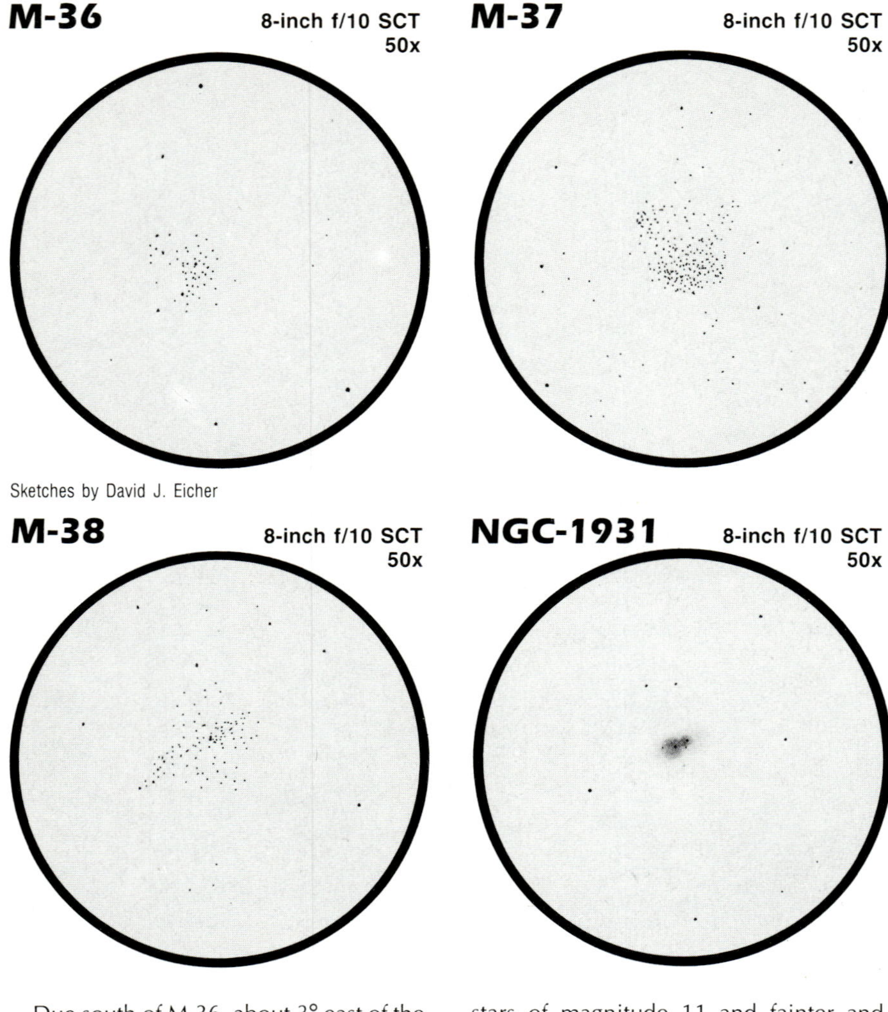

Sketches by David J. Eicher

Due south of M-36, about 3° east of the bright star Chi (χ) Aurigae, is the tiny, faint emission nebula **NGC-1985**. This object measures a mere 1′ across and appears in large telescopes as a round spot of hazy light. It certainly isn't impressive in most backyard telescopes, but it offers a challenge to those with a large telescope and dark sky.

On Auriga's border with Perseus is another large, bright star cluster designated **NGC-1664**. Containing 40 stars of magnitude 11 and fainter and measuring some 15′ across, NGC-1664 is visible in binoculars as a hazy peppering of faint light. A small refractor shows a large scattered group of two dozen stars encased in a faint haze; a 6-incher shows an area half the size of the Full Moon dotted with three dozen lights of various brightnesses and colors. NGC-1664 is a fine star cluster as viewed with virtually any instrument.

Another nice group of stars is **NGC-1893** which is located south of the midway point of a line drawn between M-36 and AE Aurigae. This group is irregularly shaped and contains 20 stars between 9th and 12th magnitude in an area spanning 12′. It is easily visible in most any telescope on a good night and quite bright in a telescope of 8-inches or more aperture. On very dark nights an 8-incher shows a faint haze pervading the cluster; in this case, it is not caused by unresolved faint stars but by the emission nebula **IC 410**. This faint cloud of gas measures a whopping 20′ across, but only the brightest few arc-minutes are visible in backyard telescopes.

The lone bright planetary nebula in Auriga lies just 3° north of Beta (β) Aur, one of the stars forming the pentagon. **IC 2149** is a bright disk of greenish light visible in any telescope, but it is so small that on nights of poor seeing it could be mistaken for a star. Its 10th-magnitude, 10″ disk requires high power and steady air to observe. Wait until a good night, find what appears to be a "fuzzy star" in the right area, and crank up the magnification to verify the disk structure. Large backyard scopes may show the 14th-magnitude central star.

Auriga contains two bizarre variable stars — **Epsilon** (ε) **Aurigae** and **Zeta** (z) **Aurigae**. Epsilon lies 3° southwest of Capella (Alpha [α] Aurigae) and is normally a third-magnitude star. Once every 27 years, however, something extraordinary happens to the Epsilon Aur system: it undergoes an eclipse, dimming by almost a magnitude for nearly two years. What causes such a strange fading? The current model is that Epsilon Aur has a large, dark flat disk companion which passes in front of it every 27 years. The next eclipse should start in January, 2010. Zeta, found 2¾° south of Epsilon, is an eclipsing variable star made up of a giant K4 star and a small B7 main sequence star. Once every two years, the magnitude 3.8 star undergoes an eclipse and fades by a tenth of one magnitude for several weeks.

Object	M#	Type	R.A. (2000)	Dec.	Mag.	Size/Sep./Per.
NGC-1664		⊙	4h 51.1m	+43°42′	7.6	18′
Epsilon (ε)		EV	5h 02.0m	+43°49′	2.9↔3.8	9892d
Zeta (ζ)		EV	5h 02.5m	+41°05′	3.7↔4.0	972d
IC 405		□E	5h 16.2m	+34°16′	—	30′ x 19′
AE		IV	5h 16.3m	+34°19′	5.8↔6.1	irr.
IC 410		□E	5h 22.6m	+33°31′	—	40′ x 30′
NGC-1893		⊙	5h 22.7m	+33°24′	7.5	11′
NGC-1907		⊙	5h 28.0m	+35°19′	8.2	7′
NGC-1912	M-38	⊙	5h 28.7m	+35°50′	6.4	21′
NGC-1931		□E	5h 31.4m	+34°15′	—	3′
NGC-1960	M-36	⊙	5h 36.1m	+34°08′	6.0	12′
NGC-1985		□E	5h 37.3m	+32°00′	—	3′
NGC-2099	M-37	⊙	5h 52.4m	+32°33′	5.6	24′
IC 2149		■	5h 56.3m	+46°07′	11.2p	8″

EV	*Eclipsing Variable*
IV	*Irregular Variable*
⊙	*Open Star Cluster*
■	*Planetary Nebula*
□E	*Emission Nebula*

CAMELOPARDALIS
URSA MAJOR
NGC-1664
M-38
NGC-1907
AE
IC 405
M-36
1931
IC 410
M-37
NGC-1985
Galactic Equator
GEMINI
TAURUS
M-35

Above: The 9th-magnitude globular NGC-5466 is elusive under poor transparency because of its faint constituent stars. Viewed from a dark, transparent sky, however, the cluster appears like a sprinkling of tiny stars embedded in a nebulous halo. Photo by Martin C. Germano.

Left: The Sc-type galaxy NGC-5248 is visible as an oval-shaped smear of nebulosity in a 4-inch telescope. Lick Observatory photograph.

Bootes

Boo
Bootis

Bootes the Herdsman is a transitional constellation. Surrounded by Milky Way star groups to the northeast, the relative emptiness of the circumpolar sky to the north, and innumerable galaxy fields to the south and west, Bootes isn't classifiable as a constellation rich in either Milky Way or extragalactic goodies. Instead, it has a sprinkling of faint galaxies floating behind a lone globular and a few standout double stars.

The single globular cluster in Bootes is **NGC-5466**, a fairly large object composed of very faint stars. This combination results in a low surface brightness that makes NGC-5466 difficult to spot in telescopes of 3 inches or less aperture. To find this object, begin at Arcturus (Alpha [α] Bootis), and then move 6° north-northwest to the 5th-magnitude star d Bootis. From here move 3° northwest and you'll see a pair of bright stars, 9 and 11 Bootis — 11 to the east and slightly brighter, 9 to the west. The cluster lies 2° northeast of the star 11 Bootis.

A 6-inch scope typically shows this cluster as a 5′-diameter hazy glow without individual stars, although on a night of exceptional seeing some of the brightest of NGC-5466's stars are visible with this aperture. A 10-inch telescope at high power shows a uniformly lit nebulous glow with a sprinkling of tiny glowing cluster members. Larger instruments show progressively more stars, but this cluster remains mostly unresolved in even the biggest backyard telescopes.

Five degrees east of NGC-5466 lies one of the many faint galaxies in Bootes. **NGC-5641** is an S(B)b-type barred spiral dimly glowing with a blue magnitude of 12.9 and measuring 2.7′ by 1.6′ across. NGC-5641 isn't terrifically impressive in backyard telescopes because it appears like a faint, oval smudge of nebulosity. It does, however, lie in an attractive field containing three bright stars arranged in a triangle, which makes for a pretty sight.

BEST VISIBLE DURING
SPRING

Forming an equilateral triangle with NGC-5466 and NGC-5641 is the faint galaxy **NGC-5523**, which lies 1° east-northeast of the star 12 Bootis. Although this galaxy shines at only blue magnitude 12.9, it is a wonderful object for observers with large telescopes because it is inclined nearly edge-on and is reminiscent of NGC-4565 in Coma Berenices. The galaxy is a type Sb object and measures 4.5′ by 1.4′ across.

Lying in the extreme southwestern corner of Bootes is the constellation's brightest galaxy, **NGC-5248**. Shining at magnitude 10.2 and spanning 6.5′ by 4.9′,

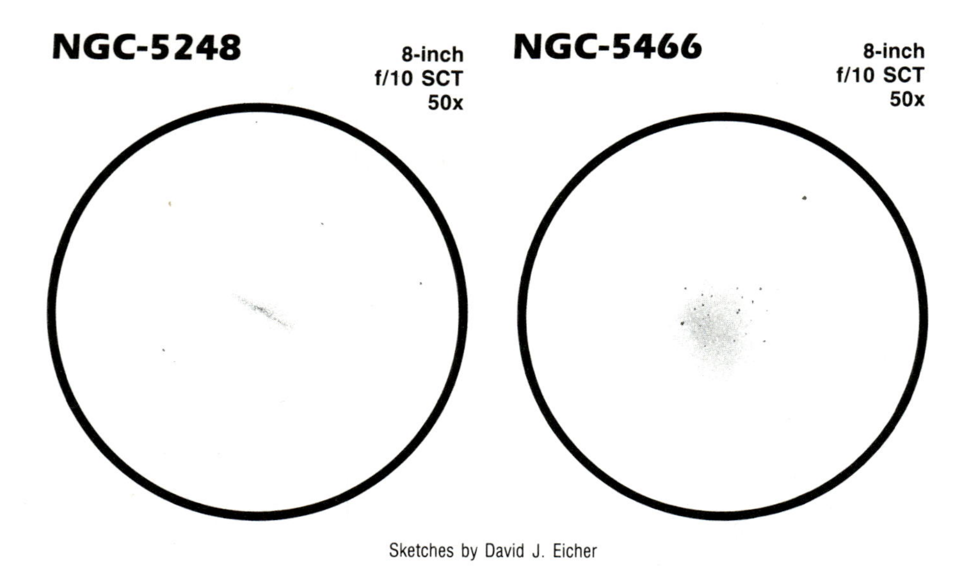

Sketches by David J. Eicher

this Sc-type galaxy is a fine target for any backyard telescope. A 2-inch scope shows the galaxy as a hazy oval patch of nebulosity without detail. NGC-5248 becomes interesting, however, when viewed with a 6-inch or larger instrument: the galaxy's spiral arms, visible as a faint, elongated haze, surround a bright, round central hub. Large backyard telescopes in the 12-inch to 18-inch class are capable of showing the delicate spiral arms of NGC-5248 on a night of excellent transparency.

About 11° east and a little south of NGC-5248 lies a prominent curved row of three 5th- and 6th-magnitude stars. Move another 3° northeast and you'll come across the amorphous galaxy **NGC-5669**. This galaxy is a type Sc object measuring 4.1′ by 3.2′ and glowing at blue magnitude 12.3. Although it's classified as an Sc galaxy, NGC-5669 appears more like an irregular: a high-power view through a 10-inch scope shows a faint wisp of light streaked over a small, diffuse knot of brighter nebulosity. This object certainly is not impressive, but its visibility may serve as a test of sky conditions.

Back in central Bootes is the spiral galaxy **NGC-5614**, located 2° east-southeast of the 5th-magnitude star A Bootis. This little galaxy glows at magnitude 11.7 and measures 2.7′ by 2.3′ in extent. In a 4-inch scope it appears like a pale disk of nebulosity some 1′ across; larger apertures don't dramatically improve the view.

A fine trio of galaxies — separated from one another by only 1° — lies in northern Bootes. **NGC-5660** lies just 30′ southeast of the 7th-magnitude star g Bootis, and both are visible in a low-power field of view. NGC-5660, an Sc-type spiral oriented more or less face-on, covers 2.8′ by 2.6′ of sky and shines at magnitude 11.8, so it is visible as a fuzzy patch in 4-inch scopes. An 8-inch telescope shows this galaxy to be a moderately bright disk of nebulosity with a brighter central condensation some 30″ across.

Almost 1° to the southeast of NGC-5660 is the much brighter galaxy **NGC-5676**, another Sc specimen. This galaxy shines at magnitude 10.9 and spans 3.9′ by 2.0′, and it appears as an elongated smear of whitish light in small scopes. A 10-inch scope reveals a bright nucleus encapsulated in a milky envelope of grayish-green nebulosity some 2′ by 1.5′ in extent.

Another degree to the southeast is the edge-on barred spiral **NGC-5689**. This galaxy measures 3.7′ by 1.2′ and glows softly at magnitude 11.2. It is visible as an indistinct smudge in a 4-inch scope at moderate power, but to see any noticeable detail in this galaxy, an 8-inch scope is required. An 8-inch instrument reveals NGC-5689's tiny, bright nuclear region and a thin strip of faint nebulosity running through it and extending for 1′ on both sides.

Northeastern Bootes holds two more galaxies for backyard telescopes. **NGC-5820** is a 12th-magnitude elliptical galaxy lying only 15′ west of a 7th-magnitude double star. The galaxy is visible as a round nebulous patch of light measuring some 2.5′ by 2.3′ across. About 12° south-southeast of NGC-5820 is **NGC-5899**, lying about 15′ southeast of a lone 7th-magnitude star. This Sb-type spiral shines at magnitude 11.8 and measures 3.0′ by 1.3′ across. It appears as an elongated blotch of nebulosity with a slightly condensed core in small backyard telescopes.

Bootes offers three impressive double stars for users of small scopes. The first of these, **Xi (ξ) Bootis**, lies some 9° east of Arcturus. Composed of magnitude 4.7 and 7.0 stars separated currently by 8.0″ in p.a. 230°, Xi Bootis is easily split by any telescope at reasonably high magnification. Its stars are, in the words of Thomas W. Webb, "clear yellow and reddish purple."

More challenging is the brighter star **Epsilon (ε) Bootis**, part of the kite-shaped asterism that forms the central figure of Bootes. Epsilon Bootis is made up of magnitude 2.5 and 4.9 stars separated by 2.8″ in p.a. 339°. The components are far enough apart that almost any telescope can split them with high power, and the colors are yellowish-orange and blue.

In northeastern Bootes lies the 5th-magnitude double star **i Bootis**, also known as 44 Bootis. This star contains magnitude 5.3 and 6.2 components currently split by a mere 1.6″ in p.a. 220°. The colors are yellow and blue.

Object	M#	Type	R.A. (2000)	Dec.	Mag.	Size/Sep./Per.	H
NGC-5248		§	13h 37.5m	+8°53′	10.2	6.5′x 4.9′	Sc
NGC-5466		•	14h 05.5m	+28°32′	9.1	11.0′	
NGC-5523		§	14h 14.8m	+25°19′	12.9_B	4.5′x 1.4′	Sb
NGC-5614		§	14h 24.1m	+34°52′	11.7	2.7′x 2.3′	S
NGC 5641		§B	14h 29.3m	+28°49′	12.9_B	2.7′x 1.6′	S(B)b⁻
NGC-5660		§	14h 29.8m	+49°37′	11.8	2.8′x 2.6′	Sc
NGC-5669		§	14h 32.7m	+9°53′	12.3_B	4.1′x 3.2′	Sc
NGC-5676		§	14h 32.8m	+49°28′	10.9	3.9′x 2.0′	Sc
NGC-5689		§B	14h 35.5m	+48°45′	11.9	3.7′x 1.2′	SBa
Epsilon (ε)		★²	14h 45.0m	+27°04′	2.5,4.9	2.8″	
Xi (ξ)		★²	14h 51.4m	+19°06′	4.7,7.0	8.0″	
NGC-5820		0	14h 58.7m	+53°33′	11.9	2.5′x 2.3′	E5
i		★²	15h 03.8m	+47°39′	5.3,6.2	1.6″	
NGC-5899		§	15h 15.0m	+42°03′	11.8	3.0′x 1.3′	Sb⁺

★²	*Double Star*
•	*Globular Cluster*
§	*Spiral Galaxy*
§B	*Barred Spiral Galaxy*
0	*Elliptical Galaxy*

H = Hubble type for galaxies

Subscript "P" denotes photographic magnitude; subscript "B" denotes blue magnitude.

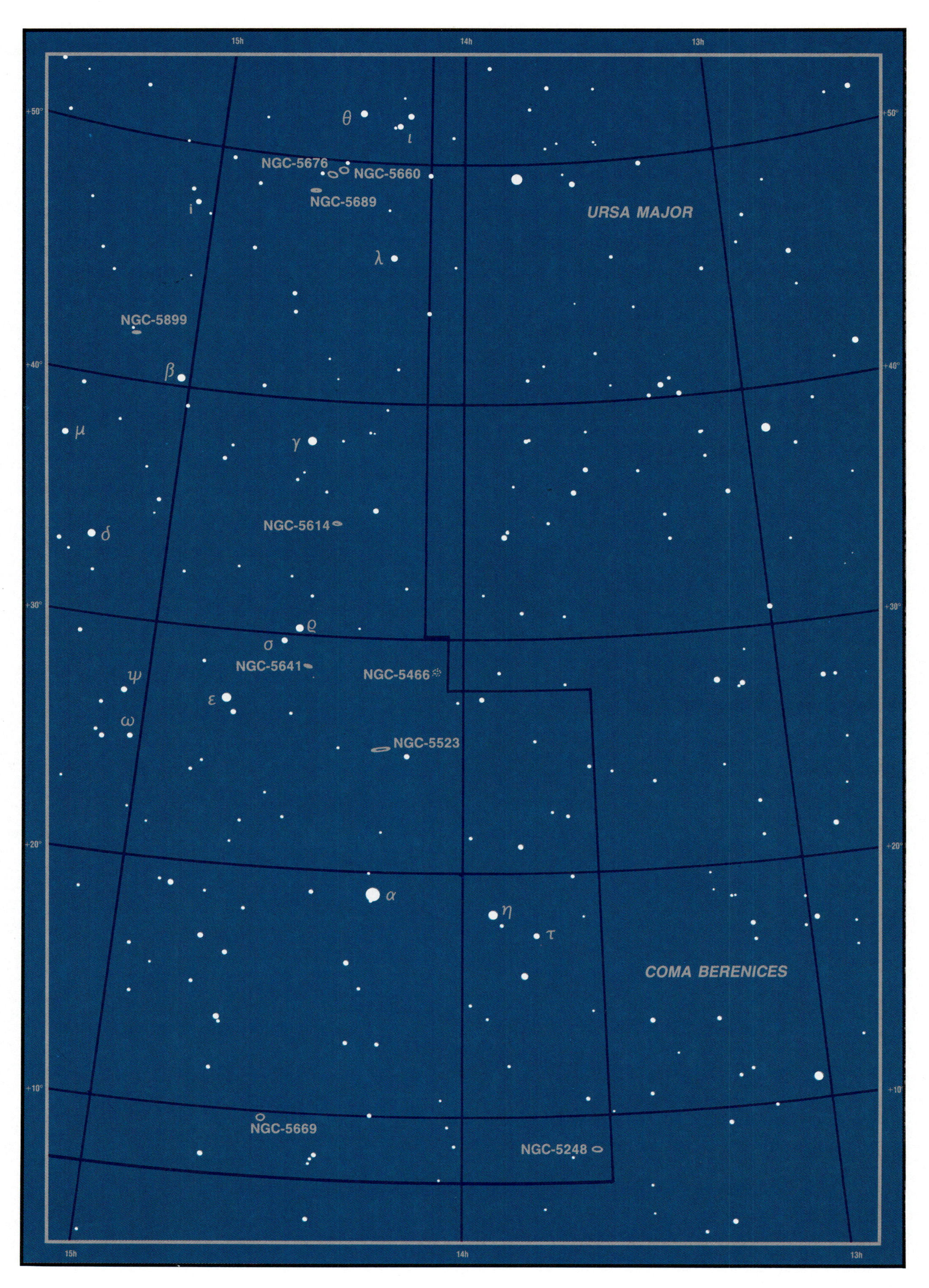
15h
14h
13h
+50°
+40°
+30°
+20°
+10°
θ
ι
NGC-5676
NGC-5660
NGC-5689
i
URSA MAJOR
λ
NGC-5899
β
μ
γ
NGC-5614
δ
ϱ
σ
NGC-5641
NGC-5466
ψ
ε
ω
NGC-5523
α
η
τ
COMA BERENICES
NGC-5669
NGC-5248

Above: NGC-2403 is one of the brightest galaxies in the northern sky. Photo by Jack Newton.

Left: Open cluster NGC-1502 contains forty-five stars in an 8' area and shines at 6th magnitude. Photo by Lee C. Coombs.

Below: NGC-1501 is a faint planetary nebula visible as a hazy smudge in medium-size telescopes. Photo by Lee C. Coombs.

Camelopardalis

Cam
Camelopardali

Lying between the rich Milky Way clusters and nebulae of Cassiopeia and the myriad spiral and elliptical galaxies of Ursa Major is the large, inconspicuous constellation **Camelopardalis** the Giraffe. This spindly constellation is far enough away from the galactic equator that it contains plentiful galaxies — including one of the finest spirals in the northern sky — yet it also offers two planetary nebulae and an unusually rich star cluster.

Placed near two 6th-magnitude stars in the southeastern corner of Camelopardalis is the magnitude 8.4 galaxy **NGC-2403**. This magnificent spiral lies only 3.5 megaparsecs away and is a member of the same group of galaxies that includes M-81 and M-82 in Ursa Major.

Because it is relatively near and appears more or less face-on to our line of sight, NGC-2403 presents intricate detail in its spiral arms. The structure of this galaxy resembles that of M-33 in Triangulum, the "Pinwheel Galaxy." Both show loose, mottled patches of bright and dark nebulosity in their arms and bright, small nuclei. Astrophotographers have discovered that recording this detail on film with long exposures and long focal length systems is rather easy. Seeing this detail in either galaxy, however, requires a large backyard telescope and a dark, transparent night.

From suburban or rural locations NGC-2403 is visible in large finder scopes as a nebulous spot of milky light. Viewed with a 4-inch telescope, the galaxy appears as an oval fuzz with a bright core. An 8-inch instrument reveals faint mottling in the galaxy's arms, and the galaxy appears as a bright oval nebulosity surrounded by a dim haze speckled with bright and dark regions visible only with averted vision. Backyard telescopes in the 16-inch class, especially when used with a Deep-Sky filter, show the patchy distribution of light in the galaxy's arms with relative ease. In any telescope the visual appearance of NGC-2403 is highlighted by its location against a rich, starry foreground.

BEST VISIBLE DURING
SPRING

NGC-2403 was the first galaxy outside the Local Group in which Cepheid variable stars were spotted — twenty-seven in all by 1960. About one hundred emission nebulae are known to habitate NGC-2403's graceful spiral arms. The largest of these nebulae — spanning 275 parsecs — are some of the fuzzy knots visible in large backyard instruments.

Nearly 4 hours of right ascension west and a little north of NGC-2403 lies the large, round barred spiral **IC 342**. Once thought to be a member of the Local Group of galaxies, this spiral lies at a distance of about 3 megaparsecs, which makes it too distant to be considered a Local Group galaxy.

IC 342 is so heavily obscured by the thick Milky Way in front of it that the spiral went undiscovered by astronomers until relatively recently. It was discovered by W. F. Denning in 1890, who reported it to J. L. E. Dreyer for inclusion in the first *Index Catalogue* in 1895. Denning and Dreyer considered IC 342 a fuzzy galactic nebula; it wasn't until 1934 that two Mount Wilson astronomers, Edwin Hubble and Milton Humason, discovered IC 342's spiral structure and identified it as a galaxy.

A large telescope is necessary for a satisfactory view of IC 342. The galaxy spans 17.8′ by 17.4′ and brightly glows at magnitude 9.1 in blue light (its visual magnitude is brighter yet). But its light is so spread out that IC 342 has a very low surface brightness and individual little pieces of it appear dim. A 4-inch scope reveals a fuzzy spot of nebulosity, the galaxy's nucleus, surrounded by a bright, colorful starfield. An 8-inch telescope shows IC 342 as a moderately bright circular fuzz surrounded by a very faint 10′-diameter haze that appears strongest using averted vision. A 17.5-inch scope shows this galaxy as a bright, compact nucleus centered in a large halo of nebulosity with patchy striations suggesting spiral arms. The entire glow is peppered with tiny stars.

Midway between NGC-2403 and IC 342 lies the little-observed spiral **NGC-1961**. This galaxy is a type-Sb peculiar system, shining at 11th magnitude and measuring 4.3′ by 3.0′ across. In backyard telescopes it appears as a roundish glow some 2′ across with a bright nucleus and faint fuzzy halo of nebulosity.

Several Camelopardalis galaxies lie at high enough declinations that they are circumpolar from the mid-northern latitudes. **NGC-2366** lies at declination +69° and is a curious irregular galaxy. Though its total magnitude is 11, which makes it technically bright enough to see in a

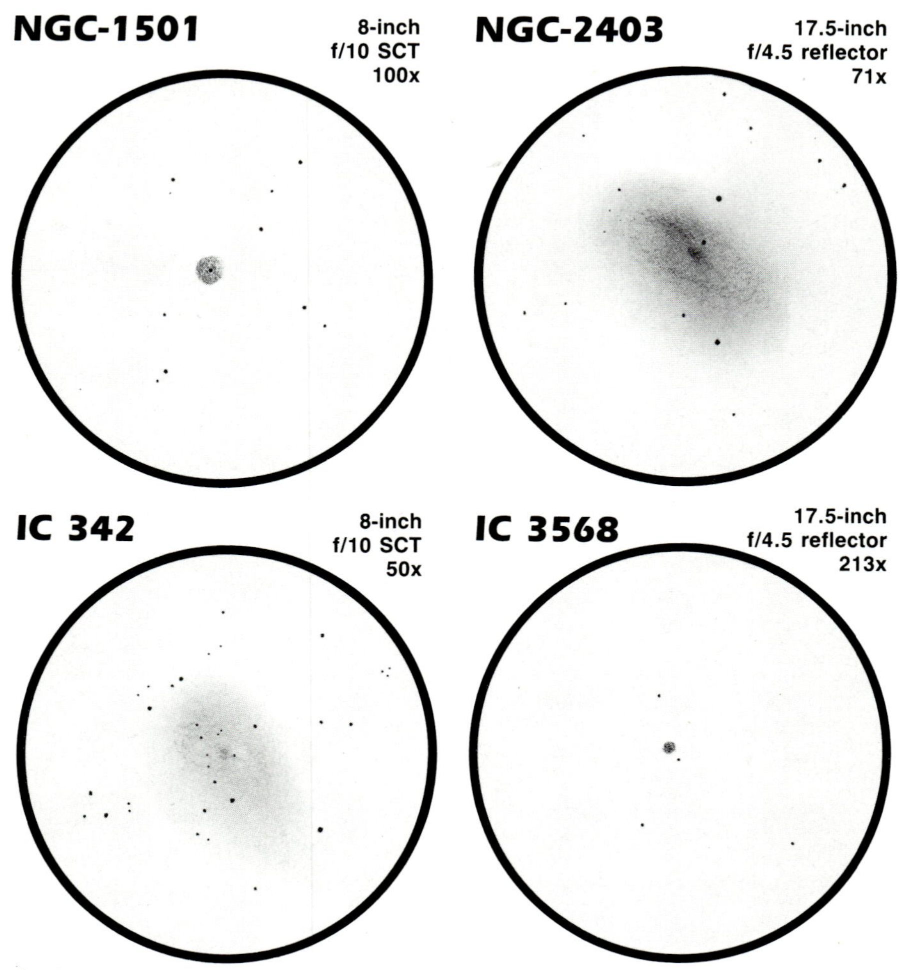

Sketches by David J. Eicher

3-inch scope, the galaxy has a rather low surface brightness. Its magnitude and overall dimensions of 7.6′ by 3.5′ make it visible as a misty patch of nebulosity in a 5-inch telescope.

NGC-2523 is a beautiful barred spiral lying at declination +73°. Although it glows at only 12th magnitude and covers a scant 3′ by 2′ of sky, this object is attractive in 10-inch and larger scopes, which show it as a small oval fuzz with a faint circular nucleus. A 16-inch telescope at high power clearly shows the galaxy's bar encased in a ring of gray-green nebulosity. If you observe with a large telescope, don't pass this object on your next night out.

Three galaxies lie at +78° declination, separated by 3 hours of right ascension. **NGC-2146** is a magnitude 10.5 barred spiral wedged between two scattered groups of bright stars in the central part of Camelopardalis. It appears as an oval smear of greenish nebulosity, measuring 6.0′ by 3.8′ in full angular extent. Large scopes reveal its bright nucleus and a mottled faint haze representing the galaxy's spiral arms. Lying only 2° apart are the galaxies **NGC-2655** and **NGC-2715** in the constellation's northeastern corner. NGC-2655 is a 10th-magnitude barred spiral measuring 5.5′ by 4.4′; NGC-2715 is an 11th-magnitude spiral measuring 5.0′ by 1.4′. Each is visible as a nebulous smudge when viewed with a 3-inch scope.

NGC-2336 is the northernmost of Camelopardalis' bright galaxies, lying at +80° declination just south of a triangle composed of two 6th-magnitude stars and one 7th-magnitude star. This object is a fine sight in backyard scopes. A spiral glowing at magnitude 10.5 and measuring 6.9′ by 4.0′, NGC-2336 appears as a bright oval nebulosity surrounding a concentrated, nearly stellar core.

The bright galactic objects in Camelopardalis consist of one open cluster and two planetary nebulae. **NGC-1502** holds forty-five stars in an area measuring 8′ across, giving it a rich, crowded appearance even in finder scopes. With a total magnitude of 5.7, NGC-1502 is technically a naked-eye object under a pitch black sky, but it is small enough to make identifying it a difficult task. The best views of this cluster come with 6- to 12-inch scopes at low power, which show a group of stars widely ranging in magnitude and color.

Camelopardalis' two planetaries offer a striking contrast to each other. **NGC-1501** is a typical planetary nebula: a small, bright, smoke ring with a faint central star. **IC 3568** is a faint, exceedingly small object that will test the limits of your equipment, your visual acuity, and the sky's steadiness.

Object	M#	Type	R.A. (2000)	Dec.	Mag.	Size/Sep./Per.	H
IC 342		§B	3h 46.8m	+68°06′	9.1_B	17.8′x 17.4′	S(B)c
NGC-1501		■	4h 07.0m	+60°55′	13.3_P	52″	
NGC-1502		⊙	4h 07.7m	+62°20′	5.7	8′	
S		SRV	5h 41.0m	+68°48′	7.7↔11.6	327.2d	
NGC-1961		§	5h 42.1m	+69°23′	11.1	4.3′x 3.0′	Sb pec
NGC-2146		§B	6h 18.7m	+78°21′	10.5	6.0′x 3.8′	SBb^- pec
NGC-2336		§	7h 27.1m	+80°11′	10.5	6.9′x 4.0′	Sb
NGC-2366		#	7h 28.9m	+69°13′	10.9	7.6′x 3.5′	Ir^+
NGC-2403		§	7h 36.9m	+65°36′	8.4	17.8′x 11.0′	Sc
NGC-2523		§B	8h 15.0m	+73°35′	12.0	3.0′x 2.0′	SBb^-
NGC-2655		§B	8h 55.6m	+78°13′	10.1	5.1′x 4.4′	S(B)a
NGC-2715		§	9h 08.1m	+78°05′	11.4	5.0′x 1.9′	Sc
NGC-2748		§	9h 13.7m	+76°29′	11.7	3.1′x 1.3′	Sc
IC 3568		■	12h 32.9m	+82°33′	11.6_P	6″	

SRV	*Semiregular Variable Star*
⊙	*Open Star Cluster*
■	*Planetary Nebula*
§	*Spiral Galaxy*
§B	*Barred Spiral Galaxy*
#	*Irregular Galaxy*

H = Hubble type for galaxies

Subscript "P" denotes photographic magnitude; subscript "B" denotes blue magnitude.

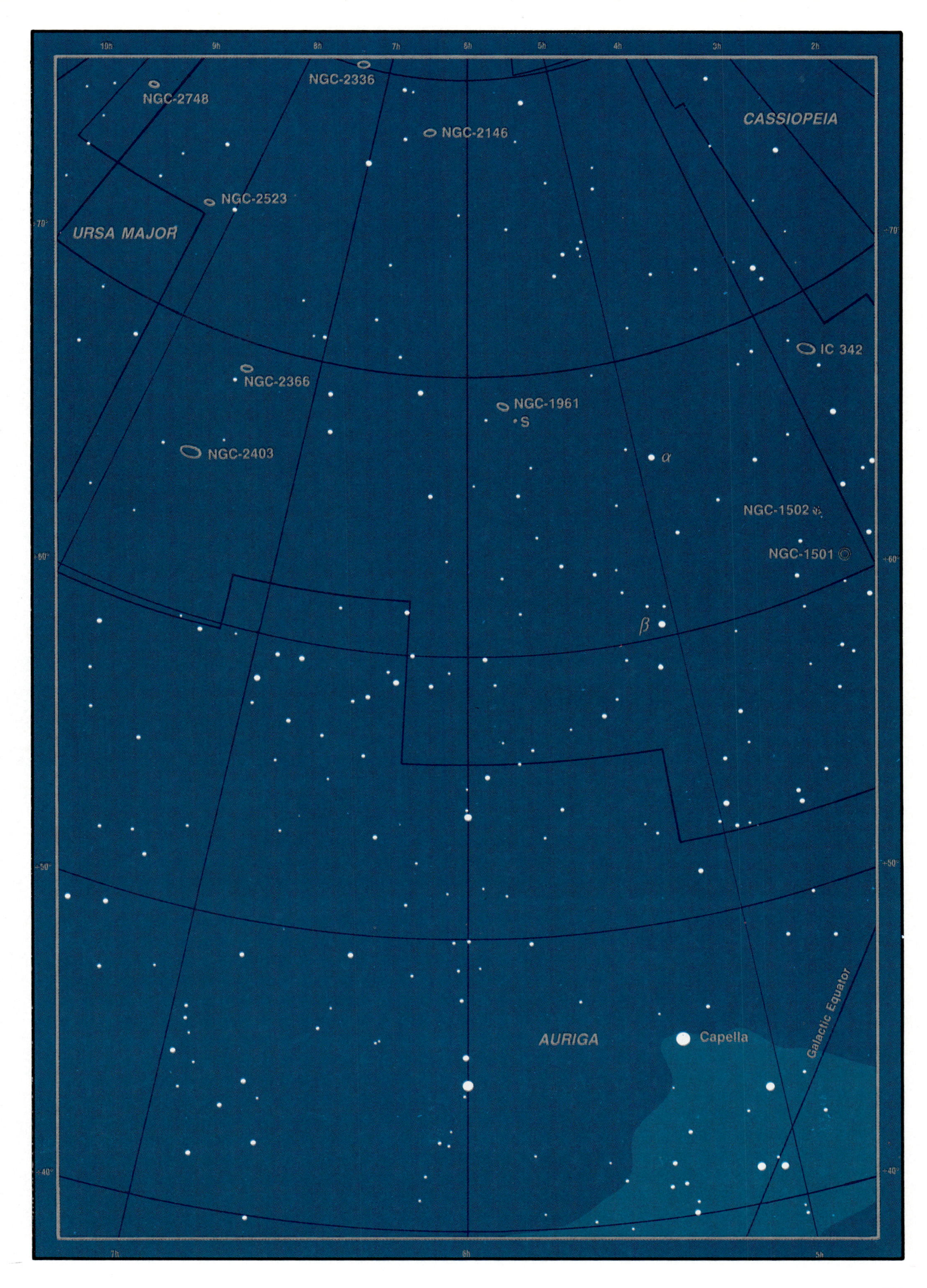
NGC-2748
NGC-2336
NGC-2146
CASSIOPEIA
NGC-2523
URSA MAJOR
IC 342
NGC-2366
NGC-1961
S
NGC-2403
α
NGC-1502
NGC-1501
β
AURIGA
Capella
Galactic Equator

Top: The Beehive cluster in Cancer (M-44) is one of the brightest clusters and holds fifty stars in an area spanning three Moon diameters. Photo by Lee C. Coombs.

Above: Hydra's star cluster M-48 glows with a collective magnitude of 5.8, just visible as a misty glow to the naked eye on dark nights. Photo by Jack Newton.

Left: M-67 in southern Cancer contains two hundred faint stars that combine for a 7th-magnitude glow. Photo by Clive Gibbons.

Cancer

Cnc
Cancri

Hydra

Hya
Hydrae

Bounded by the bright constellations Gemini and Leo to the north, the inconspicuous part of sky containing **Cancer** the Crab and western **Hydra** seems to be almost empty when viewed with the naked eye. Only a sprinkle of faint naked-eye stars and the 3rd-magnitude glow of the Beehive cluster are visible. But view Cancer and western Hydra with a backyard telescope, and you'll discover a region containing several bright, scattered star clusters, two multiple-star systems, and many galaxies.

The brightest deep-sky object in Cancer is the sprawling open cluster **M-44** (NGC-2632). Visible under a dark sky without optical aid, M-44 appears as a large, hazy patch of light containing a few bright stars. Because of its naked-eye visibility, M-44 was known in antiquity: Aratus called it the "little mist" as far back as 260 B.C., and Hipparchus referred to the cluster as a "little cloud" in 130 B.C. Aratus and Pliny used M-44 as a weather indicator: if the nebulous patch was invisible, the Greek astronomers forecast the imminent approach of a violent storm. More recently commentators have called M-44 "Praesepe," or the "Beehive," and this last name stuck, since many people agreed that when viewed with opera glasses the cluster's stars looked like a swarm of bees buzzing about their hive.

In 7x50 binoculars the Beehive's nebulous appearance vanishes and a myriad of stars becomes visible. The cluster contains fifty stars in an area measuring 95′ (nearly three Moon diameters) across. The brightest is Epsilon (ε) Cancri, a 6th-magnitude A-type star. Twelve more of its members — eight type-A stars and four K-type giants — shine brighter than magnitude 7.5. (As a standard of comparison, at M-44's distance of 160 parsecs the Sun would dimly glow at magnitude 10.9.)

BEST VISIBLE DURING
SPRING

These bright stars make M-44 a pretty area of sky to view with a small telescope. The cluster is so scattered you must use low-power eyepieces to see all of it at once. The most noticeable feature of M-44 in a 3-inch or 4-inch telescope is a trapezoid formed by magnitude 6.5 stars at the heart of the Beehive. Several arcminutes west is a bright triangle of stars, and some 20′ north lies a bright double star. Viewed with 10-inch or 12.5-inch telescopes, M-44 begins to lose its clustered appearance — these scopes have focal lengths long enough that even at low power their fields of view are rather restricted — and becomes a dull object. Perhaps the instrument best suited for viewing the Beehive cluster is a 4-inch rich-field telescope.

Eight degrees south and two degrees east of the Beehive lies Cancer's other open cluster, overshadowed by its famous neighbor to the north and completely ignored by many observers. **M-67** (NGC-2682) provides a marvelous contrast to the Beehive. Packing two hundred stars in a 30′ diameter, M-67 is a rich group of very faint stars, the brightest glowing at magnitude 9.7. The group's total magnitude is 6.9, which places it below the typical limit of naked-eye visibility.

M-67 is a beautiful object. Two- and three-inch reflectors show a misty patch of light speckled with a few tiny, gleaming stars. A 6-inch telescope at 120x resolves the cluster into dozens of faint pinpoints, but an overall haze pervades the field of view. A 10-inch scope at 100x does a better job, because it resolves the group into what is clearly an open star cluster without nebulosity.

M-67 is one of the oldest open clusters known — estimated to be 3.2 billion years old. This explains why its many stars are so faint and yellow, orange, and red in hue: they are highly evolved old stars. Only NGC-6791 in Lyra and Melotte 66 in Puppis (6.3 billion years); NGC-188 in Cepheus (5 billion); NGC-2141 in Orion, NGC-2420 in Gemini, NGC-2506 in Monoceros, and NGC-7142 in Cepheus (4 billion); NGC-2243 in Canis Major (3.9 billion); and NGC-6819 in Cygnus (3.5 billion) are older. M-67 lies 800 parsecs away, five times more distant than the Beehive.

Cancer offers more than just two star clusters. Six degrees due north of the Beehive lies a tiny, faint galaxy called **NGC-2623**. This galaxy is hardly impressive unless it is viewed with a large backyard telescope: it dimly glows at magnitude 13.8 and measures a mere 30″ by 25″ across. However, NGC-2623 is an unusual target for two reasons: it lies a whopping 70 megaparsecs away and is a peculiar galaxy that — in long-exposure photographs — looks like two galaxies in collision. NGC-2623 has two long, thin streamers of material emanating from its core and it closely resembles the brighter

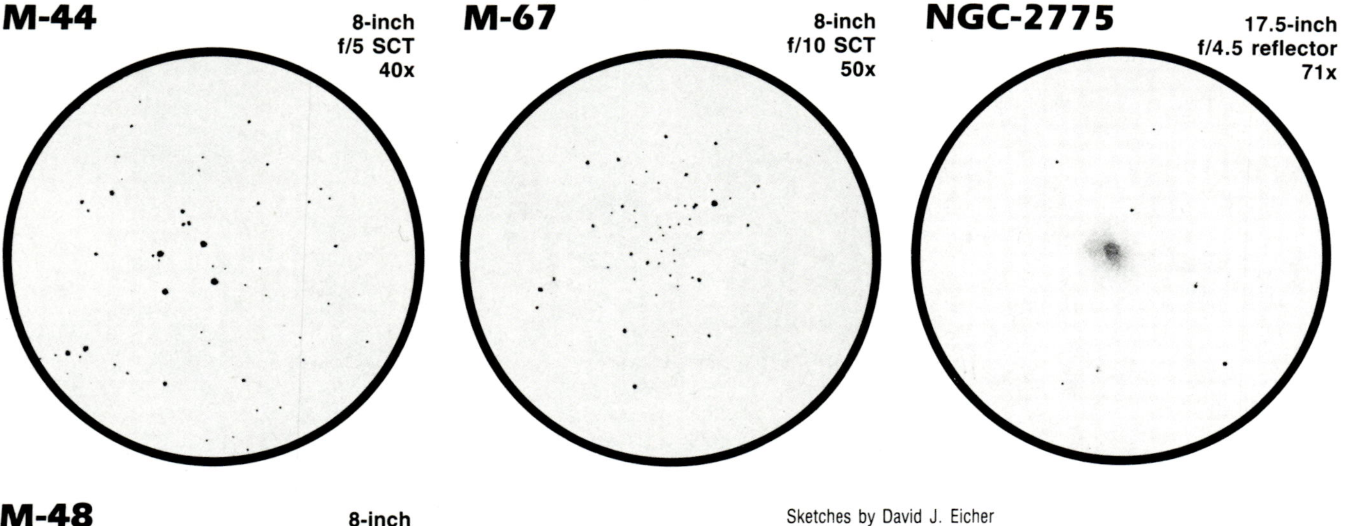

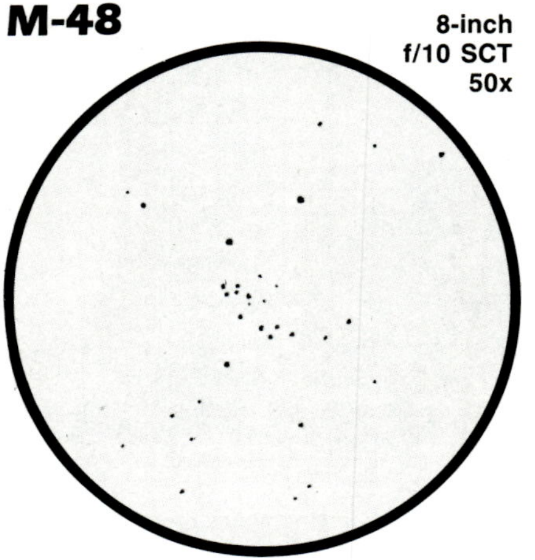

Sketches by David J. Eicher

peculiar galaxy NGC-4038/9 in Corvus, the "Ring-Tail Galaxy."

The triple star **Zeta (ζ) Cancri**, 6° east and slightly south of the Beehive, provides a challenge to long-focal length instruments. Its three components shine at magnitudes 5.6, 6.0, and 6.2, and the corresponding separations are 0.8" and 5.7". Spotting the 6.2-magnitude star and the 5.6-6.0 pair is easy, but splitting the 5.6-6.0 pair is not unless seeing is steady and high power employed. All the stars are Sunlike in color.

Near the Cancer-Hydra border, 6° southeast of M-67, is another faint galaxy. **NGC-2775** is a 10th-magnitude spiral measuring 4.5′ by 3.5′ across; the galaxy appears as a nebulous streak of pale light in a 6-inch scope at 100x.

Hydra contains objects similar to those in Cancer, except its star clusters aren't as bright as those in Cancer. **M-48** (NGC-2548), on the extreme western side of Hydra, 4° southeast of Zeta Monocerotis, is a scaled-down version of M-44: it contains eighty stars in an area 54′ across, for a combined magnitude of 5.8. The stars aren't as bright as those in M-44, and the brightest only shines at magnitude 8.2. M-48 lacks the distinctive patterns seen in the Beehive but is nevertheless a fine field for scanning with a low-power eyepiece.

Hydra contains numerous galaxies, some of them bright when viewed with backyard telescopes. **NGC-2713** is a 12th-magnitude Sb-type spiral some 3.9′ long by 1.7′ across. It shows in an 8-inch scope as a moderately bright, oval fuzz surrounding a much brighter central glow. **NGC-2784** is a 10th-magnitude lenticular galaxy some 5.1′ by 2.3′ in extent: it appears as a football-shaped nebulosity with a bright, condensed nucleus. **NGC-2835**, a spiral measuring 6.3′ by 4.4′ and shining at magnitude 11.1 in blue light, is a fine galaxy for small telescopes because of its high surface brightness.

Farther east lie three more impressive galaxies. The 10.4-magnitude (blue light) irregular system **NGC-3109** shows as a patchy cigar-shaped blob in 8-inch scopes; **NGC-3124**, a fainter barred spiral, spans 3.2′ by 2.7′ and has a bright nucleus; and the spiral **NGC-3145** is visible in 6-inch scopes as a hazy smear of light surrounding an oval nucleus.

Object	M#	Type	R.A. (2000)	Dec.	Mag.	Size/Sep./Per.	H
Zeta (ζ) Cnc		★³	8h 12.2m	+17°39′	5.6,6.0,6.2	0.8″,5.7″	
NGC-2548	M-48	⊙	8h 13.8m	−5°48′	5.8	54″	
NGC-2610		■	8h 33.4m	−16°09′	13.6_P	37″	
NGC-2623		#	8h 38.4m	+25°45′	13.8	0.6′x 0.5′	Pec
NGC-2632	M-44	⊙	8h 40.1m	+19°59′	3.1	95′	
Epsilon (ε) Hya		★⁴	8h 46.8m	+6°25′	3.8,4.7,6.8 12.4	0.2″,2.8″ 19.2″	
NGC-2682	M-67	⊙	8h 50.4m	+11°49′	6.9	30′	
NGC-2713		§	8h 57.3m	+2°55′	11.7	3.9′x 1.7′	Sb
NGC-2775		§	9h 10.3m	+7°02′	10.3	4.5′x 3.5′	Sa
NGC-2784		§L	9h 14.1m	−24°10′	10.1	5.1′x 2.3′	S0
NGC-2835		§	9h 17.9m	−22°21′	11.1_B	6.3′x 4.4′	S pec
NGC-3109		#	10h 03.1m	−26°09′	10.4_B	14.5′x 3.5′	Ir⁺
NGC 3124		§B	10h 06.7m	−19°13′	12.4_B	3.2′x 2.7′	S(B)b⁺
NGC-3145		§	10h 10.2m	−12°26′	12.4_B	3.3′x 1.7′	Sb⁺

★³	*Triple Star*
★⁴	*Quadruple Star*
⊙	*Open Star Cluster*
■	*Planetary Nebula*
§	*Spiral Galaxy*
§B	*Barred Spiral Galaxy*
§L	*Lenticular Galaxy*
#	*Irregular/Peculiar Galaxy*

H = Hubble type for galaxies

Subscript "P" denotes photographic magnitude; subscript "B" denotes blue magnitude.

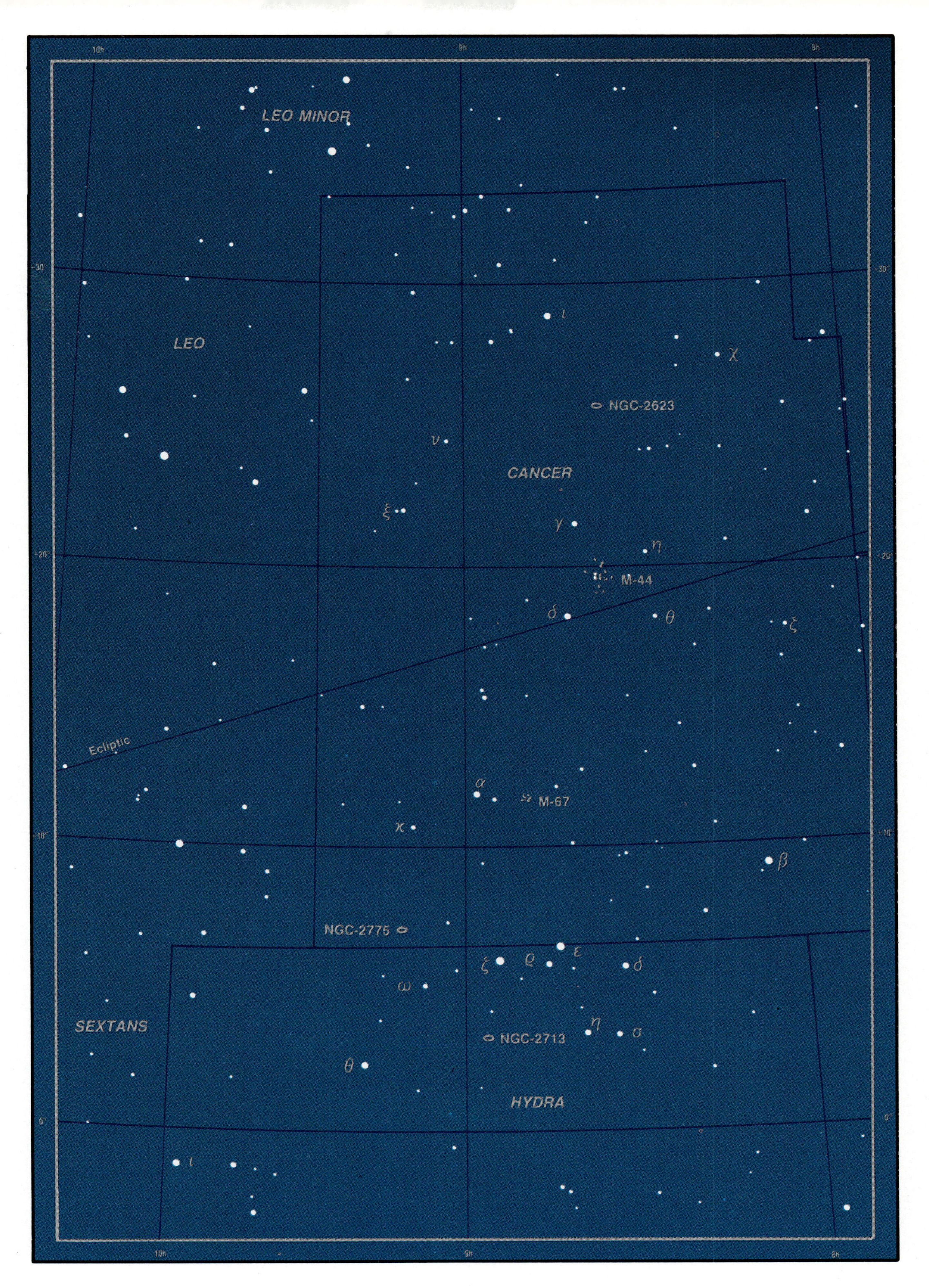
10h
9h
8h
LEO MINOR
LEO
CANCER
NGC-2623
M-44
Ecliptic
M-67
NGC-2775
SEXTANS
NGC-2713
HYDRA
30°
20°
10°
0°

Left: M-51, the Whirlpool Galaxy, is one of the finest spirals in the sky. When observed with a 4-inch or larger telescope M-51 appears as two distinctly separate patches of light in contact. Photo by Ben Mayer. Below left: NGC-4656 (left) and NGC-4631 are two distorted galaxies lying within the same low-power field. Photo by Jack Marling. Below: M-106 is one of the brightest spiral galaxies in the northern sky. Photo by Jack Marling. Below center: M-64, the Blackeye Galaxy, is well known for its easily observed dust patch. Photo by Preston Justis. Bottom: M-63 shows a tiny nucleus surrounded by knotty tightly-wound spiral arms. Photo by Jack Newton.

Canes Venatici

CVn
Canum Venaticorum

Keith Ward

Canes Venatici, the Hunting Dogs, consists of three faint naked-eye stars. It is inconspicuously tucked between the familiar constellation Ursa Major and the great cloud of galaxies spread across Virgo and Coma Berenices. In fact, Canes Venatici is richest in galaxies, most of which belong to the Virgo supercluster.

Canes Venatici's finest object — indeed one of the best galaxies in the sky — is **Messier 51** (NGC-5194). Just to the north of it is **NGC-5195**, a small irregular galaxy currently interacting with the larger spiral. Some 3½° southwest of Eta (η) Ursae Majoris, the end star in the Big Dipper's handle, M-51 is a large face-on spiral known as the Whirlpool Galaxy due to its graceful form. M-51 shines at magnitude 8.7 and covers some 10.0′ x 5.5′ of sky, making it easy to spot as a fuzzy patch with large binoculars or small telescopes. A 2-inch refractor reveals a 5′ diameter roundish patch of weak light. Large telescopes show considerable detail: a 12-inch scope reveals the subtle spiral structure across the galaxy's face, a dim star seen projected on M-51's surface, dark mottling across both galaxies, and a tenuous bridge of material arcing from the spiral over to NGC-5195.

The M-51 system lies only about 10 megaparsecs away, which accounts for its great size and brightness. For small telescopes, it is one of the finest examples of two galaxies interacting, but not the only such object in Canes Venatici.

The galaxies **NGC-4485** and **NGC-4490** also comprise an interacting system. Although visible in backyard instruments, they are much fainter than the M-51/NGC-5195 pair, so you'll need at least a 6-inch scope to find them. NGC-4490 is a 10th magnitude spiral measuring 5.0′ x 2.0′ — almost exactly half the size of M-51. NGC-4485, some 3′ to the north, glows at only magnitude 12.5 and spans a mere 1.3′ x 0.7′ of sky. This strange galaxy is variously classed as either a peculiar elliptical or an irregular system. On a dark night, a 6-inch scope will show NGC-4490 as a typically oval galaxy with a bright center; NGC-4485 appears as a tiny, fuzzy "star" with a bright center.

Almost due east of the NGC-4485/NGC-4490 pair are two bright Messier galaxies offering lots of detail for backyard telescopes. **M-94** (NGC-4736) is a bright, compact, tightly-wound spiral — unmistakable as a galaxy in any telescope. It measures 5.0′ x 3.5′ across and shines at magnitude 8.9. Upon looking at M-94 with a telescope larger than 5-inches' aperture, you should notice one obvious characteristic: the galaxy's nucleus is a brilliant, condensed pinpoint which seemingly shines right through the surrounding nebulosity. Telescopes in the 16-inch and larger range show a bright ring of nebulosity centered on the intense core, and dusty mottling inside the outer halo of light.

Continuing eastward from M-94, we come to the fine spiral **M-63** (NGC-5055). This is a multiple-arm spiral covering some 9.0′ x 4.0′ of sky and glowing with the light of a magnitude 9.8 star. Recognizable as a galaxy in telescopes as small as 2-inches' aperture, M-63 shows no detail unless viewed with medium or larger backyard telescopes. A 6-inch scope at high power reveals a bright core surrounded by a smooth envelope of nebulosity. A 10-inch scope exposes an oval shaped core, which lies in the midst of a mottled haze — the light from M-63's spiral arms. M-63 lies roughly 10 megaparsecs away — the same distance as M-51.

Another large, bright Messier galaxy lies in the northwest corner of Canes Venatici. **M-106** (NGC-4258) is a ninth-magnitude object measuring a whopping 19.5′ x 6.5′; despite being spread over such a large area, its surface brightness is remarkably high. Even when viewed with a 2-inch scope, M-106 is unmistakably nebular and *looks* like a galaxy. A 6-inch scope under dark skies reveals the nuclear region as a bright oval, with mottling around its periphery and a very faint outer halo of nebulosity. A 17.5-inch telescope easily shows the galaxy's inclined spiral structure and dusty patches in and around the nucleus.

Positioned halfway between M-106 and the interacting pair NGC-4485/NGC-4490 is an odd, box-like object designated **NGC-4449**. Aim a 6-inch or larger telescope at this object and you'll be startled: you'll see a more-or-less hard-edged rectangle of light! This 10th magnitude patch, 4.2′ x 3.0′ in diameter, is an unusually-shaped irregular galaxy. At high power with a 17.5-inch telescope,

BEST VISIBLE DURING
SPRING

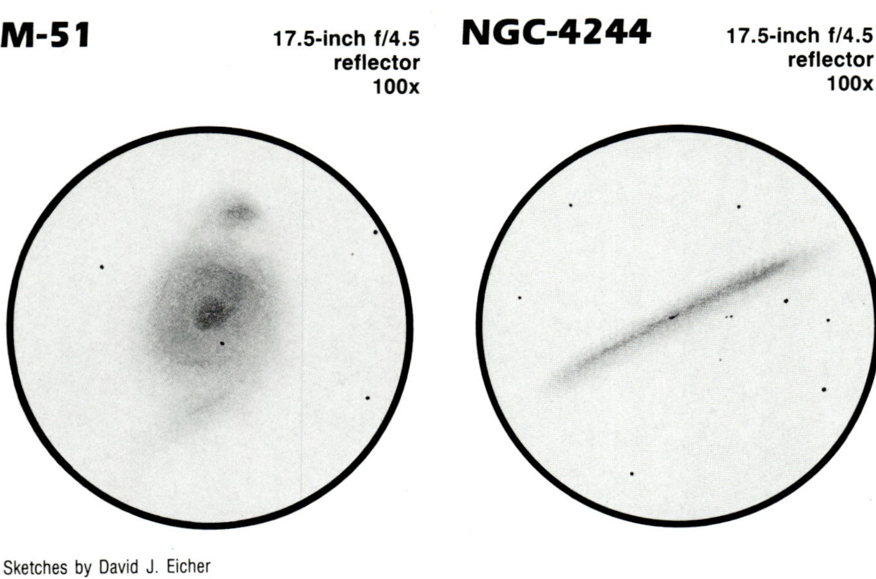

Sketches by David J. Eicher

NGC-4449 appears as a mottled rectangle with a bright center and knots of material at one edge. These knots may be bursts of recent star formation. In any case, NGC-4449 makes for a peculiar change of pace from the Messier galaxies in Canes Venatici.

Three galaxies lie in the extreme western portion of the constellation. **NGC-4151** is an 11th magnitude barred spiral galaxy with a bright nucleus. It measures 2.6′ x 1.5′ in extent and can be glimpsed with a 3-inch instrument under dark skies. Much more impressive is **NGC-4244**, an edge-on spiral brighter than 11th magnitude and measuring 13.0′ x 1.0′ across. Of the sky's bright edge-on spirals, NGC-4244 almost compares to nearby NGC-4565 in Coma Berenices, the sky's best edge-on "needle" galaxy. A 4-inch telescope shows a slender streak of greyish-green light spanning ¼ the length of the Moon. No detail is visible with larger backyard instruments, but the apparent brightness of the galaxy grows rapidly with more aperture. **NGC-4214**, south and slightly west of NGC-4244, is a magnitude 10.5 irregular galaxy. It covers 7.0′ x 4.5′ and appears as a homogeneous patch of nebulosity in a 6-inch telescope.

The southwestern edge of Canes Venatici holds three other notable galaxies. **NGC-4395** is an 11th magnitude spiral with a majestic three-armed spiral pattern. Its large, nearly circular shape spanning 10.0′ x 8.0′ gives it a fairly low surface brightness, but its bright core is visible in any telescope. Southeast of NGC-4395 are two very unusual galaxies lying within the same low-power telescopic field. These are **NGC-4631**, an edge-on spiral galaxy, and **NGC-4656**, a bizarre irregular system. NGC-4631 glows at magnitude 9.7 in an area measuring 12.5′ x 1.2′. Though it is believed to be a normal Sc-type galaxy, NGC-4631's edge-on perspective shows us immense amounts of dust and knotty clumps of material on the periphery of its disk. The largest backyard telescopes show little more than mottling, but they do show a tiny fuzzball just off NGC-4631 — this is NGC-4627, a tiny satellite galaxy. NGC-4656, a huge galaxy covering 19.5′ x 2.0′ and glowing at 11th magnitude, resembles M-106 in the smallest instruments. An 8-inch telescope shows mottling and uneven illumination across the face of this object, and also reveals that one end is "hooked." This high distortion is unexplained at the moment.

Two small galaxies in the central part of the constellation are **NGC-5005** and **NGC-5033**. NGC-5005 is an 11th magnitude spiral covering 4.7′ x 1.6′ of sky. NGC-5033 shines at the same magnitude and is the same type of galaxy, but is larger by a factor of two.

One of spring's finest double stars for small telescopes is **Cor Caroli**, also known as Alpha (*α*) Canum Venaticorum, 12 Canum Venaticorum, and Σ1692. (Cor Caroli means "Heart of Charles," named in honor of the "martyr" King Charles I of England and his son Charles II, who was restored to the throne.) This fine pair is composed of magnitude 3.0 and 5.5 stars separated by 19.6″ in p.a. 228° (it is relatively fixed). The stars appear bluish-white and white, and are easily resolved at moderate powers in virtually all telescopes.

Although galaxies are scattered all over Canes Venatici, there is at least one spectacular object located closer to home — globular cluster **M-3** (NGC-5272). This magnificent ball of perhaps a million stars lies in the southeastern part of Canes Venatici, practically on Bootes' border. Binoculars show M-3 as a hazy spot, unlike the bright stars that surround it. A 3-inch telescope under good conditions will show a hazy disk of mottled light. Larger telescopes show individual stars at the edges of the cluster, and a 12-inch instrument under black skies will resolve M-3 into its core.

Object	M#	Type	R.A. (2000)	Dec.	Mag.	Size/Sep./Per.	H
NGC-4151		§B	12h 10.5m	+39°24′	11.2	2.5′x1.6′	SB
NGC-4212		#	12h 15.6m	+36°19′	10.5	7.0′x4.5′	I
NGC-4244		§	12h 17.5m	+37°48′	10.7	13.0′x1.0′	Sb
NGC-4258	M-106	§	12h 18.9m	+47°18′	9.0	19.5′x6.5′	Sb
NGC-4395		§	12h 25.8m	+33°32′	11.0	10.0′x8.0′	S
NGC-4449		#	12h 28.2m	+44°05′	10.5	4.2′x3.0′	I
NGC-4485		#	12h 30.7m	+41°41′	12.5	1.3′x0.7′	I
NGC-4490		§	12h 30.7m	+41°38′	10.1	5.0′x2.0′	Sc
NGC-4631		§	12h 42.2m	+32°33′	9.7	12.5′x1.2′	Sc?
NGC-4656		#	12h 44.0m	+32°10′	11.0	19.5′x2.0′	Ipec
NGC-4736	M-94	§	12h 50.9m	+41°07′	8.9	5.0′x3.5′	Sb
Cor Caroli (*α*)		★^{2}	12h 56.2m	+38°19′	3.0,5.5	19.6″	
NGC-5005		§	13h 10.8m	+37°03′	10.8	4.7′x1.6′	Sb
NGC-5033		§	13h 13.5m	+36°35′	11.0	8.0′x4.0′	Sb
NGC-5055	M-63	§	13h 15.8m	+42°01′	9.8	9.0′x4.0′	Sb
NGC-5194	M-51	§	13h 29.9m	+47°12′	8.7	10.0′x5.5′	Sc
NGC-5195		#	13h 30.0m	+47°16′	11.0	2.0′x1.5′	I?
NGC-5272	M-3	●	13h 42.2m	+28°23′	6.0	18′	

H = Hubble classification type for galaxies

★^{2}	*Double Star*
●	*Globular Star Cluster*
§	*Spiral Galaxy*
§B	*Barred Spiral Galaxy*
#	*Peculiar/Irregular Galaxy*

14h
13h
12h
+60°
+50°
+40°
+30°
+20°
M-108
M-97
M-101
URSA MAJOR
M-109
NGC-5195
M-51
M-106
Y
NGC-4449
NGC-4485
M-63
M-94
β
NGC-4490
NGC-4151
α
NGC-4244
NGC-5005
NGC-5033
NGC-4214
NGC-4395
NGC-4631
NGC-4656
Coma
Star
Cluster
LEO
M-3
North
Galactic
Pole
NGC-4565
BOOTES
COMA BERENICES
M-64
M-53

Left: The great cluster M-41, lying some 4° south of Sirius, spans a Moon's diameter and shines at fifth magnitude. It is one of the sky's grandest open clusters for binoculars and small telescopes. Photo by Tom Dessert. Top: The bright, compact cluster NGC-2362 surrounds the star Tau (τ) Canis Majoris. In the same field — to the north — is the eclipsing binary UW CMa. Lick Observatory photo. Above: The rich open cluster NGC-2360 contains many stars of 9th magnitude and fainter. Lick Observatory photo.

Canis Major

CMa
Canis Majoris

Keith Ward

The brilliant constellation **Canis Major**, the Large Dog, is one of the most easily-recognized winter groups of stars. It is renowned for containing Sirius, the brightest star in the sky. Even so, Canis Major doesn't contain much variety when it comes to deep-sky objects. But those it does contain are worthy of slow, careful observing: a number of bright and impressive open clusters, a couple of complex nebulosities, a difficult planetary nebula, and two galaxies heavily obscured by the Milky Way's thick blanket of dust.

Sirius (Alpha [α] Canis Majoris) is not one of the Galaxy's most luminous stars, but appears as the brightest because it is the fifth closest at 2.7 parsecs. Sirius is an A1-type main sequence star. It shines with the light of 23 Suns, measures 1.8 solar diameters across, and contains nearly 2½ times the Sun's mass. Its apparent magnitude is −1.4, so only Venus, Jupiter, Saturn, and Mars can outshine it in the sky. As you stand under the stars and gaze up at Sirius, realize that it is over half a million times farther away than our Sun.

Sirius is also known as the Dog Star, and gains much attention from double star observers because of its faint, elusive companion. Sirius B, the "Pup," was suspected as early as 1834 by the German astronomer F.W. Bessel and discovered visually by Alvan G. Clark in 1862. Sirius B is a prime example of a white dwarf star — a tiny, incredibly dense object only about 30,000 kilometers across. A cubic inch of this star weighs over 2½ tons! This remarkable state results from the star running out of hydrogen fuel to burn, collapsing upon itself, and shining only because its temperature remains high. Such stars are "dead" and will eventually become dark as they cool off.

BEST VISIBLE DURING
WINTER

Because it is so much fainter than Sirius, the Pup is difficult to observe in backyard telescopes. It shines at magnitude 8.7, but at present lies only 9″ away from Sirius in p.a. 72°. (Its period is about 50 years; it will close to 3½″ during the 1990s.) The glare from Sirius is usually overpowering, but you may spot the companion by moving Sirius slightly out of your field of view or placing an occulting bar over the bright star's image in the eyepiece (try using a paper clip painted flat black). If you spot the Pup, you'll be among a relatively select group of backyard astronomers.

Of Canis Major's many open clusters, **Messier 41** (NGC-2287) easily stands out as the finest. It lies about 4° south of Sirius and is easily found in 7x50 binoculars or finder telescopes. M-41 measures 32′ across and shines with the light of a fifth magnitude star.

Through the telescope, M-41 fills a low-power field with shimmering stars. At higher powers, the cluster appearance becomes lost and individual star colors are more easily detectable. A 3-inch scope shows 40 or 50 stars, while an 8-inch or bigger instrument reveals a multitude of additional fainter stars painted across the background.

M-41 holds some 100 stars, the brightest being seventh-magnitude objects. The very brightest are G- and K-type giants, followed by many bright blue B-type giants with high intrinsic luminosities. M-41 lies about 750 parsecs away and has a density of 1.1 stars per cubic parsec.

Two fine open clusters lie in the Dog's hindquarters — **NGC-2354** and **NGC-2362**. The former is nearly as large as M-41, but is composed of 60 faint stars in a loose arrangement. Again, use low powers when gazing at this object. NGC-2362 is a small group of 40 stars, which surrounds the bright star Tau (τ) Canis Majoris. It measures only 6′ across, but is unusually attractive. With a finderscope, it appears as an unresolved haze blanketing fourth-magnitude Tau. A 2-inch or larger scope resolves the cluster, showing two dozen stars of seventh-magnitude and fainter. Larger telescopes show all 40 stars and make NGC-2362 satisfying viewing.

NGC-2362 is one of the youngest known open clusters, probably less than one million years old. On the H-R diagram, the cluster's stars are displaced to the right of the normal main sequence, which suggests they have not finished contracting. (Other very young clusters are NGC-2264 in Monoceros and the Double Cluster in Perseus.) NGC-2362 lies about 1.4 kiloparsecs away, with its brightest stars O- and B-type giants. As viewed from this group, our Sun shines at a mere magnitude of 15.5. Tau itself may be a member also, but if it is, it's one of the most luminous supergiants known.

Assumed to be at the cluster's distance, Tau would outshine the Sun by 50,000 times!

Undoubtedly one of the most massive and luminous stars known is the odd variable **UW Canis Majoris**, which lies only 24′ north of NGC-2362. UW CMa is an eclipsing binary consisting of two stars that orbit each other in just 4.4 days. The distance between the stars, both of which are flattened into ellipsoids by the mutual tug of gravity, is 17 million miles — around 0.2 astronomical unit. The distance to this pair is about 1.0 kiloparsec, making the system's luminosity 16,000 times that of the Sun. With a short period and small magnitude range (4.7 to 5.0), detecting light variations in this system is difficult with backyard telescopes.

Canis Major contains three nebulae observable in small telescopes. **NGC-2359**, along with its fainter partner **IC 468**, presents a dim and challenging region of emission nebulosity. Both nebulae are illuminated by a magnitude 10.4 star. Shaped like a twisted comet and measuring 10′ x 5′, NGC-2359 appears as a faint streamer of pale grey light in an 8-inch scope. A 16-inch telescope shows irregular lighter and darker patches across its surface, but not much detail due to a low surface brightness. IC 468, adjacent to 2359, measures 20′ across and on dark nights shows up as a pale circle of weak light.

Difficult to find but worth pursuing is the tiny planetary nebula **IC 2165**, which is located in the northwest corner of the constellation. At magnitude 12.5, this nebula is bright enough to see with a 4- or 5-inch telescope, but measures only 8″ across. At low powers, it appears nearly stellar and is hard to pick out from the Milky Way's rich background. Once you've determined IC 2165's approximate position, crank up the magnification and look for a slightly bluish, fuzzy "star."

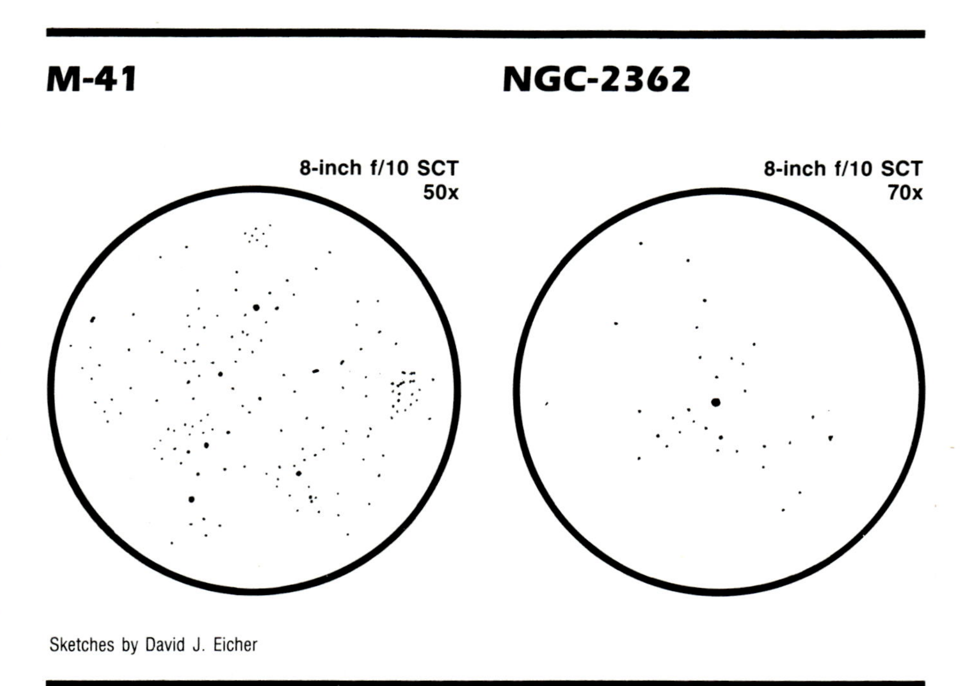

Sketches by David J. Eicher

Canis Major contains several more bright open clusters, three of which lie in the northeastern reaches near M-46 and M-47 in Puppis. These are **NGC-2360**, **NGC-2374**, and **NGC-2383**. NGC-2360 is a rich group of 50 stars up to ninth-magnitude packed into a 10′ diameter. Found about 3½° east of Gamma (γ) Canis Majoris, it is one cluster that handles higher magnifications quite well. Nearby and equally commanding is NGC-2374, which measures 15′ across and contains many widely-scattered bright stars. Lying to the south is NGC-2385 — a faint yet very rich group measuring only 2′ in diameter. On crisp, clear nights, use high magnifications and a 6-inch or larger scope to partially resolve this tight bunch of 50 stars, which at 12th magnitude and fainter appear as tiny specks of light.

Two final clusters are **NGC-2204** and **NGC-2243**. NGC-2204 is a large, unimpressive collection of 20 indistinct stars covering 10′ of sky. If it were alone in the middle of nowhere, this cluster would stand out better than it actually does — surrounded by brighter cousins in the rich starfields of Canis Major. NGC-2243 is a small, fairly rich group; its 50 constituent stars shine rather dimly in a 4′ circle.

Observing in the winter Milky Way, you wouldn't expect to find galaxies because the region is dominated by obscuring dust. Yet **NGC-2207** and **NGC-2217** — two 12th magnitude objects — are routinely visible in 6-inch scopes. NGC-2207 is a complex system measuring 2.5′ x 1.5′ across, and is perhaps a double galaxy or an interacting pair. NGC-2217 is a run-of-the-mill type-SBa barred spiral covering an area 4.0′ x 3.0′. On dark, crisp nights an 8-inch telescope shows it as a dim, even halo of light surrounding a sharply brighter, highly-condensed core.

Object	M#	Type	R.A. (2000)	Dec.	Mag.	Size/Sep./Per.	N★	Mag.★
NGC-2204		⊙	6h 15.7m	−18°36′	9.6	10′	30	11....
NGC-2207		§	6h 16.4m	−21°22′	12.3	2.5′x1.5′		
NGC-2217		§B	6h 20.8m	−27°15′	12.0	4.0′x3.0′		
IC 2165		■	6h 21.9m	−12°59′	12.5	8″		
NGC-2243		⊙	6h 29.5m	−31°17′	—	4′	50	
Sirius (α CMa)		★²	6h 45.2m	−16°42′	−1.4, 8.7	9″		
NGC-2278	M-41	⊙	6h 47.0m	−20°45′	5.0	30′	55	7....
NGC-2354		⊙	7h 14.2m	−25°43′	—	25′	60	
IC 468		□E	7h 17.5m	−13°10′	—	20′		10.4
NGC-2360		⊙	7h 17.6m	−15°38′	9.1	10′	50	9....
NGC-2359		□E	7h 17.7m	−13°12′	—	10′x5′		10.4
UW		EV	7h 18.7m	−24°34′	4.7↔5.0	4.4d		
NGC-2362		⊙	7h 18.7m	−24°58′	3.9	3′	240	4....
NGC 2374		⊙	7h 24.1m	−13°15′	—	15′		
NGC-2383		⊙	7h 24.8m	−20°56′	8.8	2′	50	12....

N★ = number of stars, Mag.★ = magnitude range of cluster or magnitude of central star, (...) indicates many fainter.

★²	*Double Star*
EV	*Eclipsing Variable*
⊙	*Open Star Cluster*
■	*Planetary Nebula*
□E	*Emission Nebula*
§	*Spiral Galaxy*
§B	*Barred Spiral Galaxy*

8h
7h
6h
CANIS MINOR
Galactic Equator
0°
MONOCEROS
-10°
10°
NGC-2374
IC 468
NGC-2359
M-47
IC 2165
NGC-2360
M-46
γ
α
Sirius
ι
β
PUPPIS
NGC-2204
-20°
NGC-2383
M-41
20°
NGC-2207
o^2
o^1
UW
M-93
NGC-2362
NGC-2354
δ
NGC-2217
η
ε
ζ
-30°
30°
NGC-2243
NGC-2451
NGC-2477
-40°
40°
8h
7h
6h

The globular cluster M-30 is bright, but small and difficult to resolve in backyard telescopes. Photo by Bill Iburg.

Capricornus

Cap
Capricorni

Keith Ward

Although it lies adjacent to the rich Milky Way in Sagittarius, the constellation **Capricornus** the Sea Goat is nearly devoid of nebulous deep-sky objects. Instead, it contains one bright globular cluster, two faint galaxies, and a great many double and variable stars.

Lying in the southeastern part of Capricornus is the fine globular star cluster **Messier 30** (NGC-7099). Measuring some 11′ across and glowing at magnitude 7.5, M-30 is one of the finest globulars in the fall sky and is easily identifiable in binoculars. It lies at a distance of 8.2 kiloparsecs, making it fairly difficult to resolve into stars. Nonetheless, a good 8 or 10-inch telescope operating under a transparent sky clearly shows M-30's edges partially resolved, while the cluster's core remains a white disk.

In the same low-power field, 25′ east and just slightly south of M-30, is the bright double star **41 Capricorni**. This double is not easy to pick out due to the great difference in magnitudes between its two components: the primary shines at magnitude 5.3, while the secondary — 5.5″ away in p.a. 205° (1954) — is a dim little star of magnitude 11.5. Try splitting this pair after you observe M-30 by centering the field of view on 41 Cap, inserting a high-power eyepiece, and — using averted vision — glancing about the field in search of the diminutive secondary.

In the opposite side of Capricornus — the northwestern part — lies the wide bright double star **Alpha**^{1,2} ($\alpha^{1,2}$) **Capricorni**. This pair of magnitude 3.6 (α^1) and 4.2 (α^2) stars, separated by over 6′, is best viewed with the naked eye or binoculars. Both stars are light yellow in color, α^2 being a spectral type G9 giant and α^1 a G3 supergiant. These stars are not physically associated, but appear "double" merely by chance alignment.

Each component of $\alpha^{1,2}$ Cap has a dimmer, closer companion as well. Alpha^2, the bright giant, has an 11th magnitude partner some 6.6″ away in p.a. 172° (1959). Alpha^1, the bright supergiant, has two physical companion stars. A ninth-magnitude star lies 45.4″ away at p.a. 221° (1932), while a 13th magnitude star lies 44.3″ distant in p.a. 182° (1960). After enjoying the wide pair of α^1 and α^2 Cap, "zoom in" on each star for some high-power observing and try to spot their smaller companions.

About 2½° south of Alpha^{1,2} Cap is another double star, **Beta** (β) **Capricorni**. Beta Cap contains stars shining at magnitudes 3.4 and 6.2; they are easily split with any telescope because of the 3½′ gap between them. And even a small telescope shows off their colors — white for the dazzling primary and blue for the secondary.

Two variable stars reside several degrees west of Beta Cap — **R Capricorni** and **TW Capricorni**. R Cap is a typical long-period variable whose range of magnitudes between 9.4 and 14.9 and period of 345 days combine to make observing it an ideal long-term project. You can follow R Cap with binoculars through much of its cycle, but at minimum it fades below the light-gathering ability of most backyard telescopes. TW Cap is a bright Cepheid variable with a smaller range of magnitudes — 9.7 to 10.5 — and a period of 28.6 days. This is a fine binocular variable, but you'll have to become adept at comparing measurements in tenths of a magnitude to check its variation. The range of less than one magnitude isn't great enough to make the star's variation obvious, so check it against nearby stars of similar brightness. Over the course of a month, you'll see it fade and return to normal brightness again.

Three degrees south of Beta Cap is the unusual variable star **RW Capricorni**, a bright eclipsing binary much like the more renowned Beta Lyrae. RW Cap varies between magnitudes 9.8 and 11.0 over 3.39 days, so you'll need two or three successive clear nights in order to view it. The range of variance — a little over a magnitude — suggests that you should observe with binoculars.

Southeast of Beta and RW Cap and near the ecliptic is a clump of two bright double stars and a faint galaxy. Unlike Alpha^{1,2} and Beta Cap, these double stars — **Pi** (π) **Capricorni** and **Omicron** (o) **Capricorni** — are best viewed with 100x to 150x. Omicron is the easiest to resolve with its magnitude 6.1 and 6.6 components separated by 21.9″ in p.a. 239° (1955). Both stars are bluish-white and stand out easily at low or medium powers. Pi's stars, on the other hand, are separated by only 3.2″ in p.a. 148° (1955). These magnitude 5.3 and 8.9 components differ enough in brightness to make splitting difficult unless seeing is steady. Lying adjacent to these stars is **NGC-6712**, a 14th magnitude barred

BEST VISIBLE DURING
AUTUMN

M-30

8-inch f/10 SCT
50x

Sketch by David J. Eicher

$\alpha^{1,2}$ Cap

4-inch reflector
35x

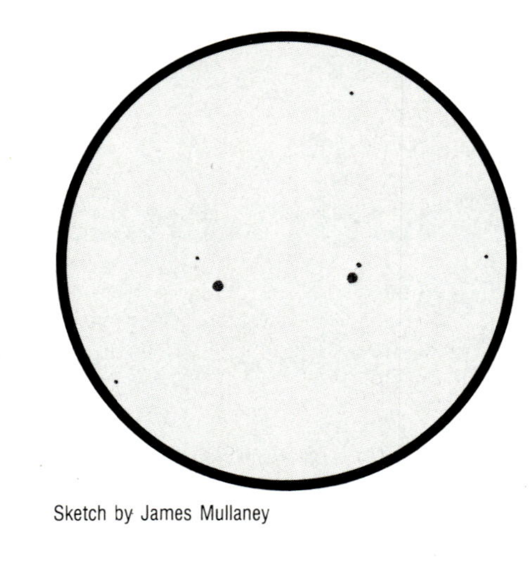

Sketch by James Mullaney

β Cap

3-inch reflector
30x

Sketch by James Mullaney

spiral galaxy measuring 1.6′ x 1.3′ across. With high magnification on a 10-inch telescope, it appears as a tiny nebulosity — a smudge of grainy light.

Lying in central Capricornus is the semi-regular variable star **RS Capricorni**, which varies between magnitudes 8.3 and 10.3 over a period spanning 345 days. Although its magnitude range isn't as dramatic as that of R Cap's, the long period provides ample time to follow the star's changes regardless of weather or other hindrances. This is another fine binocular star, but a small telescope will do just as well for comparing it with nearby stars of known constant brightness.

The eastern edge of Capricornus holds the bright double star **Delta** (δ) **Capricorni**. Appearing as a solitary star to the naked eye, modest optical aid reveals a faint companion unknown to many observers. The primary is a bluish-white magnitude 2.9 eclipsing binary around which a 12th magnitude companion orbits at a distance of nearly 2′. The primary's magnitude varies slightly over a period of less than one day. On a dark transparent night, aim your telescope at Delta Cap and insert a high-power eyepiece: you'll see the tiny dwarfish companion as a speck, nearly overpowered by the brilliant primary star.

The second brightest deep-sky object in Capricornus after M-30 is the 11th magnitude galaxy **NGC-6907**. This barred spiral measures 3.4′ x 3.0′ across and is bright enough to be viewed with a 6-inch telescope. A 10-incher shows it as a roundish patch with a bright middle, whereas large backyard telescopes reveal an evenly-illuminated, wispy halo of gray light around a bright oval nucleus.

Object	M#	Type	R.A. (2000)	Dec.	Mag.	Size/Sep./Per.	H
R		LPV	20h 10.8m	−14°16′	9.4↔14.9	345d	
TW		CV	20h 13.8m	−13°51′	9.7↔10.5	28.6d	
α		$\star^2$	20h 17.6m	−12°30′	3.6,4.2	377.7″	
RW		EV	20h 17.8m	−17°41′	9.8↔11.0	3.39d	
β		$\star^2$	20h 21.0m	−14°47′	3.4,6.2	205.3″	
NGC-6907		§B	20h 25.1m	−24°49′	11.3	3.4′x3.0′	S(B)b
NGC-6912		§B	20h 26.9m	−18°38′	14:	1.6′x1.3′	SBc
π		$\star^2$	20h 27.3m	−18°13′	5.3,8.9	3.2″	
o		$\star^2$	20h 29.9m	−18°35′	6.1,6.6	21.9″	
RS		SRV	21h 07.2m	−23°55′	8.3↔10.3	340d	
NGC-7099	M-30	●	21h 40.4m	−23°11′	7.5	11.0′	
41		$\star^2$	21h 42.0m	−23°16′	5.3,11.5	5.5″	
δ		$\star^2$	21h 47.0m	−16°08′	2.9,12.6	118.9″	

H = Hubble classification type for galaxies

$\star^2$	*Double Star*
LPV	*Long Period Variable*
SRV	*Semi-Regular Variable*
EV	*Eclipsing Variable*
CV	*Cepheid Variable*
●	*Globular Star Cluster*
§B	*Barred Spiral Galaxy*

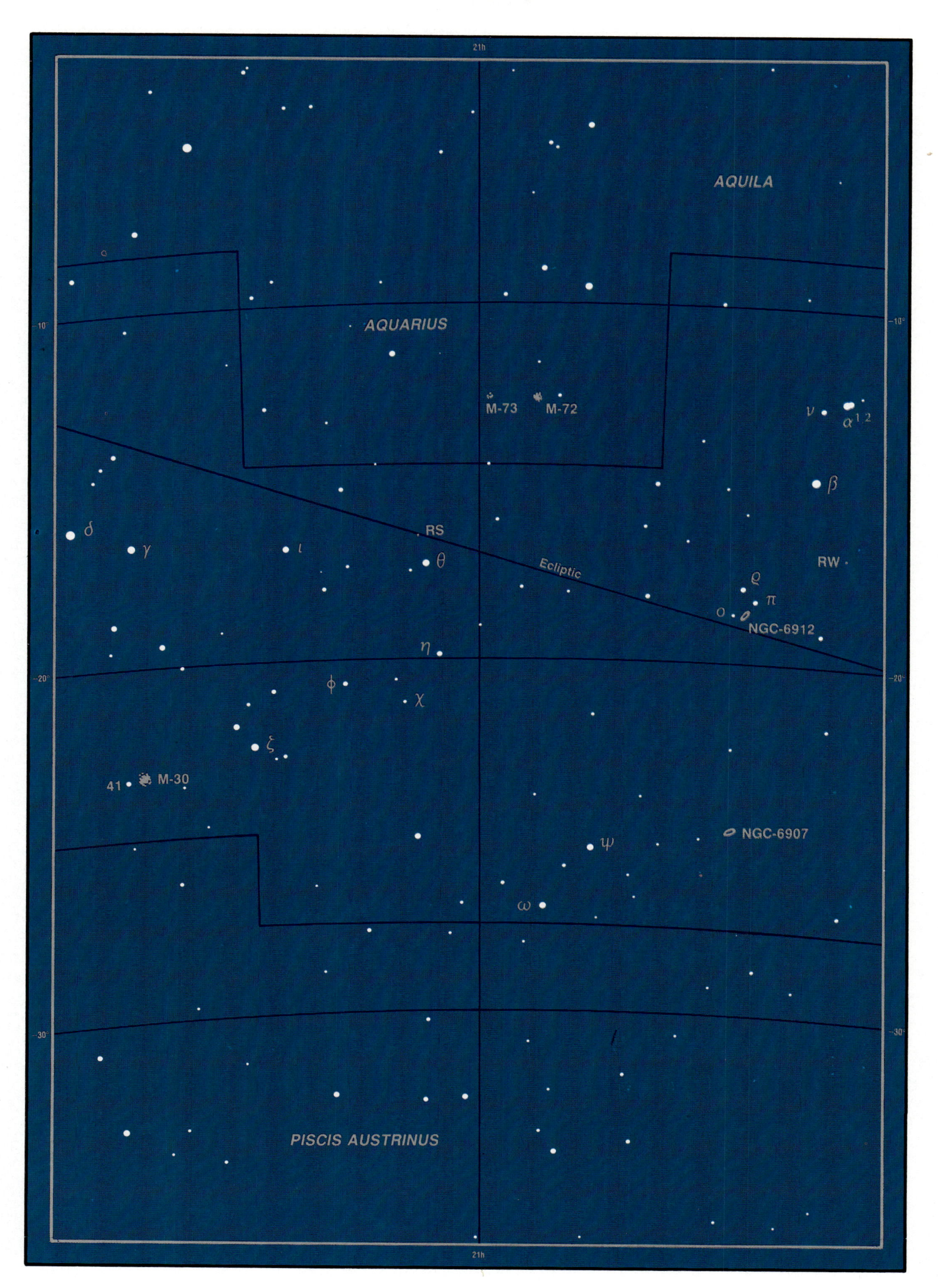

21h
AQUILA
AQUARIUS
-10°
M-73
M-72
ν
α 1 2
β
δ
γ
ι
RS
θ
Ecliptic
RW
ϱ
π
o
NGC-6912
η
-20°
ϕ
χ
ζ
41
M-30
NGC-6907
ψ
ω
-30°
PISCIS AUSTRINUS
21h

The bright, exceptionally detailed emission nebula NGC-3372 surrounds the variable star Eta Carinae. This nebula is visible as a large fuzzy patch without optical aid. Photo by Jim Barclay.

Carina
Car
Carinae

Keith Ward

Still stabilizing the old ship Argo Navis, **Carina** the Keel encompasses one of the richest areas of the southern Milky Way. Although largely invisible to most Northern Hemisphere viewers, Carina holds such a collection of deep-sky goodies that it alone warrants an observing trip to the Southern Hemisphere. Found in Carina are the huge, impressive clusters NGC-2516 and NGC-3532, the famed Eta Carinae nebula and the sky's second brightest star, Canopus.

Dazzling and unmistakable to Southern Hemisphere observers is **Canopus** (Alpha [α] Carinae), the brightest star in the sky after Sirius. Canopus lies approximately 35° south of Sirius, making it visible during the winter from latitudes as far north as the southern United States. Sirius is bright because it lies just 2.7 parsecs away, but Canopus is nearly as bright despite its distance of 60 parsecs. It shines at magnitude −.7 and is a type F0 supergiant — an intrinsically brilliant star some 30 times larger and 1,400 times more luminous than the Sun.

Another interesting object in Carina is **Eta** (η) **Carinae**, an unusual nebular variable star 500 parsecs distant with a bizarre observational history. In 1677, Edmond Halley classified Eta Car as a 4th-magnitude star. During the following century it varied erratically in brightness: in 1730 it shone at 2nd magnitude, in 1782 it was near 4th magnitude, and by 1820 it had repeated this cycle again. Eta Car then started brightening quickly, reaching 2nd magnitude in 1822 and 1st magnitude by 1827. In 1843 it was as bright as Canopus, but by 1868 it had faded out of naked-eye visibility. In 1900 the star was at 8th magnitude and has since hovered quietly around 6th or 7th.

What is going on with this star? It seems to be a type of recurring nova with a spectrum that changes dramatically during rises and subsequent falls. The unique emission features in the star's spectrum that indicate a shell of gas surrounding the star as well as ionized iron and other metals remain puzzling for astronomers. Similar features are observed in novae but Eta Carinae seems to hold a unique position, temporarily at least, among the stars in our galaxy.

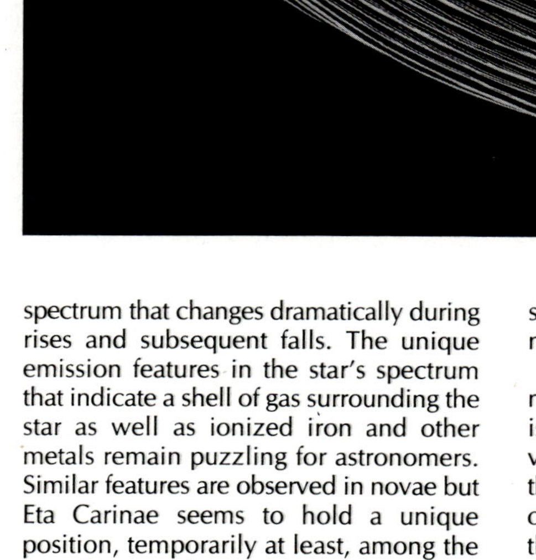

In a telescope, Eta Carinae itself isn't a terrific sight. However, surrounding the star is the bright emission nebula **NGC-3372** — sometimes called the Eta Carinae or Keyhole nebula — one of the largest and brightest deep-sky objects in the entire sky. The Eta Carinae nebula consists of four large sections of nebulosity so bright that each can be glimpsed with the naked eye. The sections are divided by a wide, opaque dust lane that provides the Keyhole nebula with its nickname. The entire object is far brighter than M-42, Orion's Great nebula, and covers a whopping 120′ of sky. It is visible as a huge, glowing, misty patch even on average nights. Binoculars trained on the nebula show bright triangle-shaped patches of nebulosity well separated by the dust lane. Since the nebula lies in the thick of the southern Milky Way, the field is strewn with hundreds of stars that are arranged in lines and pairs. A small telescope reveals an amazing amount of detail within the Eta Carinae nebula, including wisps of very faint nebulosity on the outskirts of the object and intricate jets of nebulosity projected against the dark lane.

Carina contains two other notable nebulae for small telescopes. **NGC-3581** is a small, fan-shaped nebulosity that is visible as a wisp of milky light. If you spot this nebula, look carefully around the field of view. Several detached pieces having the separate NGC designations of 3579, 3582, 3584, and 3586 lie within several arc-minutes of NGC-3581. **NGC-2867** is a bright planetary nebula that appears as an out-of-focus star in small scopes at high power. With a photographic magnitude of 9.7, it's easy to spot providing you employ medium to high power. This nebula is a mere 11″ across and has a slightly blue-green color, which may help you pick it out from the myriad field stars. The central star shines at 14th magnitude and is exceedingly difficult to spot visually.

Besides nebulae and unusual stars, Carina is loaded with bright star clusters. One of the best is **NGC-2516**, a gathering of some 80 stars of magnitude 7 and fainter over an area the diameter of the Full Moon. With a total magnitude of 3.8, NGC-2516 is quite easily visible to the naked eye and appears as a large hazy patch of unresolved light. Binoculars resolve the cluster into its constituent stars, creating a view of several dozen bright white and bluish white stars scattered across the entire field.

Also bright and large is **NGC-3114**, a cluster containing several dozen stars in

BEST VISIBLE DURING
SPRING

NGC-3372 and environs

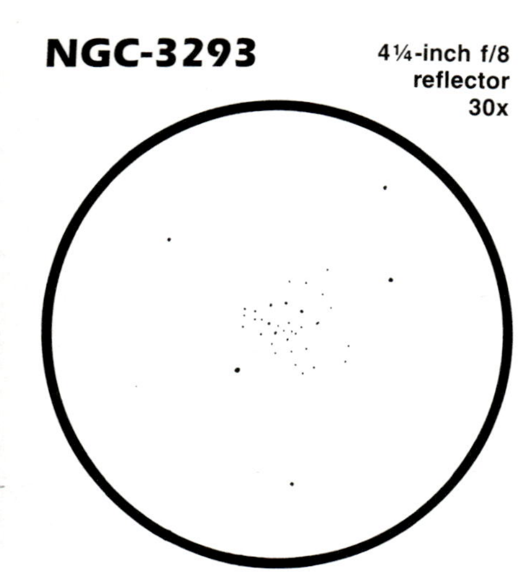

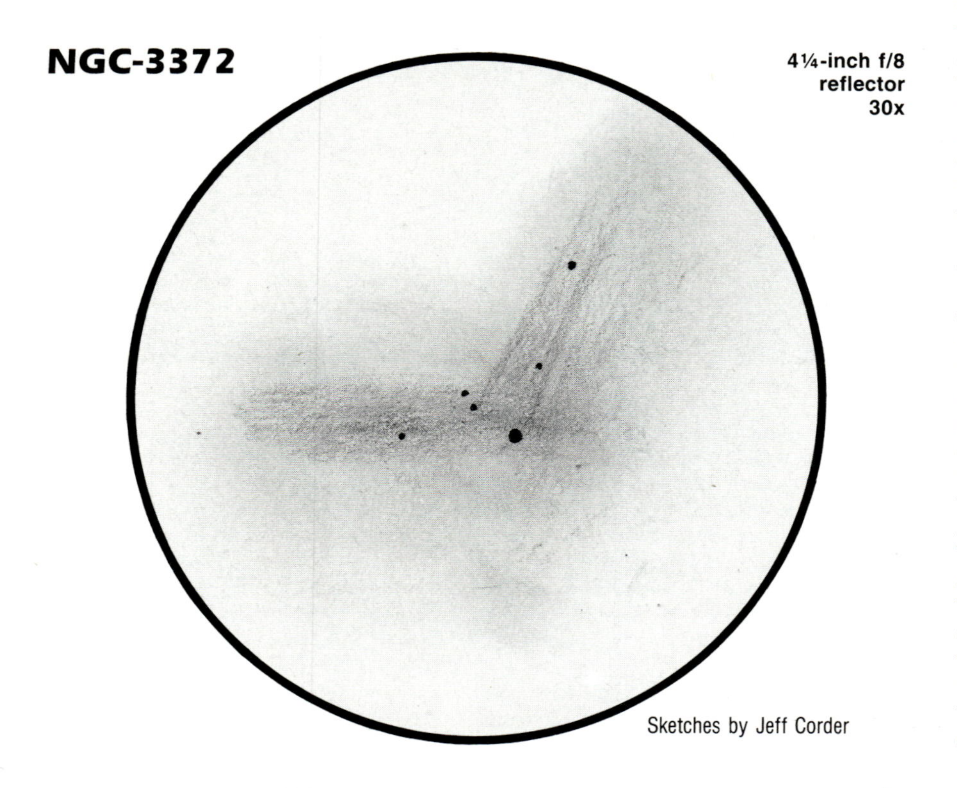

Sketches by Jeff Corder

a 35′ diameter. The group's total magnitude is 4.2 with its brightest stars shining at 7th magnitude. Compare NGC-3114 with NGC-2516 — which do you enjoy better? Which cluster has the most colorful stars? **NGC-3532** is another very bright cluster. It covers nearly a degree of sky — 55′ — and holds some 150 stars of magnitude 7 and fainter. The group's total magnitude is 3.0, making it an easy naked-eye target even on nights of poor seeing.

Carina holds four other open clusters that are ideal for small telescopes. **NGC-3293** is very small, only 6′ across, but bright; its several dozen stars give it a combined magnitude of 4.7. **IC 2581** is a cluster of 25 stars in an area spanning 8′ across. It contains a single 4th-magnitude star, which contributes greatly to its total brightness. **IC 2602**, also known as the Theta Carinae cluster, contains Theta (θ) Car and 60 other members in an area almost a degree across. Because Theta is a magnitude 2.8 star, this dazzling cluster shines at magnitude 1.9 making it visible without optics and very bright in binoculars. **IC 2714** is not so bright but very rich — it packs 100 stars within a 12′ circle.

Object	M#	Type	R.A. (2000)	Dec.	Mag.	Size/Sep./Per.
Alpha (α)		★	6h 23.9m	−52°42′	−0.7	—
NGC-2516		⊙	7h 58.3m	−60°52′	3.8	30′
NGC-2867		■	9h 21.4m	−58°19′	9.7p	11″
NGC-3114		⊙	10h 02.7m	−60°07′	4.2	35′
IC 2581		⊙	10h 27.4m	−57°38′	4.3	8′
NGC-3293		⊙	10h 35.8m	−58°14′	4.7	6′
IC 2602		⊙	10h 43.2m	−64°24′	1.9	50′
NGC-3372		□E	10h 43.8m	−59°52′	—	120′x120′
Eta (η)		IV	10h 45.1m	−59°41′	6.21v	irr.
NGC-3532		⊙	11h 06.4m	−58°40′	3.0	55′
NGC-3581		□E	11h 12.0m	−61°12′	—	—
IC 2714		⊙	11h 17.9m	−62°42′	8.2p	12′

★	*Star*
IV	*Irregular Variable*
⊙	*Open Star Cluster*
■	*Planetary Nebula*
□E	*Emission Nebula*

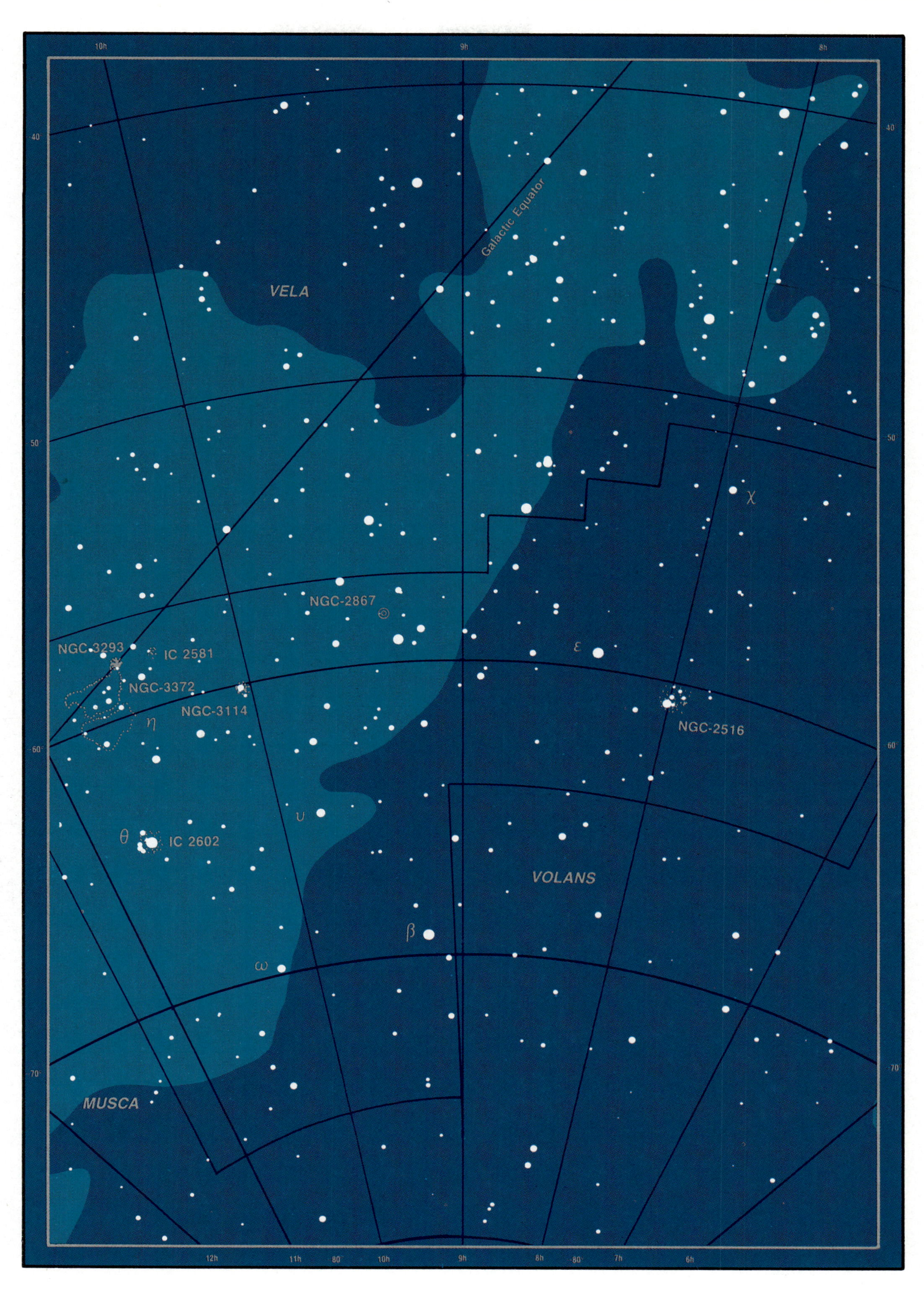

VELA
Galactic Equator
NGC-2867
NGC-3293
IC 2581
NGC-3372
NGC-3114
NGC-2516
IC 2602
VOLANS
MUSCA

Above left: One of the richest starfields in the northern Milky Way contains bright open cluster M-52 (upper left) and emission nebula NGC-7635. NGC-7635 is known as the "Bubble Nebula" because of its distinctive shape. Photo by John Kerns.

Left: Surrounding the 5th-magnitude star Phi (ϕ) Cassiopeiae is the bright, scattered cluster NGC-457. This group of eighty stars, called the "Owl Cluster," shines at magnitude 6.4. Photo by Jack Newton.

Top: Holding twenty-five stars in an area spanning 6′, the cluster M-103 has a magnitude of 7.4. It is visible in binoculars as a fuzzy patch of gray-green light some 3′ across. Photo by Jack Newton.

Above: One of the best open clusters for small telescopes is NGC-7789, a huge grouping of three hundred stars. Although none of its stars shines brighter than magnitude 10.7, the cluster shines as brightly as a magnitude 6.7 star. It is visible in a 3-inch scope as a large cloud of pale nebulosity spotted with tiny stars. Photo by Lee C. Coombs.

Cassiopeia

Cas
Cassiopeiae

Keith Ward

Smack-dab in the middle of the northern Milky Way, **Cassiopeia** the Queen is a constellation rich in galactic deep-sky wonders. Cassiopeia, a great W-shaped asterism of five bright naked eye stars, contains four bright nebulae suitable for small telescopes and nine bright examples of open star clusters. The constellation also holds three unusual galaxies, two of which are dwarf ellipticals physically connected to the M-31 system in Andromeda.

The brightest star cluster in Cassiopeia is **NGC-457**, located about 2° south-southwest of the variable star Delta (δ) Cassiopeiae. This cluster contains at least eighty stars and has a total magnitude of 6.4, which make it barely visible to the naked eye as a small patch of fuzzy light. Phi (ϕ) Cassiopeiae, a 5th-magnitude star visible to unaided eyes, appears inside this cluster but is most likely a foreground object. Another bright star, HD 7902, is a 7th-magnitude beacon immediately west of Phi. HD 7902 is probably not a cluster member either, but these two bright stars help make NGC-457 a beautiful sight in backyard telescopes.

The brightest stars in NGC-457 are arranged in distinct lines and curves, outlining the apparent form of an owl. The two bright field stars, Phi and HD 7902, represent the bird's eyes. A rectangular clump of 9th- and 10th-magnitude stars traces a body, and scattered curving rows of fainter stars outline an owl's wings, which extend roughly north-south. Because of its shape NGC-457 has acquired the nickname "the Owl Cluster."

This arrangement of stars makes NGC-457 a particularly pretty sight with just about any optical instrument. Binoculars show the cluster as a fuzzy patch of light spreading west and slightly north and south of the bright pair of stars. A 3-inch refractor at moderate power resolves the group into two dozen faint specks of white light. An 8-inch telescope shows NGC-457 as a group of several dozen stars sprinkled across a diameter of 13′. This aperture also clearly shows that the brightest cluster member, a magnitude 8.6 star, has a distinctly ruddy orange color. Although NGC-457 was not included in the Messier catalog and therefore isn't as well known as many nearby open clusters, a look at this group of stars should convince you it is one of the finest in the northern sky.

BEST VISIBLE DURING
AUTUMN

Only 1° northeast of Delta Cassiopeiae is the fine open cluster **M-103** (NGC-581). M-103 is a magnitude fainter than NGC-457 and contains twenty-five stars in an area spanning 6′. This cluster's brightest star glows dimly at magnitude 10.6, some five times fainter than the brightest member of NGC-457. Although M-103 isn't as large or as bright as the Owl Cluster, it is an impressive group of stars. A 4-inch telescope shows M-103 as a fan-shaped group containing four bright stars between 7th and 9th magnitude and one 8th-magnitude orange star. Larger telescopes show myriad fainter stars in the cluster and the surrounding starfield.

Continue in a line from Delta Cassiopeiae through M-103 and extend the line another degree to open cluster **Trumpler 1**. This cluster was named for American astronomer Robert Trumpler, who meticulously studied star clusters throughout much of the twentieth century. Trumpler 1 consists of twenty stars that are magnitude 9.6 and fainter and cover 4.5′ of sky. The cluster's total magnitude is 8.1, which makes it faint but visible in binoculars. Small telescopes trained on Trumpler 1 reveal a sprinkling of faint stars that on nights of average transparency appears as little more than an increase in density of the rich Casseiopeia Milky Way.

A more impressive cluster lies 3° southwest of the bright star Beta (β) Cassiopeiae, the westernmost star in Cassiopeia's W-shaped asterism. **NGC-7789**, one of the richest open clusters in the sky, consists of at least three hundred stars of magnitude 10.7 and fainter in an area half the size of the Full Moon. NGC-7789 can be found by sweeping southwest of Beta Cassiopeiae to Rho (ϱ) and Sigma (σ) Cassiopeiae, a pair of 5th- and 6th-magnitude stars separated by 2°. NGC-7789 lies midway between these stars.

Telescopically NGC-7789 is a treat. Binoculars show it as a gray cloud of haze of uniform brightness. A 6-inch telescope shows a cloud of nebulosity over 10′ across peppered with dozens of faint stars. A large backyard telescope in the 12- to 18-inch range shows the cluster as a nebulous disk of light with over one hundred visible stars that appear like a sparsely populated globular cluster.

About 3° northeast of Beta Cassiopeiae is the open cluster **NGC-103**. This group of thirty stars glows at magnitude 9.8 and

NGC-147

NGC-185

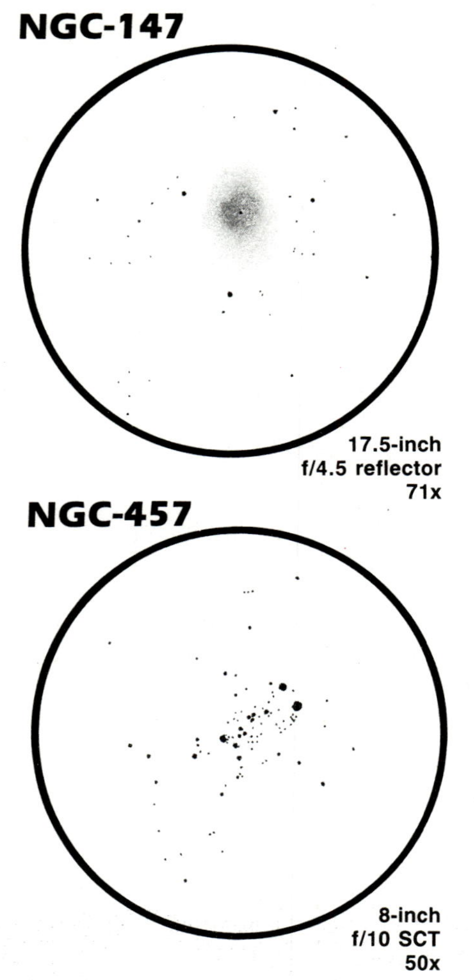

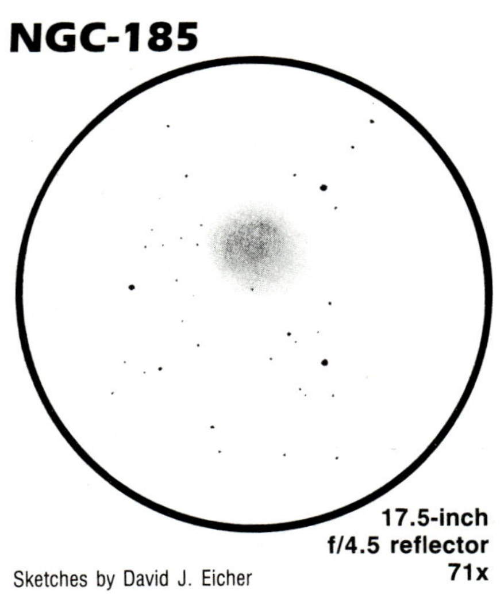

17.5-inch
f/4.5 reflector
71x

Sketches by David J. Eicher

17.5-inch
f/4.5 reflector
71x

NGC-457

8-inch
f/10 SCT
50x

covers 5′ of sky. NGC-103 is unimpressive in binoculars because its brightest stars are so faint. In a 6-inch reflector, however, the group appears as a misty cloud sprinkled with a number of 12th- and 13th-magnitude stars. One-and-a-half degrees south-southeast of NGC-103 is the more impressive cluster **NGC-129**. This sparkling array of thirty-five bright stars offers a fine contrast to NGC-103. NGC-129's diameter of 21′ and total magnitude of 6.5 make the cluster a beautiful sight in any telescope.

Two large, scattered clusters lie in the southeastern corner of Cassiopeia about 4° north of the Double Cluster in Perseus. **IC 1805** is a bright group of forty stars spread across 22′. The brightest stars of this group shine at magnitude 8, which makes the region a fine area for scanning with a pair of binoculars. About 2.5° southeast of IC 1805 is the smaller cluster **IC 1848**. This object has the same magnitude as IC 1805 but contains fewer stars and covers only half as much sky. Both of these clusters are surrounded by a network of emission nebulosity barely visible to observers with large backyard telescopes.

At the opposite end of the constellation lies **M-52** (NGC-7654), a 7th-magnitude group of one hundred stars located 1° south of the bright star 4 Cassiopeiae. Along with NGC-457 and NGC-7789, M-52 is among the best clusters in the northern Milky Way. Its stars shine at magnitude 8.2 and fainter and cover 13′ of sky, which is about one-third the size of the Moon's disk. Small telescopes show M-52 as a rich, bright mass of dozens of stars.

One degree southwest of M-52 is the unusual emission nebula **NGC-7635**, the Bubble Nebula. This object is an extremely faint emission nebula measuring 15′ by 8′ across. The brightest part of this object is bubble-shaped, hence the nickname. Observers with 8-inch scopes may see traces of the Bubble nebulosity on exceptionally dark nights.

Another faint emission nebula lies some 2.5° east of the star Alpha (α) Cassiopeiae. **NGC-281** is a large cloud of low surface brightness covering an equilateral triangle of 7th-magnitude stars separated by about 30′. To see this nebula you'll need to use a low-power eyepiece and perhaps a nebula filter on a 10-inch or 12-inch telescope.

Fainter yet are two nebulae surrounding the bright star Gamma (γ) Cassiopeiae. Both **IC 59** and **IC 63** have extremely low surface brightnesses. And 3rd-magnitude Gamma Cassiopeiae, which is invariably in the field of view while you try to observe them, doesn't help viewing. Try using a large-aperture scope, a good nebula filter, and an occulting bar — a small piece of metal — over the eyepiece to block the bright star but not the field of view around it.

At the extreme southern end of Cassiopeia lie three galaxies, two of which are members of our Local Group of galaxies. Located about 8° north of the Andromeda Galaxy, **NGC-147** and **NGC-185** are distant satellite galaxies revolving around the Andromeda spiral. Each is a dwarf elliptical galaxy, and although both are low-surface brightness objects, each is visible with a 6-inch telescope under good conditions. NGC-147 shines at magnitude 9.3 and measures 12.9′ by 8.1′. NGC-185 is a magnitude 9.2 galaxy spanning 11.5′ by 9.8′. Both galaxies appear as ghostly patches of gray light in backyard telescopes. Three degrees southeast of NGC-185 is the compact elliptical **NGC-278**. This 11th-magnitude galaxy measures 2.2′ by 2.1′ and has a bright, compact core.

Object	M#	Type	R.A. (2000)	Dec.	Mag.	Size/Sep./Per.	N★	Mag.★
NGC-103		⊙	0h 25.3m	+61°21′	9.8	5′	30	12.3
NGC-129		⊙	0h 29.9m	+60°14′	6.5	21′	35	8.6
NGC-147		⬯	0h 33.2m	+48°30′	9.3	12.9′x 8.1′		
NGC-185		⬯	0h 39.0m	+48°20′	9.2	11.5′x 9.8′		
NGC-278		⬯	0h 52.1m	+47°33′	10.9	2.2′x 2.1′		
NGC-281		□E	0h 52.8m	+56°36′	—	3.5′x 30′		
IC 59		□ER	0h 56.7m	+61°04′	—	10′x 5′		
IC 63		□ER	0h 59.5m	+60°49′	—	10′x 3′		
NGC-457		⊙	1h 19.1m	+58°20′	6.4	13′	80	8.6
NGC-581	M-103	⊙	1h 33.2m	+60°42′	7.4	6′	25	10.6
Trumpler 1		⊙	1h 35.7m	+61°17′	8.1	4.5′	20	9.6
IC 1805		⊙	2h 32.7m	+61°27′	6.5	22′	40	7.9
IC 1848		⊙	2h 51.2m	+60°26′	6.5	12′	10	7.1
NGC-7635		□E	23h 20.7m	+61°12′	—	15′x 8′		
NGC-7635	M-52	⊙	23h 24.2m	+61°35′	6.9	13′	100	8.2
NGC-7789		⊙	23h 57.0m	+56°44′	6.7	16′	300	10.7

⊙ *Open Star Cluster*
□E *Emission Nebula*
□ER *Emission and Reflection Nebula*
⬯ *Elliptical Galaxy*

N★ = number of stars, Mag.★ = magnitude range of cluster or magnitude of central star, (...) indicates many fainter.

4h
3h
2h
1h
23h
22h
21h
+70°
+60°
+50°
+40°
+30°
CEPHEUS
IC 1848
IC 1805
Galactic Equator
Trumpler 1
IC 59
IC 63
M-103
NGC-103
NGC-129
M-52
NGC-7635
The Double Cluster
NGC-457
NGC-7789
NGC-281
PERSEUS
NGC-147
NGC-185
NGC-278
The Andromeda Galaxy
ANDROMEDA
2h
1h
23h

Above: Omega Centauri, the fines globular star cluster in the sky, is visibl to the naked eye as a small fuzzy "star and is apparent in 3-inch telescopes as mass of partially resolved stars. Photo b Jack Newton. Far left: The galaxy NGC 5128, located some 4° north of Omega, i one of the sky's best active galaxies. Sma scopes show its dust lane separating tw enormous lobes of fuzzy light. Photo b Jack B. Marling. Left: NGC-4945 is a fin edge-on barred spiral that is visible i binoculars. Photo by Jim Barclay.

Centaurus

Cen
Centauri

Keith Ward

Lying between Hydra — rich in galaxies — and Crux — rich in Milky Way clusters and nebulae — is the large and bright constellation **Centaurus** the Centaur. Because it is located on the fringe of the southern Milky Way, this group of stars contains a diverse collection of both clusters and nebulae belonging to our home galaxy. It also contains a number of distant spiral and elliptical galaxies that are bright enough to shine through the Milky Way's blanket of obscuring dust. Particularly notable are the globular cluster Omega (ω) Centauri — the brightest such object in the sky — and the bright active galaxy NGC-5128. Centaurus also contains Alpha (α) Centauri, the closest star system to our Sun.

Several of Centaurus' best galaxies lie in the constellation's central region. One of these is **NGC-4945**, a large, edge-on barred spiral that is visible as a streak of wispy light even in a pair of large binoculars. To find NGC-4945, first locate the 5th-magnitude bluish star Xi^1 (ξ^1) Centauri and then move your telescope 30′ northeast. This galaxy has a blue magnitude of 9.5 (implying its V magnitude is a half magnitude or more brighter); its dimensions are an extensive 20.0′ x 4.5′. In comparison, the well-observed edge-on galaxy NGC-4565 in Coma Berenices shines at magnitude 9.6 and covers only 16.2′ x 2.8′ of sky.

NGC-4945 appears dramatically different through telescopes as the aperture increases. Large binoculars and small scopes show only a sliver of gray light spanning some one-third of the Moon's diameter. A 10-inch telescope reveals a tiny knot of a nucleus surrounded by the faint, mottled haze of the galaxy's spiral arms. An even larger scope, one in the 16-inch or bigger class, displays NGC-4945 as a highly inclined dinner plate some 15′ long with a bright stellar nucleus.

BEST VISIBLE DURING
SPRING

Just 4° northeast of NGC-4945 is the popular, impressively bright globular cluster **Omega** (ω) **Centauri**, which is catalogued by J.L.E. Dreyer as NGC-5139. This cluster is so large and bright that early astronomers believed it was a star. As early as A.D. 140, Ptolemy included Omega in his star catalogue, the *Almagest*. In the seventeenth century, Johannes Bayer designated it one of the brightest "stars" in Centaurus. The first to recognize Omega as a star *cluster* was Edmond Halley in 1677.

To the naked eye, Omega appears as a hazy star or a comet without a tail, but telescopes show an oblate disk with rich outlying sections that cover as much sky as the Full Moon (its total visual diameter is 36′). Individual stars belonging to the cluster are observed out to 90′, or some 150 parsecs.

Telescopically, Omega Centauri is far and away the most dazzling globular. A 3-inch scope shows a large, fuzzy disk appearing mottled on its edges whereas a 6-inch scope under dark skies resolves the cluster into myriad stellar points. The best view of Omega comes with 10-inch or larger scopes at low power: the huge globular literally overflows the field of view with thousands of resolved, yellow-white, crisp points of light.

About 4° due north of Omega is **NGC-5128**, one of the finest examples of an active galaxy. In large finderscopes and binoculars it appears as a matched pair of semicircles that are separated by a dark gap, which is a broad equatorial dust band. The galaxy measures 18.2′ x 14.5′ across and shines at magnitude 7.0. It is close enough at 4 megaparsecs that large backyard telescopes show structural detail in NGC-5128's halo and dust lane.

A 16-inch telescope at 100x shows NGC-5128 as a bright circular patch of fuzzy gray light, intensely bright toward the middle and bisected by the dust lane, which is thicker and twisted at opposite sides of the galaxy's edge. The bright edges of the dust lane show tiny clumps and knots of nebulosity, and an extensive network of very faint nebulosity lies inside the dark lane. This material, seen projected against the dust band, is composed of gigantic chains of hot O- and B-type stars. These stars are grouped in associations and are visible only because they lie in front of the dust as viewed from our perspective.

Astronomers have classified NGC-5128 as an active galaxy, one demonstrating powerful, chaotic disruptions within its core. NGC-5128 is an intense emitter of radio energy, which is observed along two enormous lobes on either side of the galaxy's center. This suggests that NGC-5128, also known as Centaurus A (due to its radio emission), may house a supermassive black hole in its center.

About 5° southeast of Omega is the far

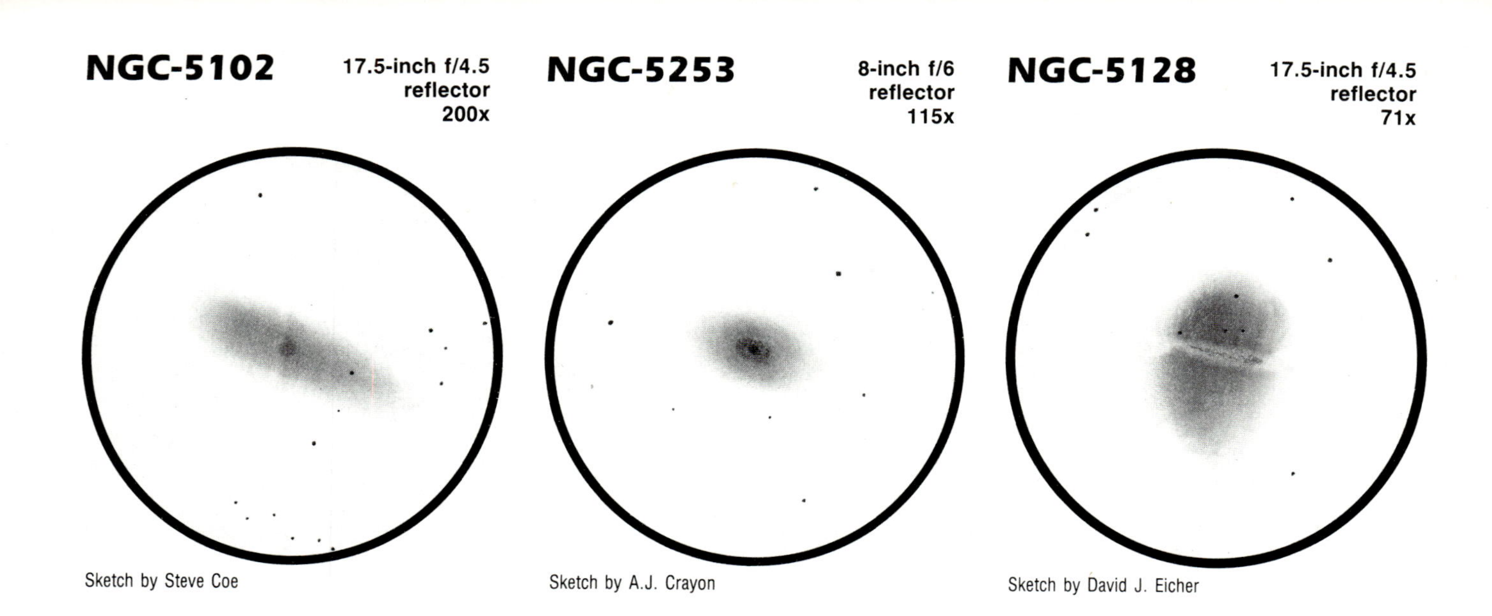

Sketch by Steve Coe

Sketch by A.J. Crayon

Sketch by David J. Eicher

smaller globular **NGC-5286**, a pretty sight in medium- and large-aperture scopes. This cluster is a magnitude 7.5 object measuring some 9.1′ across. It lies within the same low-power field of view as the 5th-magnitude orange star M Centauri. Although it is large and bright, this cluster is considerably more difficult to resolve than Omega. A 6-incher at high power shows only the outer edges of the cluster resolved into a misty haze although the brilliant core remains an unresolved disk.

Three bright galaxies lie north of NGC-5128, the best of which is **NGC-5102**. This galaxy also lies near a bright star — 17′ northeast of the 3rd-magnitude double Iota (ι) Centauri — and is an impressive sight in most telescopes. Measuring 9.3′ x 3.5′ and glowing at magnitude 9.6, this spiral appears as a bright oval in 3-inch telescopes; with 6-inch scopes it becomes detailed, showing a condensed nucleus encapsulated inside a uniformly bright halo.

Some 4° northeast of Iota and NGC-5102 is the more diminutive galaxy **NGC-5161**, a faint spiral. This galaxy shines at magnitude 12.0 and covers 5.4′ x 2.3′ of sky, appearing as a bright knot of a nucleus surrounded by a large, faint, featureless haze.

A much brighter galaxy lies 3° to the northeast near the Hydra border: **NGC-5253**. This is a large, nearby, irregular galaxy spanning 4.0′ x 1.7′ and glowing at magnitude 10.6. A small telescope shows it as a smooth, featureless glow without central condensation whereas a 12- or 16-inch scope shows a markedly brighter center in an oval-shaped grayish nebulosity. Curiously, NGC-5253 has produced two brilliant 7th-magnitude supernovae within the past century: one in 1895 and another in 1972. NGC-5253 is a member of the Centaurus group of galaxies, which also includes M-83 in Hydra, NGC-5128, NGC-4945, and NGC-5102.

In the far southern reaches of Centaurus are two dazzlingly bright stars that are observational favorites. **Beta (β) Centauri**, the tenth brightest star in the sky, is a difficult double star for small scopes; its magnitude 0.6 and 3.8 stars are separated by a mere 1.3″ in p.a. 251° (1960). Beta Cen is some 150 parsecs distant, meaning that the component stars are 200 astronomical units — some 30 billion kilometers — apart. These two stars are truly luminous: the primary is some 10,000 times brighter than the Sun and the secondary is 440 times brighter than Sol.

Four-and-a-half degrees southeast of Beta is **Alpha (α) Centauri**. Lying a mere 1.33 parsecs distant, Alpha Cen is celebrated as the primary sun in the closest star system to our own. Sometimes called Rigel Kentaurus, Alpha Cen is the third brightest star in the sky (after Sirius and Canopus), shining at magnitude −0.4. It is bright enough to be visible during twilight, and is a pretty object to view as the sky darkens. The Alpha Cen system is a beautiful triple star composed of magnitude −0.04, 1.2, and 11.0 members with separations of 21.2″ (p.a. 212°) and 131′, respectively. The stars' colors create a pretty contrast: the primary is a dazzling yellow, the secondary a dull orange, and the distant companion (called Proxima) a weak burnt red. Proxima, incidentally, is a physical member of the system and is actually the closest to our solar system, hence its name.

Object	M#	Type	R.A. (2000)	Dec.	Mag.	Size/Sep./Per.	H
NGC-3766		⊙	11h 36.4m	−61°36′	5.1	10′	
NGC-3918		■	11h 50.3m	−57°11′	8.4	10″	
NGC-3960		⊙	11h 50.9m	−55°42′	9.0	6′	
NGC-4945		§B	13h 05.3m	−49°29′	9.5B	20.0′x4.4′	SB(s)cd:
NGC-5102		§	13h 21.9m	−36°39′	9.6	9.3′x3.5′	SA0⁻
NGC-5128		§L	13h 25.3m	−43°01′	7.0	18.2′x14.5′	S0pec
NGC-5139		●	13h 26.8m	−47°29′	3.7	36.3′	
NGC-5161		§	13h 29.1m	−33°09′	12.0	5.4′x2.3′	SA(s)c:
NGC-5253		#	13h 39.9m	−31°39′	10.6	4.0′x1.7′	I0pec
NGC-5286		●	13h 46.2m	−51°22′	7.5	9.1′	
NGC-5316		§	13h 53.9m	−61°52′	8.4	12′	
Beta (β)		★2	14h 03.8m	−60°22′	0.6,3.8	1.3″	
Alpha (α)		★3	14h 39.6m	−60°50′	0.0,1.3, 10.7	8.7″,132″	

H = Hubble classification type for galaxies

★2	*Double Star*
★3	*Triple Star*
⊙	*Open Star Cluster*
●	*Globular Star Cluster*
■	*Planetary Nebula*
§	*Spiral Galaxy*
§B	*Barred Spiral Galaxy*
§L	*Lenticular Galaxy*
#	*Irregular Galaxy*

13h
14h
12h
HYDRA
M-83
-30°
NGC-5253
NGC-5161
NGC-5102
θ
-40°
ν
μ
NGC-5128
NGC-5139
ω
ξ
τ
γ
NGC-4945
-50°
NGC-5286
ε
NGC-3960
CRUX
NGC-3918
β
NGC-5316
-60°

Above: The faint emission nebula IC 1396 is a difficult target for backyard scopes — its light is spread over a diameter measuring nearly 3°. Photo by Ronald Royer. Far left: Although millions of parsecs apart, star cluster NGC-6939 and galaxy NGC-6946 are apparent neighbors lying in the same low-power field of view. Photo by Jack Marling. Top left: The cluster NGC-7510 is bright and easy to spot in binoculars. Photo by Lee C. Coombs. Bottom left: Planetary nebula NGC-40 requires high power to distinguish it from background stars. Photo by Bill Iburg.

Cepheus

Cep
Cephei

Lacerta

Lac
Lacertae

Keith Ward

Lying northeast of the brilliant Milky Way in Cygnus, the constellations **Cepheus** the King and **Lacerta** the Lizard offer numerous clusters and nebulae — and even a strange active galaxy — for backyard telescopes.

The largest object in Cepheus is a region of emission nebulosity catalogued as **IC 1396**, which covers nearly six square degrees of sky. Although first observed by E.E. Barnard with a 6-inch refractor, this faint cloud of nebulosity is normally difficult to spot with small telescopes. The nebula's total light is great, but so spread out that IC 1396 appears little more than a brightening in the Milky Way. Be sure to use a low-power eyepiece when hunting this nebula.

Just at the northern tip of IC 1396 is the unusual ruddy star **Mu** (μ) **Cephei**, often called "Herschel's Garnet Star." Mu Cep is the reddest star visible to the naked eye in the northern hemisphere, appearing distinctly deep orange or red to most observers. It is a red giant star which varies in light output semi-regularly, making it fun to follow from week to week. The magnitude range is about 3.7 to 5.0, and its approximate period is about 755 days. (Astronomers have identified additional possibly overlapping periods of 700, 900, 1100, and 4,500 days.)

Mu Cep may well vary in color as well as brilliance. Observers have occasionally mentioned seeing a peculiar purple tint. Others have seen it as deep orange or yellowish orange, while still others find it golden yellow. When you observe Mu Cep, compare it to a known white star in the area, such as Alpha (α) Cephei.

Lying to the southeast of Mu is **Delta** (δ) **Cephei**, another unusual star. This is the prototypical Cepheid variable — a class of stars which has contributed enormously to our understanding of the universe. First noticed by English astronomer John Goodricke in 1784, Delta Cep's variations have been meticulously recorded for 200 years. The star is a supergiant, and varies between magnitudes 3.6 and 4.3 in about 5.4 days.

BEST VISIBLE DURING
AUTUMN

Between Delta Cep and the bright star cluster M-52 in Cassiopeia are two unusual objects — NGC-7510 and IC 1470. **NGC-7510** is a bright, compact open star cluster made up of some 60 stars fainter than magnitude 9.7. Spanning only 4′ across, the group appears as a tiny diamond-studded haze in finderscopes and small telescopes, but becomes a fully-resolved cluster in 6-inch scopes. NGC-7510's total brightness is 7.9, making it easy to locate as a myriad of mostly white and bluish-white stars. **IC 1470**, a couple of low-power telescope fields to the southwest, is a tiny emission nebula bathed in an optically-invisible molecular cloud. It measures 70″ x 45″ across and is embedded with a 12th magnitude star which is visible in most telescopes.

NGC-7762 is a fine open star cluster lying in east-central Cepheus. Containing 40 stars, the brightest of which shine at 11th magnitude, this cluster is easily visible in any telescope as a sparse congregation of white and bluish-white stars some 11′ in diameter. Use a low-power eyepiece when first observing this cluster to note the colorful stars filling the area. Then substitute a higher power eyepiece and "zoom in" on individual stars for close inspection.

Farther north in Cepheus is the planetary nebula **NGC-40**, a bright bluish-green disk easily distinguishable from field stars at medium or high power. About 40″ across and with a photographic magnitude of 10.7, NGC-40 is one of the finest of a number of planetaries at high declinations. A 4-inch telescope at 100x shows it as a fuzzy "star," while an 8-incher at 100x clearly reveals its nebular character. At magnitude 11.6, the nebula's central star is easily spied using most telescopes.

Continuing on a wide arc from M-52 in Cassiopeia to NGC-7762 and through NGC-40, we come to the old open cluster **NGC-188** way up near the north celestial pole. Having carefully studied NGC-188's color-magnitude diagram, astronomers believe it is an astonishing five *billion* years old — immensely ancient for an open star cluster. Of the well-studied clusters, only NGC-6791 in Lyra and Melotte 66 in Puppis appear older.

In the eyepiece, NGC-188 is a fairly rich group of 120 stars of magnitude 12 and fainter packed into 14′. Many of its stars are so faint that small scopes show NGC-188 as an unresolved haze; a 6-incher at high power partially resolves the group but still provides the impression of nebulosity. In the largest backyard telescopes, this cluster is a dazzling sight with dozens of faint stars forming unending chains.

Farther south, back in the west-central part of Cepheus, is the faint reflection nebula **NGC-7023**. As are most reflection nebulae, this is a difficult object to observe because it is fairly large and diffuse. Measuring 18′ across and dimly lit by a seventh-magnitude type B5e star, NGC-7023 demands a 6-inch telescope for viewing even under dark skies. With a 10-inch or larger scope, the nebulosity is fairly easy to see but appears completely featureless.

A galaxy and an open cluster — both in the same low-power field but separated in space by five megaparsecs — lie near the far western part of Cepheus. **NGC-6939** contains 80 stars in an area spanning a mere 8′ across, making it one of the richest, most stunning open clusters in the area. Its brightest stars shine slightly brighter than 12th magnitude, giving it the overall glow of an eighth-magnitude star. It is easily visible in finderscopes and a fine object in any telescope. Large scopes at high power provide a field absolutely packed with stars. Nearby is the galaxy **NGC-6946**; at five megaparsecs' distance, it is one of the nearer galaxies outside the Local Group. It is face-on to our line of sight and appears as a tight, bright core surrounded by a faint haze representing its spiral arms. Large backyard scopes show a mottled structure in the haze, suggesting patches of dark dust.

The planetary nebula **NGC-7139** in the central part of Cepheus is large but dim. Measuring over one arc-minute across, it is large enough to avoid confusion with surrounding stars. Its photographic magnitude of about 13 leads you to expect a very low surface brightness. Nonetheless, a 12-inch scope at high power shows it as a smooth fuzzy disk devoid of a central star.

Much smaller and sandwiched between bright Cygnus and Cassiopeia, Lacerta contains several clusters, a planetary nebula, and a strange galaxy-like object. The clusters — **NGC-7209**, **NGC-7243**, and **NGC-7245** — are large and bright. NGC-7209 shines at seventh magnitude and measures a full Moon diameter across, making it obvious in binoculars. NGC-7243 is only slightly smaller and a bit brighter, so that it too is best observed with a wide field. NGC-7245, on the other hand, is 5′ across, shines at ninth magnitude, and contains only a handful of stars.

The tiny planetary nebula **IC 5217** is a difficult object to identify against the rich Milky Way in Lacerta. It is practically stellar — 6″ across — and glows at magnitude 12.6. Its bluish color may help give it away; try finding the appropriate area and using high power on "suspect" stars.

The most bizarre object in the area is **BL Lacertae** — originally thought to be a variable star. A detailed look at its spectrum, however, erased such notions — it is an unusual variable galactic nucleus, throwing off high energy radiation much like a quasar. BL Lac is faint — it varies by half a magnitude from 14.7 — and star-like, but if you observe it you'll be seeing a high-energy active galaxy some 270 megaparsecs away.

NGC-6946 8-inch f/10 SCT 50x

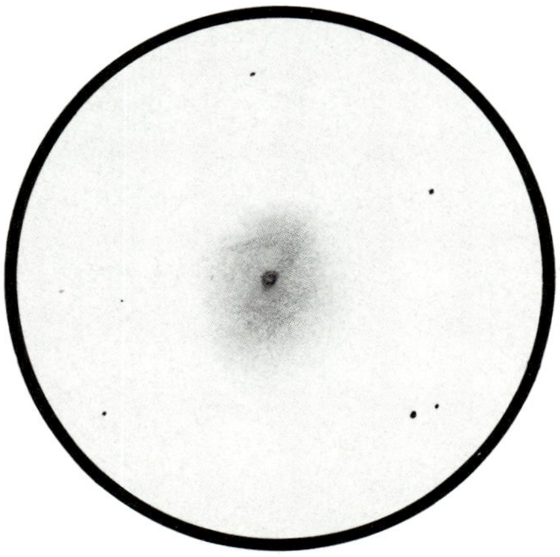

NGC-7023 8-inch f/10 SCT 50x

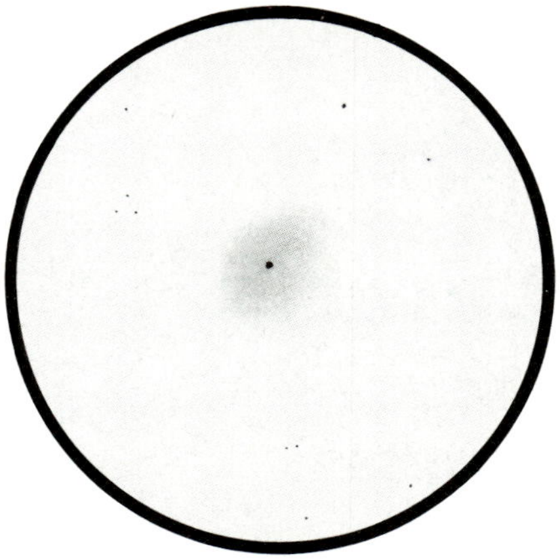

NGC-7139 8-inch f/10 SCT 50x

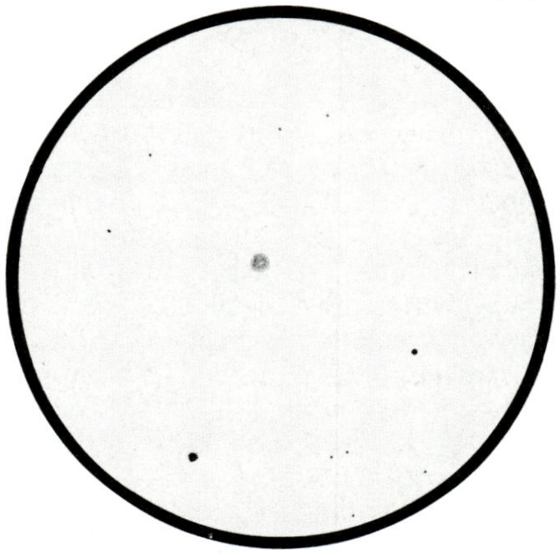

Sketches by David J. Eicher

Object	M#	Type	R.A. (2000)	Dec.	Mag.	Size/Sep./Per.	N★
NGC-40		■	0h 13.0m	+72°32′	10.7p	40″	
NGC-188		⊙	0h 44.4m	+85°20′	8.1	14′	120
NGC-6939		⊙	20h 31.4m	+60°38′	7.8	8′	80
NGC-6946		§	20h 34.8m	+60°09′	8.9	11.0′ x 9.8′	
NGC-7023		□R	21h 01.8m	+68°12′	—	18′	
IC 1396		□E	21h 39.1m	+57°30′	—	170′ x 140′	
NGC-7139		■	21h 45.9m	+63°39′	~13p	78″	
BL Lac		QSO	22h 02.7m	+42°16′	14.7v	stellar	
NGC-7209		⊙	22h 05.2m	+46°30′	6.7	25′	25
NGC-7243		⊙	22h 15.3m	+49°53′	6.4	21′	40
NGC-7245		⊙	22h 15.3m	+54°20′	9.2	5′	3
IC 5217		■	22h 23.9m	+50°58′	12.6p	6″	
IC 1470		□E	23h 05.2m	+60°15′	—	70″ x 45″	
NGC-7510		⊙	23h 11.5m	+60°34′	7.9	4′	60
NGC-7762		⊙	23h 49.8m	+68°02′	10p	11′	40

N★ = number of stars

⊙	*Open Star Cluster*
■	*Planetary Nebula*
□E	*Emission Nebula*
□R	*Reflection Nebula*
§	*Spiral Galaxy*
QSO	*Quasar*

NGC-188
DRACO
CEPHEUS
NGC-40
NGC-7023
NGC-7762
NGC-7139
NGC-6939
NGC-6946
M-52
NGC-7510
IC 1470
IC 1396
CASSIOPEIA
NGC-7789
NGC-7245
IC 5217
NGC-7243
CYGNUS
NGC-7209
LACERTA
BL

Left: The bright galaxy M-77 is one of the best-known examples of a Seyfert galaxy, a system whose tiny core emits intense ultraviolet radition. The core, surrounded by a hazy envelope of nebulosity, is easily visible in small telescopes. Photo by Jack Newton. Below left: NGC-247 is a low-surface-brightness galaxy that is difficult to see in telescopes smaller than 6 inches in aperture. Photo by Martin C. Germano. Below right: One of the largest planetary nebulae in the sky, NGC-246 appears as a ghostly incomplete ring of nebulosity; its central star glows at 12th magnitude. Photo by Bill Iburg.

Cetus

Cet
Ceti

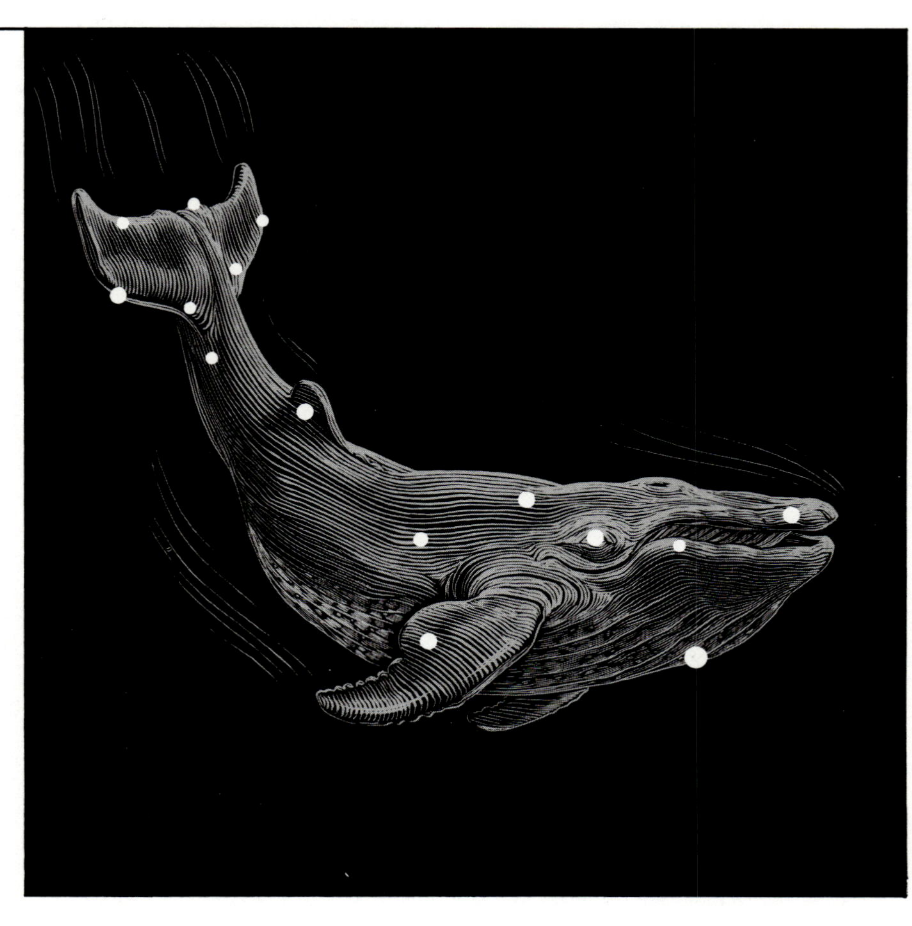

Cetus the Whale, a relatively barren constellation lying south of Pisces in the great void of the autumn sky, is home to a collection of some of the finest and most unusual galaxies in the sky. Although it is primarily known for the variable star Mira, and otherwise appears dull to the naked-eye viewer, Cetus offers a nice selection of telescopic objects for those who wish to seek them out.

A first stop in Cetus is the famed star **Mira**, Omicron (ο) Ceti, the prototype of the so-called long-period variables. Mira means "the wonderful," an appropriate name since it was the first variable star to be discovered (although the name predates the discovery); the feat was accomplished by the German astronomer David Fabricius in 1596. Mira normally fluctuates between about ninth magnitude and third magnitude over a period of about 331 days; sometimes it reaches second magnitude at maximum and the period is slightly longer or shorter, but the star is bright and predictable enough that its changes are easy to follow in a telescope or binoculars.

Mira lies at a distance of about 70 parsecs and, like all long-period variables (LPVs), is an old red giant star believed to pulsate, since its physical characteristics seem to change throughout the cycle of its variations. Astronomers believe the pulsations arise from changes in the aging star's energy-producing core.

A close, bright double star in Cetus is **Gamma** (γ) **Ceti**, whose components shine at magnitudes 3.6 and 6.2 with a mere 2.7 arc-second separation. A good 3-inch refractor will split the two stars without too much trouble, providing the seeing is steady. The primary is a type A2 main sequence star, while the companion is a type F3 dwarf; they are sometimes described as being bluish and yellowish in the telescope — perhaps simply a contrast effect. A third physically connected companion lies some 14′ to the northwest — it is an M-type dwarf called BD +2° 418, and shines at 10th magnitude.

BEST VISIBLE DURING
AUTUMN

Located roughly midway between Gamma and Omicron Ceti are a few notable galaxies. The compact galaxy **M-77** (NGC-1068) — found about 1° southeast of Delta (δ) Ceti — shines at magnitude 9.6 and measures 2.5′ x 1.7′ across, making it an easy target for small telescopes. In a 2.4-inch refractor it appears as a condensed ball of fuzzy light. A 4-inch scope at high power shows a clearly defined circular "disk" as well as a much dimmer outer halo of nebulosity. M-77 is a Seyfert galaxy, one of the curious systems discovered by the American astronomer Carl Seyfert and catalogued for their intense and variable ultraviolet emission. A 10-inch telescope at high power shows a tiny bright nucleus — the source of the high-energy radiation — embedded within the galaxy's center. M-77 is classed as an Sb-type spiral, holds around 100 billion solar masses, and measures 30 kiloparsecs across. M-77 was one of two galaxies in which large redshifts were first detected, by Vesto M. Slipher in 1913. (The other galaxy was M-104, the "Sombrero.")

Lying at a distance of about 11.8 megaparsecs, M-77 is the chief member of a small group of galaxies that also includes **NGC-1055**. This is an Sb-type system much like the Milky Way, but seen edge-on; its broad equatorial dust lane is visible in large amateur telescopes. Small backyard telescopes show NGC-1055 as a slender streak of misty light, spanning some 5.0′ x 1.0′ and glowing at magnitude 11.4. A possible member of the M-77 group is the face-on barred spiral system **NGC-1073**, which shines at magnitude 11.5 but has a low surface brightness because its light is spread over a circular area 4′ across. Photographs of this galaxy show unusual knotty structure in the arms; with the aid of a nebular filter, a large telescope may just be capable of revealing some of this detail. Most telescopes show NGC-1073 as a large glow with a slightly brighter middle, and little definition.

Lying in the western part of Cetus is the large, irregular galaxy **IC 1613**. This is a difficult object for small telescopes, as its integrated magnitude of 12.0 is spread over the large area of 11.0′ x 9.0′, resulting in a very low surface brightness. Under perfect skies, large binoculars or RFTs show IC 1613 as a dim wispy glow — but under average conditions a 6-inch or 8-inch scope is the minimum needed to show the object. IC 1613 lies at nearly the same distance as the M-31 group —

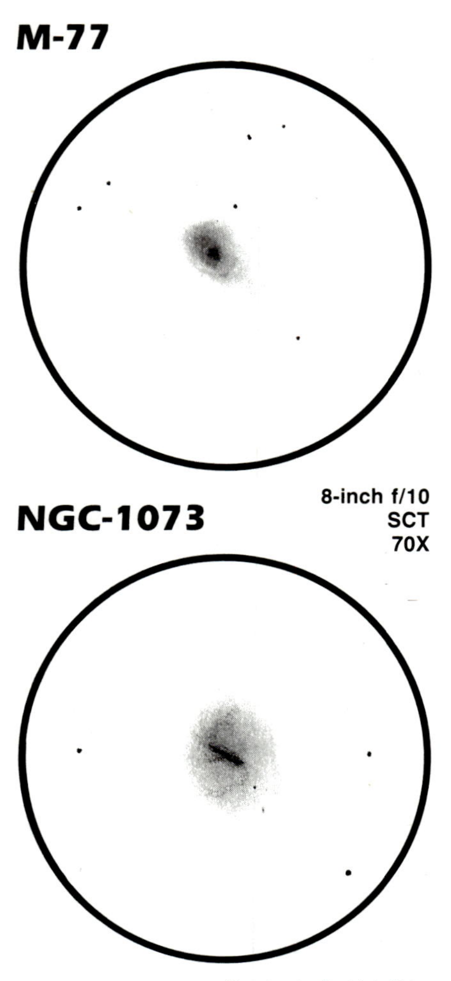

Sketches by David J. Eicher

680 kiloparsecs — comfortably earning its membership in the Local Group of galaxies, which extends out to at least 1 megaparsec from the Milky Way.

South of IC 1613 are a variety of objects, one of which is the small spiral galaxy **NGC-157**. This galaxy is one of the "many-armed spirals": Sc-type loosely wrapped galaxies with a proliferation of spiral arms (it has five or six). NGC-157 measures 2.8′ x 2.1′ across and shines at 11th magnitude, making it easy to spot as a nebulous patch in a 3-inch telescope. Large backyard scopes begin to show a knotty structure amidst the halo, but the detail is pretty faint and elusive.

A few degrees southeast of NGC-157 is **NGC-246**, one of the largest planetary nebulae in the sky: this nebula sports an apparent diameter of 240″ x 210″, over which it spreads a photographic magnitude of 8.0. It has a relatively low surface brightness, but it takes magnification well enough to show considerable structure over its face. NGC-246 is basically ring-shaped with localized spots of brighter and fainter light that give it a strangely uneven appearance. Within NGC-246's ring of nebulosity lie four bright stars; the most centrally located is the illuminating star, which shines at 12th magnitude. An 8-inch reflector at 100x easily shows all of the stars inside, as well as the ring, which appears incomplete since it is faintest toward the southeast.

The largest galaxy in Cetus is the highly inclined spiral **NGC-247**, which spans 18.0′ x 5.0′ and shines at magnitude 9.5. To find NGC-247, which is of very low surface brightness, look 3° south-southeast of the bright star Beta (*β*) Ceti. This galaxy is classed as a type Sc spiral, though the arms show weakly and the outer envelope appears to be made up of irregular patches of stars and gas. NGC-247 is a member of the Sculptor group of galaxies, which also includes the bright objects NGC-45, NGC-55, NGC-253, NGC-300, and NGC-7793. At a distance of 2.4 megaparsecs, these galaxies are the closest to us with the exception of those in our own Local Group. **NGC-45**, a spiral galaxy that measures 8.0′ x 5.5′ and glows at magnitude 11.1, also lies in Cetus. It is rather more difficult than NGC-247, but is nevertheless visible in small apertures as an oval-shaped blob of light with a bright middle. An eighth-magnitude star lies 4.5′ southwest of this galaxy.

A few more bright galaxies lie in Cetus, out in the central region of the Whale. One such object is the 12th magnitude Sc-type spiral **NGC-578**, easily visible as a roundish patch of mottled haze in telescopes of 4- or 5-inch aperture. The object measures 4.5′ x 2.5′ in diameter and has a relatively condensed nuclear region. NGC-578's integrated magnitude is about 11.5, making it visible in large finderscopes as a small nebulous patch. **NGC-908** is another late-type spiral, which shows prominently loose spiral arms — visible as a mottled area of uneven texture in telescopes in the 10-inch to 20-inch aperture range. This galaxy has a magnitude of 10.9 and measures 4.0′ x 2.5′ across, making it just a bit brighter than NGC-578. The galaxy **NGC-936** is a fine barred spiral, showing a nebulous knot in its center surrounded by a faint haze. Its magnitude of 11.2 is spread over an area of 3.0′ x 2.0′, keeping its surface brightness relatively high. The Sc-type spiral **NGC-941**, a 13th magnitude object, lies 12′ to the west.

Let's finish our tour of Cetus with a look at one of the closest stars to the Sun. At a distance of 2.8 parsecs, **UV Ceti** is a system comprised of two of the smallest and faintest stars known. Discovered by the American astronomer W.J. Luyten in 1949, the system consists of magnitude 12.4 and 13.0 stars, and has the unusually high proper motion of 3.35 arc-seconds per year, in position angle 80°. Each of these two stars has only about eight percent of the Sun's mass. The variable star designation applies to the fainter of the two, whose violent radio and optical flares are typically 100 to 1000 times more powerful than those from our Sun.

Object	M#	Type	R.A. (2000)	Dec.	Mag.	Size/Sep./Per.	H
NGC-45		§	0h 13.9m	−23°10′	11.1	8.0′x5.5′	Scd
NGC-157		§	0h 34.7m	− 8°23′	11.0	2.8′x2.1′	Sc
NGC-246		■	0h 47.2m	−11°53′	8.0	240″x210″	
NGC-247		§	0h 47.2m	−20°45′	9.5	18.0′x5.0′	Sc
IC 1613		§I	1h 05.1m	+01°18′	12.0	11.0′x9.0′	I
NGC-578		§	1h 30.4m	−22°41′	11.5	4.5′x2.5′	Sc
UV		IV	1h 38.8m	−17°58′	7↔12	Irr	
Omicron (o)		LPV	2h 19.3m	−02°58′	3↔9.5	331d	
NGC-908		§	2h 23.1m	−21°13′	10.9	4.0′x1.3′	Sc
NGC-936		§B	2h 27.6m	−01°09′	11.2	3.0′x2.0′	SB0
NGC-1055		§	2h 41.8m	+00°29′	11.4	5.0′x1.0′	Sbc
NGC-1068	M-77	§	2h 42.7m	−00°01′	9.6	2.5′x1.7′	Sb
Gamma (γ)		★²	2h 43.3m	+03°15′	3.5,6.0	2.7″	
NGC-1073		§B	2h 43.8m	+07°23′	11.5	4.0′x4.0′	SBc

★²	*Double Star*
LPV	*Long Period Variable*
IV	*Irregular Variable*
■	*Planetary Nebula*
§	*Spiral Galaxy*
§B	*Barred Spiral Galaxy*
§I	*Irregular Galaxy*

H = Hubble classification type for galaxies

ARIES
Ecliptic
PISCES
CETUS
NGC-1073
NGC-1055
M-77
IC 1613
NGC-936
Mira
UV
NGC-908
NGC-578
FORNAX
SCULPTOR

Glowing at magnitude 7.3 and spanning 11', globular cluster NGC-1851 is Columba's brightest and largest deep-sky object. Courtesy National Optical Astronomy Observatories.

Columba
Col
Columbae

Caelum
Cae
Caeli

Pictor
Pic
Pictoris

Keith Ward

Located between the bright forms of Canis Major and Carina to the east and the dim, winding figure of Eridanus to the west are the faint southern constellations Caelum, Columba, and Pictor. **Caelum** the Chisel is one of a group of faint constellations named by Nicolas Louis de Lacaille in the eighteenth century. It holds one faint galaxy. **Columba** the Dove is not difficult to recognize under a dark sky: its brightest stars shine at magnitude 3. This star group contains a bright globular star cluster, a handful of galaxies, and one unusual star. **Pictor** the Painter's Easel, another of Lacaille's creations, is a faint constellation holding a star that became a nova in 1925 and a couple of galaxies.

The brightest deep-sky objects of these three constellations are in Columba. **NGC-1851**, a magnitude 7.3 globular cluster, lies in an otherwise barren area in Columba's southwestern corner. Covering a full 11′ of sky, NGC-1851 is easily visible in binoculars as a small, misty clump of light. A 4-inch telescope shows this cluster as a pale disk spanning 5′ with a slight central condensation. Under a dark sky a 6-inch telescope just resolves the cluster's brightest stars, revealing its nature. Large backyard telescopes show speckles of faint stars strewn across an 8′-diameter soft, unresolved glow of light.

Three degrees northwest of NGC-1851 — halfway between the globular and the bright double star Gamma (γ) Columbae — lies a pair of galaxies easily accessible to small telescope users. Lying in a bright starfield, **NGC-1792** is an Sb-type spiral that glows at magnitude 10.2 and measures 4.2′ by 2.1′. Because it is relatively bright, NGC-1792 is rather easy to pick up in a small telescope. A 4-inch reflector shows the object as a pale smear of light some 2′ long. An 8-inch telescope shows the galaxy as a 3′ by 1′ elongated oval of light without much central condensation. The surface brightness of this galaxy rises slowly toward the center, but there is no sharply defined nucleus, which is typical of Sb-type spirals.

One degree northeast of NGC-1792 — and visible in the same low-power field — is the 10th-magnitude barred spiral **NGC-1808**. At low power or on nights of average transparency NGC-1808 appears very much like its neighbor. However, view NGC-1808 on a dark night — especially at high power — and you'll see some subtle differences. NGC-1808 appears longer and broader than NGC-1792 and has a bright, condensed, ball-like nucleus. Large telescopes at high power may show slight mottling over NGC-1808's spiral arms, which is indicative of dust lanes and patches.

Seven degrees north of NGC-1792 lies a group of faint galaxies visible in small telescopes. **NGC-1800** is a tiny elliptical galaxy located 4° north and slightly east of Gamma Columbae. (A helpful check on the location is provided by a 6th-magnitude star lying 1° to the west.) This little galaxy is visible as a pale spot of light equaling a magnitude 12.6 star in brightness. NGC-1800 measures 1.6′ by 0.9′ but appears nearly round in backyard telescopes. Even large backyard instruments — 12 to 16 inches in aperture — don't show any detail in this object but only a slight concentration toward the center.

Equally faint is the tiny face-on spiral **UGC A103**, found 1° northeast of NGC-1800. UGC A103, the 103rd object in the addendum of the *Uppsala General Catalogue* of galaxies, glows softly at blue magnitude 13.1 and measures 2.5′ by 2.4′. This galaxy appears similar to NGC-1800, showing no detail except for a slight central condensation. About 1.5° southeast of UGC A103 is another galaxy, **UGC A106**. This faint irregular galaxy is very different from UGC A103. UGC A106 has a magnitude of 12.6 and dimensions of 3.4′ by 2.5′. The galaxy has a very low surface brightness, so its small — almost stellar — nuclear bulge is visible with a faint, uniformly illuminated halo of light surrounding it.

Located 2.5° northeast of UGC A106, **NGC-1879** is a faint face-on barred spiral. With a blue magnitude of 13.2 and diameter of 2′, NGC-1879 appears as a pale round patch of light in medium-size telescopes. Large telescopes show a slight central condensation. A faint star lies on the galaxy's northeastern side, and a 9th-magnitude star is located some 10′ to the northwest.

Appearing as a small oval patch with a bright, condensed center, the galaxy **NGC-2090** lies in central Columba, midway and slightly north of a line drawn between the 3rd-magnitude stars Alpha (α) and Beta (β) Columbae. This galaxy is a

BEST VISIBLE DURING
WINTER

NGC-1851

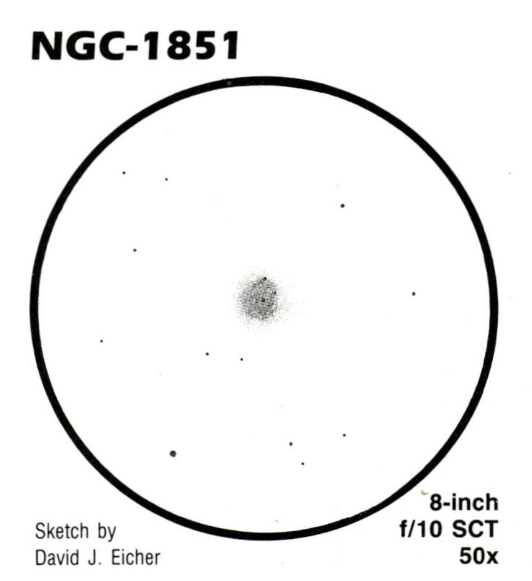

Sketch by David J. Eicher

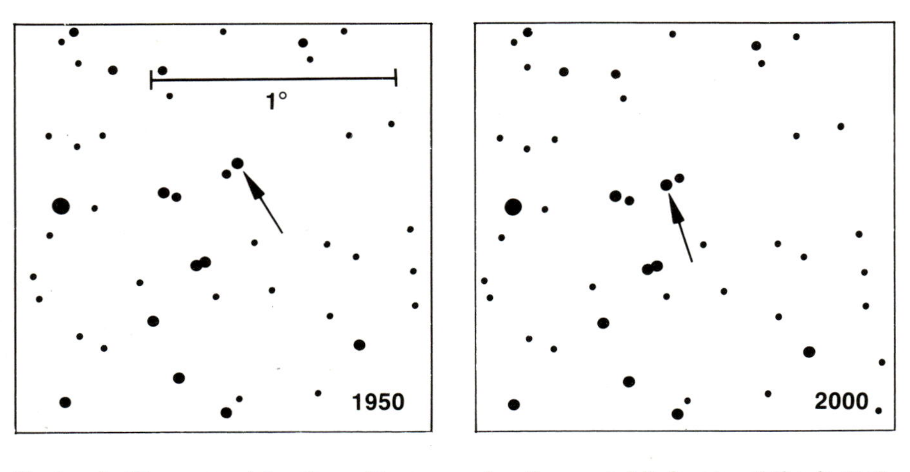

Kapteyn's Star, named for the astronomer who discovered it, is one of the fastest-moving stars in the sky. The star's proper motion will have carried it 7.3′ relative to field stars between 1950 and 2000.

blue magnitude 11.8 Sc-type spiral measuring 4.5′ by 2.3′.

On the eastern side of Columba, 2.5° northwest of the 4th-magnitude star Kappa (κ) Columbae, lies **NGC-2188**. NGC-2188 is fun to observe because it is bright enough to be visible in small scopes and has a thin, spindlelike appearance. This object is a barred spiral spanning 3.7′ by 1.1′ and glowing at magnitude 11.8. Some 2.5° north of the galaxy NGC-2090, forming a right triangle with Alpha and Beta Columbae, is the 5th-magnitude star **Mu (μ) Columbae**. Visually Mu Columbae appears to be a normal blue-white star, but photographs made of the Mu Columbae region over many years show an odd situation. The star is moving away from the Orion association — a large group of young stars in Orion — at a rate of 1′ every 2,400 years. The stars 53 Arietis and AE Aurigae also appear to be moving away from the Orion association in different directions. Tracing the paths of these three stars backward in time shows the stars met at the Orion association several million years ago. Astronomers don't know what threw these three stars outward from the Orion complex, but some speculate that such a force could have been supernova-related.

Across Columba's western border in the tall, narrow, dim constellation Caelum lies a lone galaxy. **NGC-1679** is a magnitude 13.5 galaxy that measures 1.2′ by 0.8′ and is thought to be an Sc-type spiral. To find this object, first find the 6th-magnitude star Zeta (ζ) Caeli and then move 2.5° southeast. NGC-1679 lies in a rich starfield but is easy to pick up in an 8-inch scope under a dark sky. The galaxy appears as a circular haze 1′ across with a clearly defined bright center.

Situated too far south for many Northern Hemisphere viewers, Pictor offers three treats for those who can observe it above the murk and glow of the horizon. Just 1° northeast of the 5th-magnitude star Iota (ι) Pictoris is the lenticular galaxy **NGC-1705**. Spanning 1.8′ by 1.4′ and glowing at magnitude 12.3, this object appears as a dim oval of light in small scopes.

Kapteyn's Star, named for the Dutch astronomer Jacobus C. Kapteyn, is the second "fastest" star in the sky. Kapteyn's Star has the highest proper motion, or movement relative to other stars, of any star except Barnard's Star in Ophiuchus. This star is a red dwarf with a magnitude of 8.8, located 8.5° northwest of Beta Pictoris. Its motion across the sky carries it 8.7″ per year, or one degree every 414 years. Kapteyn's Star is a mere 3.9 parsecs away, which in part accounts for its rapid motion; mostly, however, the star's proper motion is large because the star is moving through space at 280 kilometers per second. (By comparison, the Sun plods along at a modest 20 kilometers per second.) To find this star, use the accompanying chart.

Way down at −63° declination, just 2° southwest of Alpha Pictoris and 6° northeast of the Large Magellanic Cloud, lies Pictor's other famous star. A century ago **RR Pictoris** shone at 12th-magnitude, but on May 25, 1925, the South African astronomer R. Watson noticed it gleaming at magnitude 2.3. RR Pictoris became a brilliant nova that spring and summer. By June 9, the star brightened to magnitude 1.2 and then faded to 4th magnitude by July 4. After fluctuating and attaining three maxima, the star faded rapidly and it now glows at its original 12th magnitude. For a time, however, RR Pictoris captured the eyes of all astronomers far enough south to see it.

Object	M#	Type	R.A. (2000)	Dec.	Mag.	Size/Sep./Per.	H
NGC-1679		§	4h 49.8m	−31°59′	13.5	1.2′x 0.8′	Sc:
NGC-1705		§L	4h 54.2m	−53°22′	12.3	1.8′x 1.4′	S0 pec.
NGC-1792		§	5h 05.2m	−37°59′	10.2	4.2′x 2.1′	Sb⁺
NGC-1800		§L	5h 06.4m	−31°57′	12.6	1.6′x 0.9′	E6
NGC-1808		§B	5h 07.7m	−37°31′	9.9	7.2′x 4.1′	S(B)a
UGC A103		§	5h 10.7m	−31°36′	13.1_B	2.5′x 2.4′	S⁺
Kapteyn's Star		★	5h 11.4m	−44°56′	8.8	—	
UGC A106		#	5h 12.0m	−32°58′	12.6	3.4′x 2.5′	Ir⁻
NGC-1851		●	5h 14.1m	−40°03′	7.3	11.0′	
NGC-1879		§B	5h 19.8m	−32°09′	13.2_B	2.2′x 1.9′	S(B)⁻
Mu (μ) Col		★	5h 45.9m	−32°18′	5.2	—	
NGC-2090		§	5h 47.0m	−34°14′	11.8_B	4.5′x 2.3′	Sc
NGC-2188		§B	6h 10.1m	−34°06′	11.8	3.7′x 1.1′	SBm
RR Pic		N	6h 35.6m	−62°39′	1.2↔12.4	—	

★	*Star*
N	*Nova*
●	*Globular Star Cluster*
§	*Spiral Galaxy*
§B	*Barred Spiral Galaxy*
§L	*Lenticular Galaxy*
#	*Irregular or Peculiar Galaxy*

H = Hubble type for galaxies

Subscript "P" denotes photographic magnitude; subscript "B" denotes blue magnitude.

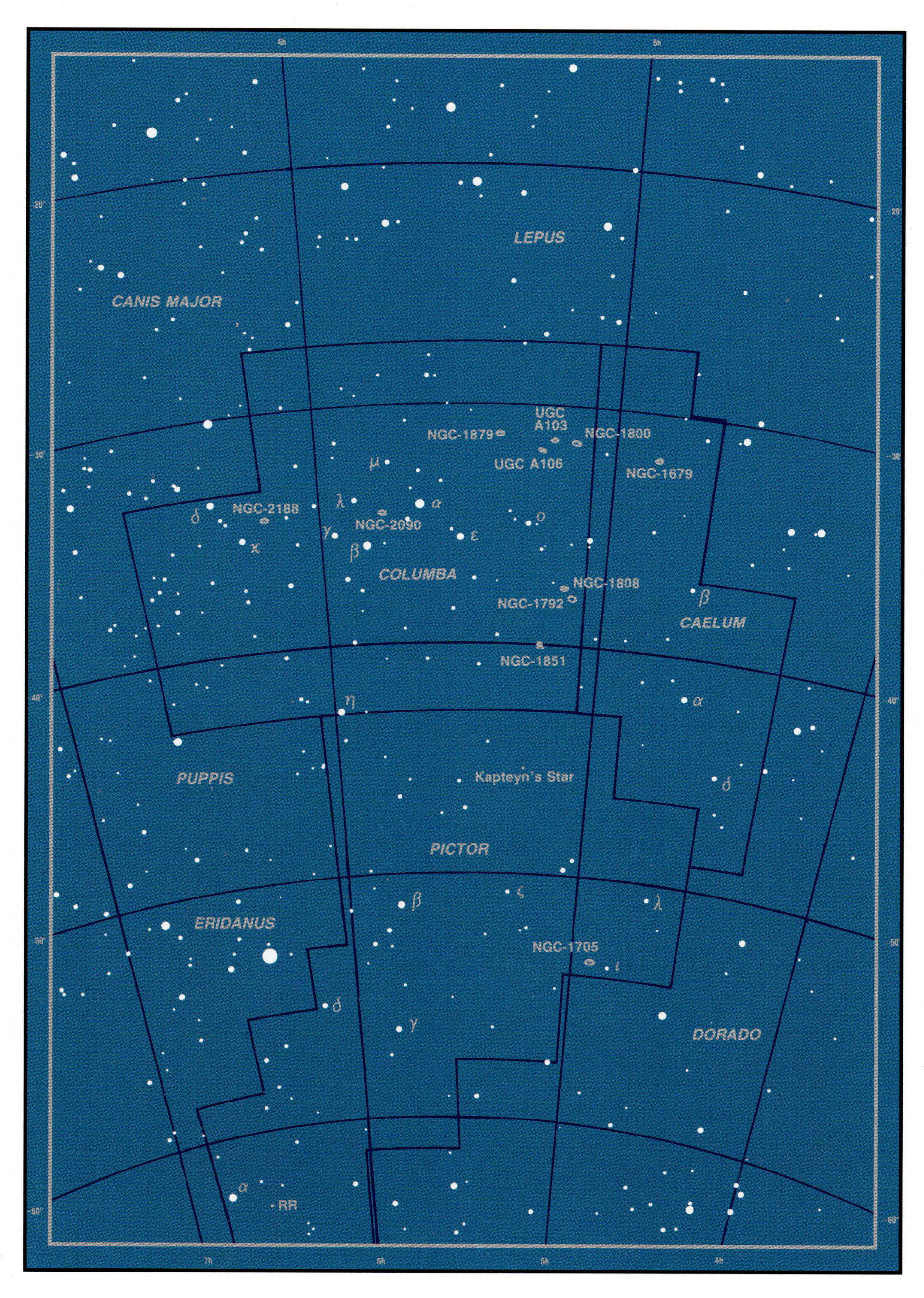

6h
5h
-20°
-30°
-40°
-50°
-60°
LEPUS
CANIS MAJOR
UGC
A103
NGC-1879
NGC-1800
UGC A106
NGC-1679
NGC-2188
NGC-2090
COLUMBA
NGC-1808
NGC-1792
CAELUM
NGC-1851
PUPPIS
Kapteyn's Star
PICTOR
ERIDANUS
NGC-1705
DORADO
RR
7h
6h
5h
4h

Above: M-64, the Blackeye Galaxy, is the brightest galaxy in Coma Berenices. Its dust patch is visible to 4-inch scopes on good nights. Photo by Jack Newton. Top right: Another Newton photograph shows NGC-4565, the finest example of an edge-on spiral in the entire sky. Its broad dark lane is visible in 6-inch telescopes. Middle right: Tom Dessert photographed the sharp stellar nucleus and faint outer halo of spiral M-88, one of the best sights in the Virgo cluster. Bottom right: The delicate three-arm spiral shape of M-99 is captured in this Lee Coombs photo. Ten-inch telescopes on dark nights show this spiral structure. Below: M-53, an eighth-magnitude globular, is partially resolved at high power in backyard scopes. Photo by Jack Newton.

Coma Berenices

Com

Comae Berenices

Keith Ward

In the direction of the small irregular group of stars known as **Coma Berenices** — Berenice's Hair — lies one of the densest concentrations of external galaxies visible to us. Looking in this direction allows us to see toward the Virgo cluster of galaxies, the heart of which lies across the border in Virgo. Several dozen bright examples of spirals, ellipticals, lenticulars, and peculiar galaxies are visible in small telescopes, many within the same low-power field of view. Coma Berenices also holds the north galactic pole, the imaginary point in the sky positioned 90° above all points on the galactic equator.

The constellation's outstanding feature is not a galaxy, but rather a large, loose open star cluster called **Mel 111** — for the astronomer P.J. Melotte, who catalogued it in 1915 — or the Coma star cluster. It is more obvious to the naked eye than the Pleiades or the Beehive cluster, as it measures a whopping 6° in diameter and glows at magnitude 2.9. The Coma cluster is one of the finest areas for scanning with binoculars, although telescopes don't show much of anything due to their limited fields of view. This group is composed of 47 stars totaling some 100 solar masses; the brightest are the fifth magnitude suns 12, 13, 14, 16, and 21 Comae Berenices. At a distance of 75 parsecs, the Coma cluster is the third nearest open star cluster after the Hyades in Taurus and the Ursa Major Moving Group.

Only 3° from the north galactic pole and 1.7° east of the star 17 Comae Berenices (close to the Coma cluster) lies **NGC-4565**, the best and brightest example of an edge-on spiral galaxy in the sky. With a 2-inch telescope, you'll see a small thin streak of light; a 4-inch at high power reveals a thicker needle of nebulosity and a hint of the central bulge about the galaxy's nucleus. NGC-4565 measures 15.5′ x 2.1′ in diameter and has a magnitude of 10.7; on good nights an 8-inch at high power shows a long nebulosity with a broad dust band running alongside its edge, and a bright starlike nucleus embedded in the roundish central bulge. This Sb-type galaxy is probably 10 megaparsecs distant, and is possibly an outlying member of the Virgo cluster, although it lies some 13° north of the cluster's heart.

BEST VISIBLE DURING
SPRING

Just 2° north of NGC-4565 is the bright Sc-type spiral **NGC-4559**, a galaxy measuring 10.0′ x 3.0′ and glowing at magnitude 10.5. This galaxy is composed of several coarse knotty arms; on dark nights large backyard telescopes may show a mottled condensation within the galaxy's halo, corresponding to dust lanes and bright gas clouds in the galaxy's arms. A 4-inch shows a bright oval smudge with a condensed nucleus.

On the other side of the north galactic pole from the NGC-4565 area, about 2.3° west of Beta [β] Comae Berenices, lies the tiny dim elliptical galaxy **NGC-4889**. This little speck of light shines at magnitude 13.2, measures 1.0′ x 0.6′, and is classed as an E4 galaxy. Through small telescopes NGC-4889 looks like a small, dim planetary nebula; its great distance of 130 megaparsecs masks its great size and causes it to appear unimpressive. This is the principal member of the Coma Berenices cluster of galaxies, a group much like the Virgo cluster but 20 times farther away. At least 1000 galaxies are members of this group, scattered across about 3° and centered on NGC-4889, but these systems are very faint, most being 15th magnitude and fainter.

In the eastern part of the constellation, 1° northwest of the binary Alpha [α] Comae Berenices, lies the globular cluster **M-53** (NGC-5024). This cluster has a total magnitude of 8.7 and a diameter of 14.0′, making it a nice example of a globular under low-power viewing — a 6-inch at 50x shows a crisply defined ball of stars, grainy at the edges and milky white in its center. Only 1° southeast is the peculiar globular cluster **NGC-5053**, a sparse system containing only 3,400 stars, compared to the Hercules cluster's one million stars. This loose group measures 8.9′ across and glows at magnitude 10.9, making it visible in a 6-inch as a grainy patch of weak light.

Midway between α Comae and the Coma cluster lies the bright star 35 Comae Berenices. One degree (one low-power telescopic field) east-northeast of this star is the beautiful eighth-magnitude system **M-64** (NGC-4826), the "Blackeye Galaxy." One of the brightest and most easily observed galaxies anywhere in the sky, it received its nickname for a dust patch superimposed across its face. This black eye is visible in even a 4-inch telescope at a good site, and is comfortably visible in a 6-inch from most observing sites. A good 8-inch scope shows the dark patch, a sharp condensed

M-64

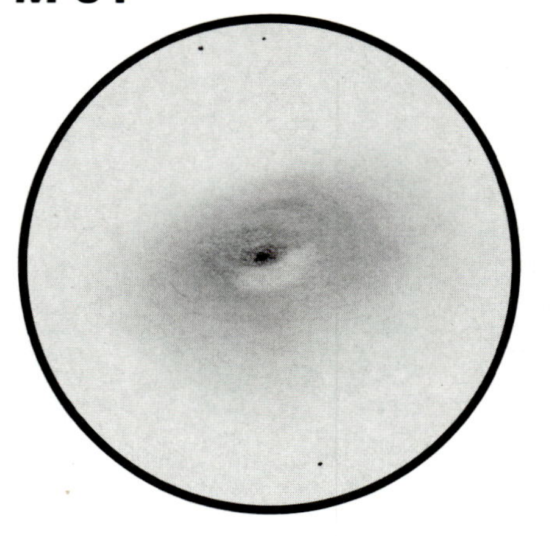

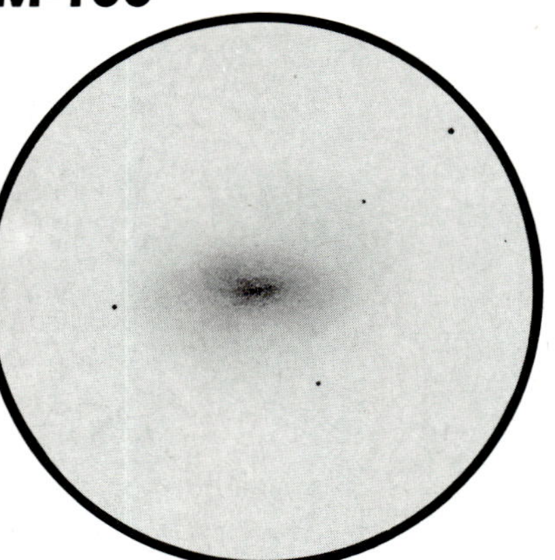

8-inch SCT
170x

M-99

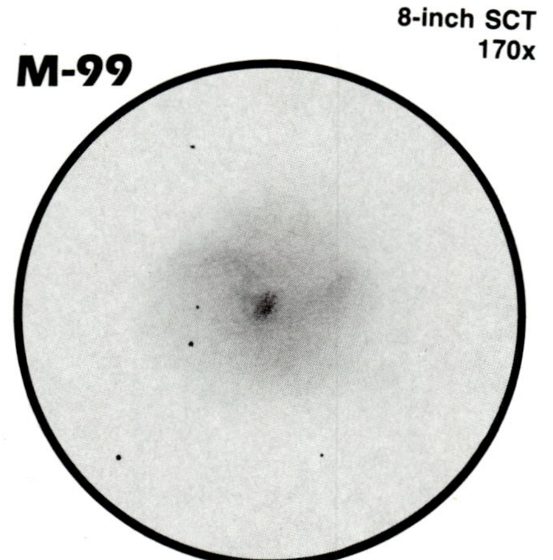

Sketches by David J. Eicher

nucleus, bright ringlike core, and a large outer envelope of faint gas. This Sb galaxy (some astronomers say Sa) lies roughly 9 megaparsecs away and shines with the luminosity of 13 billion Suns. Curiously, not one supernova has ever been recorded in this galaxy.

In the western part of the constellation, near the Leo border, lies the bright star **2 Comae Berenices**. This is a fine, relatively fixed binary system composed of magnitude 6.0 and 7.5 stars separated by 3.6″ in position angle 237°. As the distance between the spectral type A8 and F2 components is pretty close, find this star and then "crank up" to the highest usable power that delivers a smooth, steady image. You should be able to split the stars handily when the seeing is good. Another double in the constellation is **24 Comae Berenices**, an exquisite wide pair composed of magnitude 5.0 and 6.5 stars separated by 20.3″ in p.a. 271°. The K2- and A7-type stars show as yellow and bluish suns, the K2 star being the brighter. Perhaps Coma's finest variable star is the long-period variable **R Comae Berenices**, which changes in brightness from magnitude 7.5 to fainter than 14 in a period of 362 days. Since its period is almost exactly a year, estimating the brightness of this star once a week will give you a good prospect for an annual observing program.

Now we move south, approaching the border with Virgo; this is the direction of the main body of the Virgo cluster, the largest concentration of galaxies in our part of the universe. About 5° north of the cluster's heart, we find **M-85** (NGC-4382), a bright lenticular system measuring 3.0′ x 2.0′ across and shining at magnitude 10.5. In small telescopes it appears as a ball of fuzz showing little structure save for a condensed nuclear region. The distance to this system is about 13 megaparsecs, and it holds about 100 billion solar masses. The barred spiral galaxy NGC-4394 lies 7.8′ east of M-85.

The loose Sc-type spiral **M-100** (NGC-4321), the largest spiral in the Virgo cluster, is a much better eyepiece target. Its dimensions of 5.2′ x 5.0′ and magnitude of 10.4 — suggesting a roughly face-on inclination — provide a view of the nucleus and a faint hazy envelope of matter encapsulating it. With large amateur scopes some dust structure is visible, but only on supremely good nights. In 1979 a bright (12th magnitude) supernova erupted in this system, offering backyard telescopists a good view for over two weeks.

M-98 (NGC-4192) is an elongated nearly edge-on type Sb spiral, measuring 8.2′ x 2.0′ and shining at magnitude 11.0. This galaxy's surface brightness is rather low, making it a tricky object at high power. Backyard telescopes show this galaxy as a thin streak of greenish light, slightly curved, showing a faint envelope of gas and a sharp nucleus. Its neighbor **M-99** (NGC-4254), on the other hand, is a face-on type Sc spiral showing a bright hub, sharp nucleus, and faint outer haze suggesting spiral arms in a 6-inch on dark nights. A 10-inch at high power easily shows spiral structure in M-99 under dark skies. This system is a magnitude 10.4, 4.5′ x 4.0′ roundish glow.

The multiple arm galaxy **M-88** (NGC-4501), a magnitude 10.5 system measuring 5.7′ x 2.5′, is a fine sight in backyard instruments. It is inclined 30° from edge-on with respect to our line of sight, giving it a nice high surface brightness and a willingness to withstand high magnification. One of the better objects in the Virgo cluster for small scopes, it shows a broad elongated central glow and a large outer envelope of nebulosity.

Object	M#	Type	R.A. (2000)	Dec.	Mag.	Size/Sep./Per.	H
R CrB		LPV	12h 04m	+18°47′	7.5-14.0	362d	
2 CrB		★²	12h 04m	+21°27′	6.0,7.5	3.6″	
NGC-4192	M-98	∮	12h 15m	+14°54′	11.0	8.2′x2.0′	Sb
NGC-4254	M-99	∮	12h 19m	+14°25′	10.4	4.5′x4.0′	Sc
NGC-4321	M-100	∮	12h 23m	+15°49′	10.4	5.2′x5.0′	Sc
Melotte 111		⊙	12h 25m	+25°43′	2.9	360′	
NGC-4382	M-85	0L	12h 25m	+18°11′	10.5	3.0′x2.0′	S0
NGC-4501	M-88	∮	12h 32m	+14°26′	10.5	5.7′x2.5′	Sb
24 CrB		★²	12h 35m	+18°23′	5.0,6.5	20.3″	
NGC-4559		∮	12h 35m	+27°57′	10.5	10.0′x3.0′	Sc
NGC-4565		∮	12h 36m	+26°00′	10.5	15.0′x1.1′	Sb
NGC-4826	M-64	∮	12h 57m	+21°41′	8.6	7.5′x3.5′	Sa/Sb
NGC-4889		0	13h 00m	+27°59′	13.2	1.0′x0.6′	E4
NGC-5024	M-53	●	13h 13m	+18°10′	8.0	10.0′	
NGC-5053		●	13h 16m	+17°41′	10.5	8.0′	

H = Hubble classification type for galaxies

★²	*Double Star*
LPV	*Long Period Variable*
⊙	*Open Cluster*
●	*Globular Cluster*
∮	*Spiral Galaxy*
0	*Elliptical Galaxy*
0L	*Lenticular Galaxy*

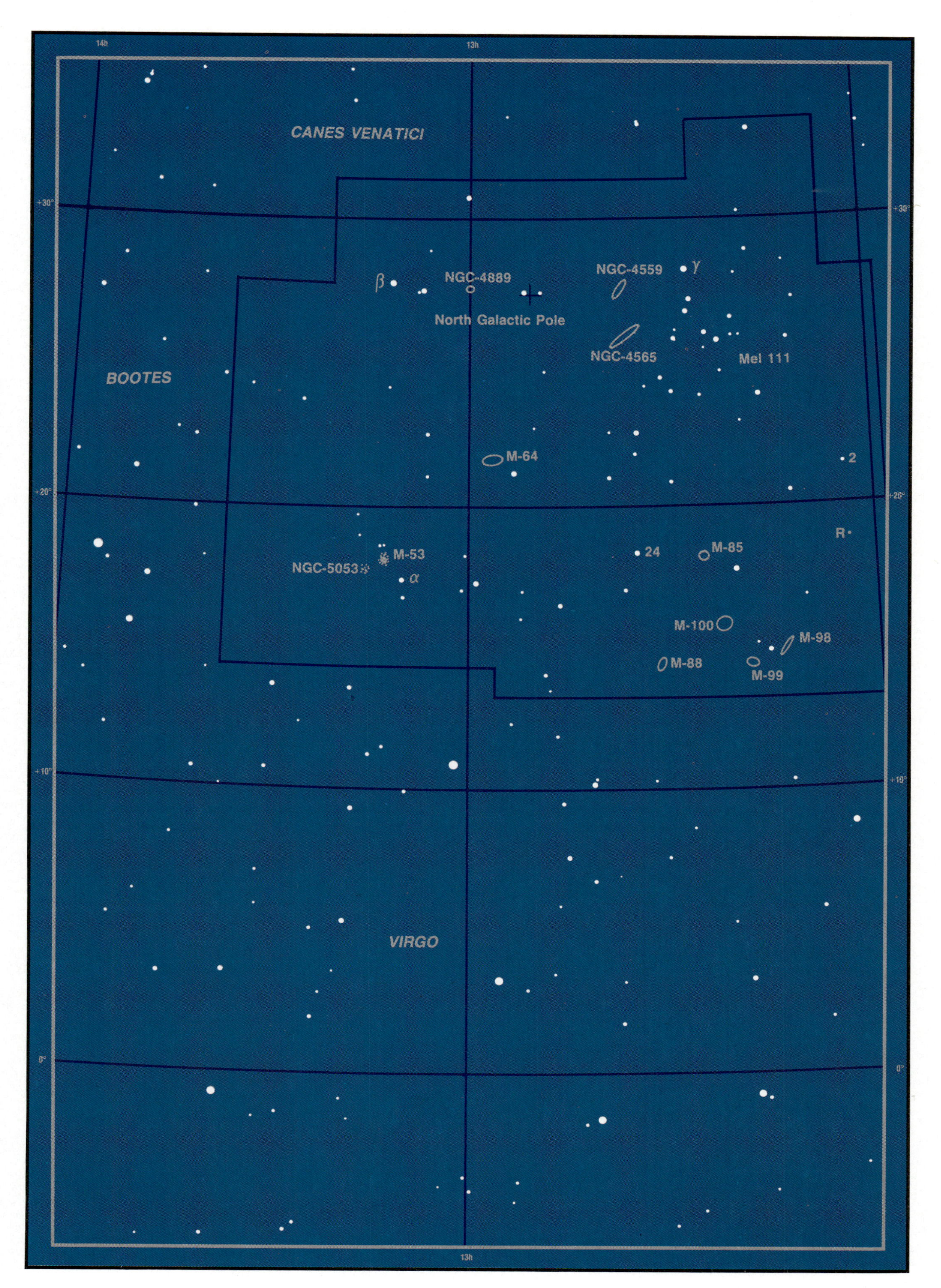
14h
13h
CANES VENATICI
+30°
β
NGC-4889
North Galactic Pole
NGC-4559
γ
NGC-4565
Mel 111
BOOTES
M-64
2
+20°
R
M-53
NGC-5053
α
24
M-85
M-100
M-98
M-88
M-99
+10°
VIRGO
0°
13h

This pair of photographs shows the changing brightness and dimensions of the nebulae designated NGC-6726/7 and NGC-6729. The area photographed on June 14, 1917 (above) shows NGC-6726/7 as a double knot of bright nebulosity and NGC-6729 as a comet-shaped blob; the same area photographed on July 11, 1936, reveals a different structure in the nebulae. Both photographs were made by Carl O. Lampland using the 13.1-inch astrographic camera at Lowell Observatory in Flagstaff, Arizona. Lowell Observatory photos.

Corona Australis

CrA
Coronae Australis

Telescopium

Tel
Telescopii

Keith Ward

Ten degrees below the bright Teapot of Sagittarius lies the diminutive constellation **Corona Australis** the Southern Crown. Its dim stars form a graceful arc that grazes the treetops for most Northern Hemisphere viewers but is sufficiently high for observing by telescope owners south of about 30° north latitude. To its south lies the even more inconspicuous constellation **Telescopium** the Telescope. It is difficult to see north of about 20° north latitude.

Despite their southerly declinations, these two constellations offer a handful of bright deep-sky objects for viewers with medium- or large-size telescopes. Corona Australis holds several fine globular clusters and an unusual complex of emission and reflection nebulosity. Telescopium, which lies farther away from the plane of the Milky Way, contains a number of elliptical and spiral galaxies.

The best globular in Corona Australis is **NGC-6541**, a magnitude 6.6 object that lies within the same low-power field as a 6th-magnitude star. To locate this group of stars, start at Shaula (Lambda [λ] Scorpii), the bright star marking the Scorpion's stinger. Now move southeast to the equally bright Kappa (κ) Scorpii, another star in the tail of Scorpius. Continuing in a straight line, pass by the wide unequal double formed by Iota^1 (ι^1) and Iota^2 (ι^2) Scorpii and continue 3° to a lone 6th-magnitude sun. Then move another 3° along the line to a second 6th-magnitude star; you're now aimed at the field containing NGC-6541.

NGC-6541 spans 13.1′, although in binoculars and finderscopes you'll probably see it as a glow some 3′ in diameter. Small telescopes reveal a nebulous disk measuring 5′ across located in a rich starry field. An 8-inch telescope at high power partially resolves this globular's faint outer edge into stars while the nucleus remains bright and unresolved.

About 2° southwest is a 6th-magnitude star that forms an isosceles triangle with the star accompanying NGC-6541 and the equally bright star located 3° southeast of the Iota^{1,2} pair. Within the same low-power field as this third 6th-magnitude star is another globular, **NGC-6496**. (The star is actually in Scorpius, but NGC-6496 lies just within the border of Corona Australis.) This cluster is far smaller and dimmer than its neighbor but is nevertheless a fine sight for small telescope users.

Shining at about magnitude 9, NGC-6496 covers nearly 7′ of sky. This globular is visible as a hazy patch of light in binoculars under a dark sky, though it may be difficult to see when it lies near the horizon if there is even slight light pollution or an unsteady atmosphere. A 6-inch telescope typically reveals NGC-6496 as a milky glow some 3′ across peppered with pinpoint stars. The surrounding field is so rich that these stars are likely foreground Milky Way stars. NGC-6496's rather great distance of 19.0 kiloparsecs makes it doubtful that such a small instrument could resolve the cluster's constituent stars.

Some 9° southeast of these two clusters lies Telescopium's sole bright globular, **NGC-6584**. This cluster appears remarkably similar to NGC-6496, measuring 7.9′ in diameter and glowing at around 9th magnitude. A 4-inch scope shows NGC-6584 as a featureless disk of gray light some 3′ across, and an 8-incher resolves the cluster's edge into a sprinkling of faint stars. Several lanes of these resolved stars emanating outward from the cluster are reminiscent of the much larger and brighter rows of stars visible in M-13, the Hercules Cluster.

Corona Australis' three brightest stars, all shining at 4th magnitude, form the brightest segment of the constellation's unmistakable arc. Alpha (α) CrA, Beta (β) CrA, and the wide double Gamma (γ) CrA are also markers for finding one of the strangest fields of nebulosity in this part of the sky. Set your telescope cross hairs 1° west of Gamma and you'll come upon the 7th-magnitude double star Epsilon (ε) CrA. To the northeast in this same low-power field is the complex of nebulosity dubbed **NGC-6726/7** and **NGC-6729**. NGC-6726/7, a double nebulosity that earned two separate NGC numbers, is a reflection nebula shaped like a figure eight some 2′ in extent. The nebula's southwestern section contains a 7th-magnitude pure white star and the northeastern part holds the erratic variable star **TY Coronae Australis**. This unusual star fluctuates irregularly between magnitudes 8.8 and 12.6; its magnitude can be checked against that of the 7th-magnitude star located nearby.

About 4.7′ southeast of the double nebula is the even stranger nebulosity NGC-6729. This patch of gray light is 1′

BEST VISIBLE DURING
SUMMER

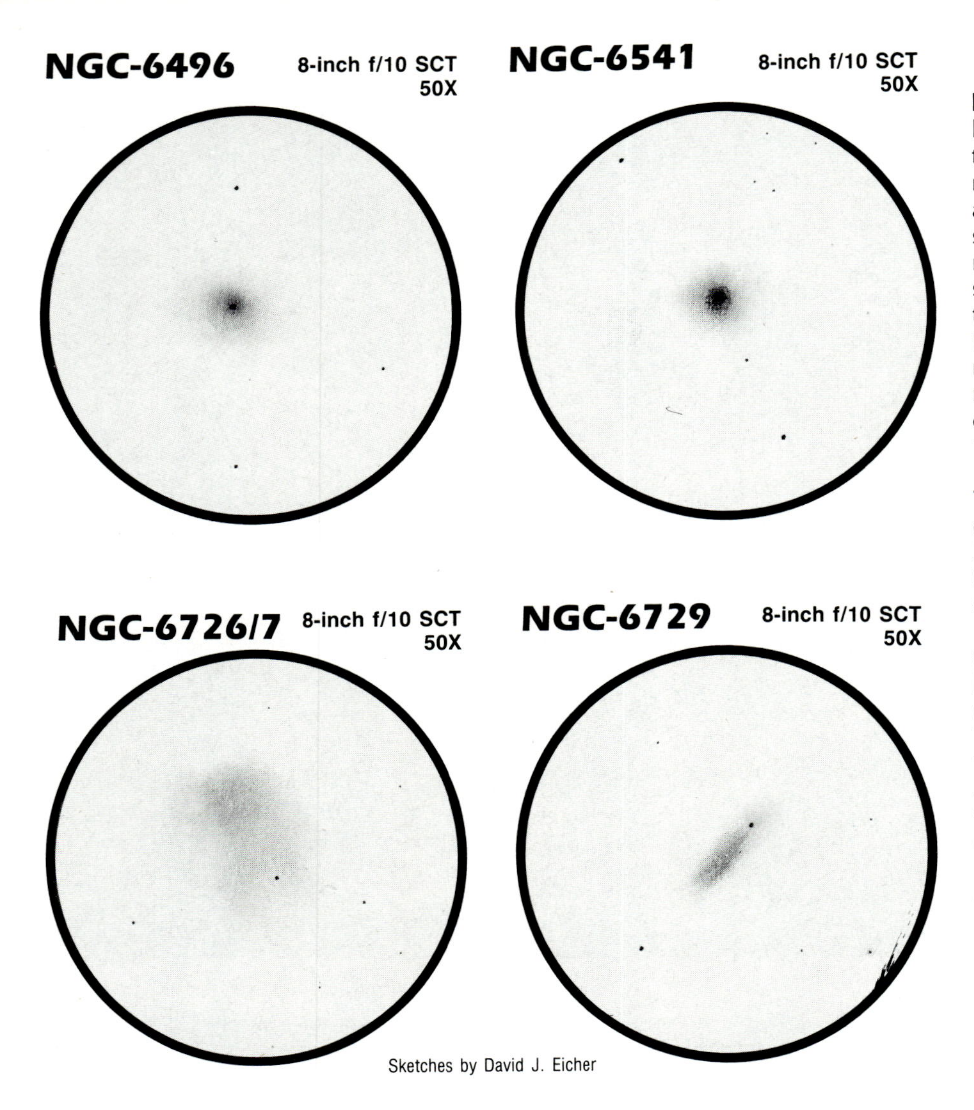

Sketches by David J. Eicher

long and appears as a tiny comet complete with a fuzzy tail. Embedded within this object is the bizarre nebular variable R Coronae Australis. This star's fluctuations in light change the visible size of the nebulosity, much like Hubble's Variable Nebula in Monoceros. The star nominally shines at 12th magnitude but can flare to 10th in just a few days, noticeably changing the nebula's shape. Inside the "tail" of this emission and reflection nebulosity is yet another variable, T Coronae Australis. This and another star in the field, S Coronae Australis, fluctuate much like R CrA but are about two magnitudes fainter. (Less than a degree north of this nebulosity lies the globular NGC-6723, across the border in Sagittarius.)

Corona Australis and Telescopium each hold one bright planetary nebula. **IC 1297** lies wedged between 7th- and 8th-magnitude stars 2° southeast of Beta CrA. This nebula appears as a pale blue disk 7″ across in a field swimming with stars. It shows no detail within the disk and reveals no central star to backyard telescopes. **IC 4699** lies equidistant between the stars Alpha and Epsilon Telescopii and is visible in small scopes as a dim circular nebulosity some 10″ across. An 11th-magnitude star lies about 1′ north of the disk, and there is a faint pair of stars some 1′ to the southwest.

Telescopium's oddest star is **RR Telescopii**, a variable found in the southeastern corner of the constellation a mere 3° northwest of Alpha Pavonis. This star is an extreme case of a "slow nova," with a freakish history of behavior. During the 1930s, the star varied between magnitudes 12 and 15 over a period of 387 days, but in the early 1940s it fell to several minima of magnitude 16.5. In 1944, RR Tel went nova and exploded in brightness to magnitude 7, remaining that bright for a number of years. By 1951 the star had faded to 8th magnitude and it has continued fading since.

A number of bright galaxies lie scattered across Telescopium. **IC 4797** is an 11th-magnitude elliptical galaxy found 2° south of Lambda Tel. It measures 2.8′ x 1.3′ across and appears as an oval patch of light in 4-inch or larger scopes. **IC 4889** lies 7° due east of IC 4797 and appears quite similar. It, too, is elliptical; its magnitude is 12.4 in blue light and its dimensions are 2.6′ x 1.4′. **NGC-6861** and **NGC-6868** lie in the same wide telescopic field in the northeastern part of Telescopium. The former is an 11th-magnitude lenticular galaxy; the latter is an elliptical of comparable size and brightness. **NGC-6887**, 5° southeast of the NGC-6861/NGC-6868 pair, is a relatively large, faint spiral appearing in small scopes as a hazy mass of milky light.

Object	M#	Type	R.A. (2000)	Dec.	Mag.	Size/Sep./Per.	H
NGC-6496		●	17h 59.0m	−44°16′	9.2:	6.9′	
NGC-6541		●	18h 08.0m	−43°42′	6.6	13.1′	
NGC-6584		●	18h 18.6m	−52°13′	9.2:	7.9′	
IC 4699		■	18h 18.6m	−45°59′	11.9_P	10″	
IC 4797		()	18h 56.5m	−54°18′	11.39	2.8′x1.3′	E5
NGC-6726-7		□R	19h 01.7m	−36°53′	—	2′x2′	
TY CrA		IV	19h 01.7m	−36°53′	8.8↔12.6	irr.	
NGC-6729		□ER	19h 01.9m	−36°57′	—	1′	
IC 1297		■	19h 17.4m	−39°37′	—	7″	
IC 4889		()	19h 45.3m	−54°20′	12.4_B	2.6′x1.4′	E5
RR Tel		IV	20h 04.2m	−55°43′	6.5↔16.5	irr.	
NGC-6861		§L	20h 07.3m	−48°22′	11.1	2.7′x1.4′	S0
NGC-6868		()	20h 09.9m	−48°23′	12.0_B	2.7′x2.2′	E2
NGC-6887		§	20h 17.2m	−52°47′	12.5_B	4.1′x1.7′	Sb+:

H = Hubble classification type for galaxies
subscript "P" denotes photographic magnitude; subscript "B" denotes blue magnitude.

IV	*Irregular Variable*
●	*Globular Cluster*
■	*Planetary Nebula*
□ER	*Emission/Reflection Nebula*
□R	*Reflection Nebula*
()	*Elliptical Galaxy*
§L	*Lenticular Galaxy*
§	*Spiral Galaxy*

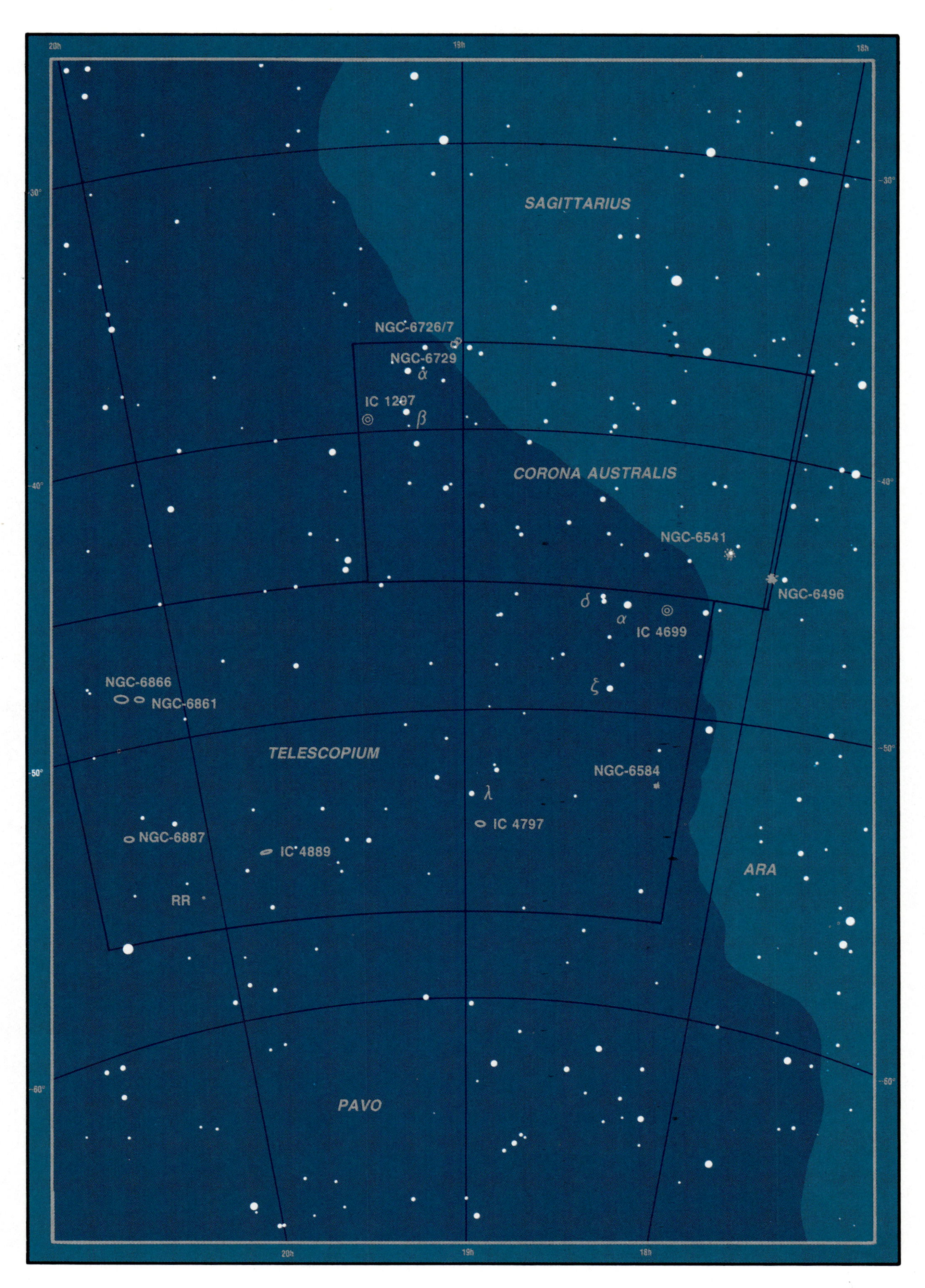
20h
19h
18h
-30°
-40°
-50°
-60°
SAGITTARIUS
NGC-6726/7
NGC-6729
α
IC 1287
β
CORONA AUSTRALIS
NGC-6541
NGC-6496
δ
α
IC 4699
NGC-6866
NGC-6861
ζ
TELESCOPIUM
NGC-6584
λ
IC 4797
NGC-6887
IC 4889
ARA
RR
PAVO
20h
19h
18h

Above: M-83, the brightest galaxy in Hydra, appears as a large circular halo of nebulosity around a bright core. Under dark skies, backyard telescopes show its spiral shape and patches of dust within the arms. Photo by Jack B. Marling. Below: The colliding galaxies NGC-4038 and NGC-4039 appear as a single nebulous arc in a small scope. Photo by Jack Newton. Below right: Although diminutive, the planetary nebula NGC-7662 is bright enough to withstand high magnification which reveals its inner detail. Photo by Joe Liddell. Bottom right: Globular cluster M-68 appears as a fuzzy ball of stars in small telescopes, requiring high power to resolve its edges. Photo by Martin C. Germano.

Corvus

Crv
Corvi

Crater

Crt
Crateris

Eastern Hydra

Hya
Hydrae

Keith Ward

The region of sky containing **Corvus**, **Crater**, and **Eastern Hydra** — southwest of the great Virgo cloud of galaxies and far east of the winter Milky Way — appears barren. However, it contains many small galaxies visible in backyard telescopes and a handful of other sights, including a globular cluster, a planetary nebula, and several unusual variable stars.

Corvus' most unusual deep-sky object is the pair of interacting galaxies **NGC-4038** and **NGC-4039**. These galaxies lie some 12 megaparsecs away and shine with the combined light of an 11th magnitude star. Together, NGC-4038 and NGC-4039 measure 2.5′ x 2.5′ across. Their fuzzy, lopsided shapes — one a distorted spiral, the other a peculiar eruptive galaxy — are clearly visible in a good 8-inch telescope on a dark night. A larger telescope reveals mottled areas of detail across the galaxies' surfaces, resembling a backward question mark. These objects undoubtedly comprise one of the easiest-to-observe pairs of physically-connected galaxies.

One of the nicest planetary nebulae in the spring sky is **NGC-4361**, a magnitude 10.5 bright blue disk measuring 80″ across. When viewed with small telescopes, NGC-4361 appears as a fuzzy "star" at low power. A good 8-inch telescope at high power, however, reveals the nebula's faint outer halo which envelopes an inner bright disk. The central star shines at 13th magnitude and is rather difficult to spot, but a 10-inch or larger telescope will reveal it shining through the fluorescing ball of gas.

BEST VISIBLE DURING SPRING

Delta (δ) **Corvi**, also known as Algorab, is a third-magnitude double star suitable for viewing with any size telescope. Discovered by John Herschel and James South in 1823, Algorab consists of a magnitude 3.0 primary with a magnitude 8.4 secondary 24.2″ away in position angle 214°. (The p.a. and distance have remained fixed for the last 160 years.) The primary is a spectral type B9 main sequence star, whereas the secondary is a dwarf K2-type star. Their colors are often described as yellowish and pale lilac, or simply white and orange. This system lies some 40 parsecs away, which means that the primary star shines with the luminosity of 75 Suns. In spite of the primary's brilliance, this double star is easily visible in virtually any telescope because of the secondary's considerable distance.

The pattern of stars known as Crater is somewhat ill-defined, but worth seeking out because of three galaxies which reside within its boundary. **NGC-3511** is a 12th magnitude Sc-type spiral with a bright nucleus surrounded by faint haze. A 6-inch telescope shows it as a tiny core — perhaps one arc-minute across — with a larger faint halo of nebulosity spanning perhaps 3′ across. **NGC-3672** is an Sb-type spiral, nearly as bright though somewhat smaller than NGC-3511. Observatory photos show that NGC-3672 is a highly-inclined multiple-arm spiral with numerous intertwining dust patches. But this detail is impossible to spot visually: 10-inch scopes show only a 1′ diameter bright nucleus and a smooth halo of greyish-green nebulosity. The Sc-type spiral **NGC-3887** is brighter and rounder than its neighbors, shining at magnitude 11.6 and measuring 2.6′ x 2.0′ across. A 6-inch telescope shows it as a bright, round, featureless spot.

The sprawling eastern half of the constellation Hydra contains a variety of celestial targets. Among its galaxies is **Messier 83** (NGC-5236), the finest face-on Sc-type spiral in the sky. With a combined magnitude of 8, this magnificent object is visible in binoculars as a bright nebulous disk, sharply brighter toward its center and spanning one-third the Moon's diameter. In telescopes, M-83 really shows itself: a 6-incher at high power reveals a condensed, brilliant nucleus within a weakly bar-shaped outer envelope of low surface brightness nebulosity. In between the faint spiral-structured arms are dark patches blocking the starlight, which represent huge areas of dust in the galaxy's disk.

M-83 was discovered by Nicholas Louis de Lacaille in 1752. The distance to M-83 is about 4.5 megaparsecs, making it one of the closest spiral galaxies outside our Local Group — which includes M-31, M-33, NGC-6822, and the Milky Way. A total of five supernovae have occurred within M-83 during the last 60 years, one in 1923, 1950, 1957, 1968, and 1983. This unusually high rate of producing supernovae makes M-83 a suitable target for backyard supernova hunters. Next time you're out observing M-83, take along a photo showing stars around the galaxy. If you see a star that doesn't appear on the photo, you may have made a discovery!

Three other notable galaxies are tucked away in eastern Hydra — **NGC-3109**, **NGC-3145**, and **NGC-3621**. NGC-3109 is a patchy irregular galaxy with low surface brightness. Its large dimensions of 11.0′ x 2.0′ and bright magnitude of 11 belie a tricky object, since the light is so widespread. Use a low-power ocular under dark skies, if possible; with a 6-inch scope, this will show a long, thin spike of faint nebulosity. A 10-inch telescope reveals a mottled, uneven texture along the galaxy's surface along with several faint stars in and around the nebulosity. NGC-3145 is a more typical object — a magnitude 12.5 barred spiral covering 2.4′ x 1.0′. A 10-inch scope displays it as a small, oval core enveloped in a very faint haze. A 17.5-inch telescope's view brightens the nebulosity and shows it to be mottled, but does not expose much more detail. NGC-3621, an Sc- or Sd-type spiral, measures 5.0′ x 2.0′ and shines with the light of a magnitude 10.6 star. A 6-inch telescope shows this galaxy as a large and bright oval nebulosity containing a condensed nucleus.

Eastern Hydra contains two fine variable stars — **R Hydrae** and **V Hydrae**. R Hya is a Mira-type long period variable whose period of 386 days carries the star between magnitudes 4 and 10. Although it is visible to the naked eye for only several weeks during each cycle, this star can be followed with binoculars or a small telescope. V Hyd is a semi-regular variable with a period of 533 days. It varies between magnitudes 6.5 and fainter than 12, so you'll need a 6-inch scope to follow its complete cycle.

M-83 8-inch f/10 SCT 50x

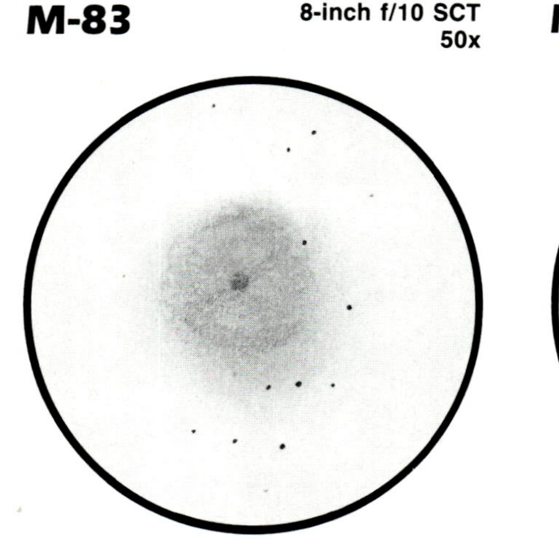

M-68 8-inch f/10 SCT 100x

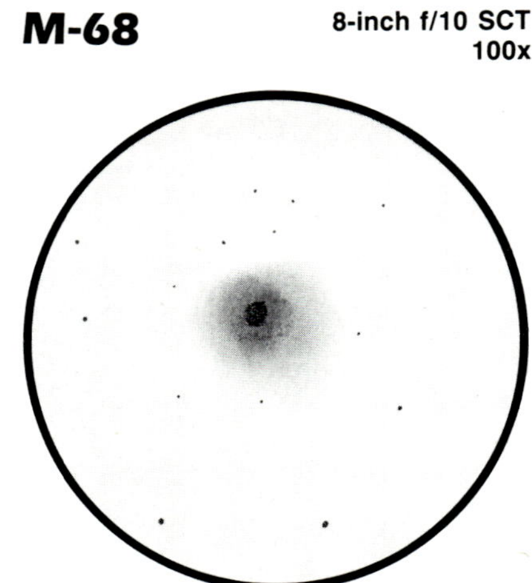

Sketches by David J. Eicher

NGC-3242 17.5-inch f/4.5 reflector 100x

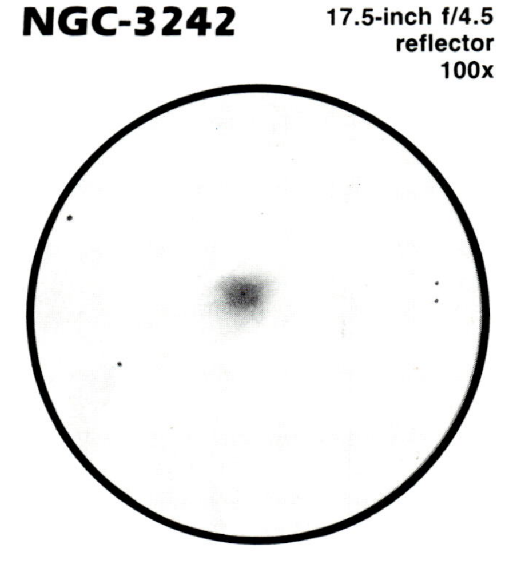

Another M-object in Hydra is **Messier 68** (NGC-4590), a bright, medium-sized globular cluster of stars. Although it isn't as spectacular as M-13 in Hercules or perhaps M-22 in Sagittarius, M-68 measures a full 9′ across and glows at magnitude 8; it is obviously nonstellar even in binoculars. Under dark skies, a 4-inch scope at high power shows M-68 as a large, fuzzy ball of unresolved light. An 8-inch telescope at high power begins to resolve M-68's edges, showing it to be a cluster of stars. **NGC-5694** is an 11th magnitude globular measuring only 2′ across — it is one of the most distant globulars and impossible to resolve with backyard telescopes.

Tiny, yet often observed, planetary nebula **NGC-3242** has much to offer for small telescope viewers. It shines at 9th magnitude, measures 40″ across, and contains an obvious 11th magnitude central star. Binoculars show NGC-3242 as a bluish "star," but small telescopes at medium power display its disk and may even reveal the central star as a tiny dot inside the nebula. Large telescopes — in the 10-inch to 20-inch range — show that NGC-3242 is composed of a bright central star, a bright inner disk, and a fainter outer halo of greenish-blue nebulosity.

Object	M#	Type	R.A. (2000)	Dec.	Mag.	Size/Sep./Per.	H
NGC-3109		#	10h 03.1m	−26°10′	11.2	11.0′x2.0′	I
NGC-3145		§B	10h 10.1m	−12°25′	12.5	2.4′x1.0′	SBb
NGC-3242		■	10h 24.8m	−18°38′	8.9	40″	
V Hya		SRV	10h 51.6m	−21°15′	6.5↔12...	533d	
NGC-3511		§	11h 03.2m	−23°06′	11.9	4.2′x1.5′	Sc
NGC-3621		§	11h 18.3m	−32°48′	10.6	5.0′x2.0′	Sc/Sd
NGC-3672		§	11h 25.0m	− 9°48′	11.8	3.5′x1.4′	Sb
NGC-3887		§	11h 47.1m	−16°52′	11.6	2.8′x2.0′	Sc
NGC-4038/9		#	12h 01.9m	−18°52′	11.0	2.5′x2.5′	S?pec
NGC-4361		■	12h 24.5m	−18°46′	10.5	80″	
δ Crv		$\star^2$	12h 29.9m	−16°31′	3.0,8.5	24.2″	
NGC-4590	M-68	●	12h 39.5m	−23°16′	8.0	9.0′	
R Hya		LPV	13h 29.7m	−23°16′	4.0↔10.0	386d	
NGC-5236	M-83	§	13h 37.1m	−29°52′	8.0	10.0′x8.0′	Sc
NGC-5694		●	14h 39.6m	−21°32′	11.0	2′	

H = Hubble classification type for galaxies

$\star^2$	*Double Star*
LPV	*Long Period Variable*
SRV	*Semi-Regular Variable*
●	*Globular Cluster*
■	*Planetary Nebula*
§	*Spiral Galaxy*
§B	*Barred Spiral Galaxy*
#	*Peculiar/Irregular Galaxy*

12h
11h
Ecliptic
0°
VIRGO
NGC-3672
-10°
CORVUS
NGC-3887
NGC-4361
NGC-4038
-20°
CRATER
NGC-3511
HYDRA
M-68
-30°
NGC-3621
ANTLIA
-40°
13h

Above: With the naked eye, the North America Nebula is easy to see, appearing as a large fuzzy patch, but its light is so spread out that it is difficult to detect with a telescope. Photo by Joseph Glowicki and Andrew Toth. Above right: One of the least flashy Messier objects, M-29 is nevertheless a bright, compact cluster composed of ninth magnitude and fainter stars. Photo by Lee C. Coombs. Right: The eastern half of the Veil Nebula is NGC-6992/5, which is visible as a ghostly arc in 8-inch scopes. Photo by Jack Newton. Below right: M-27, the Dumbbell Nebula, is one of the most impressive of the planetary nebulae. Photo by Ben Mayer. Below: The region of Gamma (γ) Cygni is filled with the huge emission nebula IC 1318. Photo by Al Lilge.

Cygnus

Cyg
Cygni

Vulpecula

Vul
Vulpeculae

Keith Ward

Among groupings of stars in the northern Milky Way, the constellation **Cygnus** the Swan is the richest, as it contains multitudes of bright, young open star clusters, patchy regions of dark nebulae, wispy, elusive emission and reflection nebulosity, many planetary nebulae, and a handful of double and variable stars. The small constellation **Vulpecula**, the Little Fox, adjacent to Cygnus, also has a worthwhile telescopic target among its dim stars.

One of the brightest objects in Cygnus is the large, scattered open cluster **M-39** (NGC-7092). This group contains at least 28 stars, the brightest of which shine at seventh magnitude, in a diameter approximately half that of the Moon, resulting in a total magnitude of 4.6. This makes M-39 one of the finest large open clusters for binocular users, as most telescopic fields of view do not contain the entire group. The cluster lies about 245 parsecs distant, and has a linear diameter of 2.2 pc. The other Messier object in Cygnus is another cluster, **M-29** (NGC-6913). This little group of 81 stars of ninth magnitude and fainter is contained — in stark contrast to M-39 — in a diameter of only 7′. The combined magnitude of M-29 is 6.6, so it is easily visible in finder scopes, but it is not impressive as open clusters go.

More enthralling but much harder to observe is the complex of nebulosity surrounding the M-29 area and centered on the bright star Gamma (γ) Cygni, or Sadr. **IC 1318**, a vast region of emission nebulosity, is excited by young, hot blue-white stars and separated by dark nebulae into five major parts. The total magnitude of this nebulosity is probably fairly high, but its light is spread out enough to make small parts of it dim. Under pitch black skies, a 5-inch RFT will show the wispy outlines of this large nebulosity. Embedded within the IC 1318 complex, on the northeastern side of γ Cyg, is the fine open cluster **NGC-6910**. This group consists of 66 stars of 10th magnitude and fainter in an area spanning 8′, with a combined magnitude of 7.4.

Southwest of the γ Cyg region lie a number of unusual objects, one of which is the faint supernova remnant **NGC-6888**. This is a young shell of ionized gas that expanded outward from a hot central Wolf-Rayet star. Since it is one of only a handful of observable supernova remnants, it is a worthwhile target on nights of excellent transparency. Its surface brightness is extremely low — something like 14 — so you'll need either a very transparent night or large optics coupled with a red-transmitting nebular filter. Measuring 18′ x 12′, the nebula is visible in large binoculars under superb skies.

The open cluster **NGC-6871** lies in the "downtown" section of the Cygnus Milky Way — an area strewn with star clouds and cluttered with thousands of stellar images. The group stands out enough to be an obvious binocular target, however, because it measures 40′ across, contains 66 stars of seventh magnitude and fainter, and shines with the light of a magnitude 5.2 star.

Also abundant in unique objects is the area around Deneb (Alpha [α] Cygni). One of the most sought-after low-surface-brightness objects is the **North America Nebula**, NGC-7000. This giant emission region measures 120′ x 100′; under dark skies, it is visible as a large hazy patch of light just southeast of Deneb. The integrated magnitude of NGC-7000 is perhaps 4.5, but it is so large that in the telescope little bits of it appear as 13th magnitude nebulosities; a 17.5-inch with nebular filter easily shows its outlines, but it fills several fields of view. Once again, you'll find binoculars often provide the best view. Due west of NGC-7000, separated by a broad band of dust, is the even more elusive **Pelican Nebula** (IC 5067-70). This is a part of the same region, measures 80′ across, and has a surface brightness of something like 14. Faintly visible with large binoculars or RFTs, it becomes harder to pick out with large telescopes.

NGC-7027 is a strange object near the North America Nebula. It was originally believed to be a star, then a planetary nebula, and now it appears to be a bizarre emission object showing a complex spectrum. It measures only 18″ x 11″ across and has a photographic magnitude of 10.4, so it isn't always easy to pick out of a rich starfield at low power. It displays a slight blue-green color, so use medium power and look for a colorful "star" whose disk exceeds in size the stars that surround it. Also nearby are two

BEST VISIBLE DURING
SUMMER

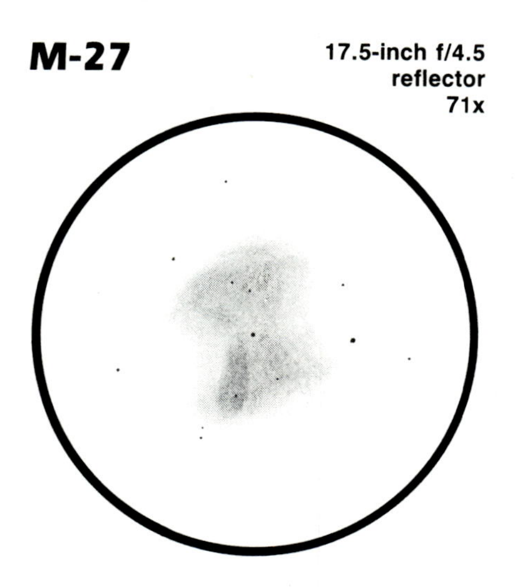

Sketch by David J. Eicher

planetaries: **NGC-7026**, a 12th magnitude object measuring 28″ x 13″, and **NGC-7048**, an 11th magnitude planetary whose disk spans 60″ x 50″ of sky. NGC-7026 can be difficult to find, but NGC-7048 is easily spotted within the low-power field of any small telescope.

Two more planetaries in Cygnus are **NGC-6826** and **NGC-7008**. NGC-6826 shines at ninth magnitude and measures 27″ x 24″. It is often called the Blinking Planetary, as it displays a curious effect that neatly demonstrates the technique of averted vision. After you find 6826 using averted vision, center it in the field and look directly at it. The nebula will fade, leaving only the pinpoint 11th magnitude central star. Look to the edge of the field, and the nebula reappears. If you switch rapidly back and forth between averted and direct vision, the nebula appears to blink on and off. NGC-7008 is large but faint, measuring over one arc-minute in diameter but glowing dimly at 13th magnitude. Telescopes 10 inches in aperture may show it as a large fuzzy glow, devoid of detail.

The third and final grand complex of nebulosity in Cygnus is the sky's brightest supernova remnant, the **Veil Nebula**. Consisting of more than four distinct bright units of nebulosity, each of which has a separate NGC number, the Veil Nebula is the result of a supernova explosion that occurred some 30,000 years ago, about 450 pc away toward the Cygnus Arm of the Galaxy. The brightest section, designated NGC-6992 and NGC-6995, comprises the eastern half of the main loop, and is visible as a ghostly arc in an 8-inch telescope at low power. Its angular extent is 78′ x 8′, so it appears as a long, thin twisting spike, cutting through an entire field of view. NGC-6960, the western half, is easier to locate because it passes through the bright star 52 Cygni — a star that makes spotting the nebulosity harder than it should be. NGC-6960 is nonetheless visible in a 6-inch scope, especially with a nebular filter. NGC-6979, the central part of the Veil, is markedly fainter than the two main halves. A large telescope is capable of showing sharply defined, intricate detail throughout the Veil complex — detail reminiscent of photographs.

Beta (β) Cygni, known as **Albireo**, is one of the most observed double stars in the sky. Its magnitude 3.1 and 5.1 components are separated by a generous 34.3″, making the system visible in small scopes. The most pleasing aspect of observing Albireo is the color contrast: a brilliant yellowish primary coupled with a dazzling soft-blue secondary. Albireo lies 120 pc distant; its stars are separated by 4400 astronomical units, or about 400 billion miles. Another notable double in Cygnus is **61 Cygni**, a pair composed of 5.3 and 5.9 magnitude stars separated by 28.4″; the distance to this pair, which was the first star to have its distance determined (by F.W. Bessel in 1838), is 3.4 pc.

A beautiful object is the bright and famous planetary M-27 (NGC-6853), the **Dumbbell Nebula** in Vulpecula, whose high-surface-brightness disk is easily visible in any telescope. This nebula measures 480″ x 240″ across and has an integrated magnitude of 7.3; binoculars show it to be a hazy little patch located inside a circular grouping of bright stars. A 6-inch scope under dark skies shows the dumbbell shape and perhaps even the "ears" of faint low-surface-brightness gas that extend outward. Larger telescopes show an extensive halo of nebulosity around the Dumbbell and the nebula's 13th magnitude central star, one of several stars floating inside the nebulosity (the others are foreground objects).

Object	M#	Type	R.A. (2000)	Dec.	Mag.	Size/Sep./Per.	N★	Mag.★
Beta (β) Cyg		★²	19h 30.7m	+27°57′	3.1, 5.1	34.3″		
NGC-6826		■	19h 44.9m	+50°31′	8.7	27″ x 24″		10.6
Delta (δ) Cyg		★²	19h 44.9m	+45°07′	3.0,6.5	2.2′		
Chi (χ) Cyg		LPV	19h 50.5m	+32°55′	5.1-14.3	407d		
NGC-6853	M-27	■	19h 59.6m	+22°43′	7.3	480″ x 240″		12.0
NGC-6866		⊙	20h 03.7m	+44°00′	7.6	8.0′	129	10....
NGC-6871		⊙	20h 06.7m	+35°51′	5.2	40.0′	66	7...
NGC-6888		SNR	20h 12.5m	+38°25′	11	18′ x 12′		
IC 1318		□E	20h 16.5m	+41°48′		24″ x 17″		2.3
IC 1318		□E	20h 18.8m	+40°39′		50′ x 25′		2.3
IC 1318		□E	20h 19.8m	+40°07′		45′ x 25′		2.3
NGC-6910		⊙	20h 23.1m	+40°47′	7.4	8.0′	66	10...
IC 1318		□E	20h 23.8m	+40°45′		85′ x 50′		2.3
NGC-6913	M-29	⊙	20h 24.0m	+38°31′	6.6	7.0′	81	9...
IC 1318		□E	20h 28.5m	+39°57′		70′ x 20′		2.3
NGC-6960		SNR	20h 45.6m	+30°43′	7	70′ x 6′		
IC 5067-70		□E	20h 48.6m	+44°22′	9	85′		
NGC-6979		SNR	20h 51.5m	+31°01′	11			
NGC-6992/5		SNR	20h 56.4m	+31°42′	7	78′ x 8′		
NGC-7000		□E	20h 58.7m	+44°20′	4.5	120′ x 100′		
61 Cyg		★²	21h 06.3m	+38°40′	5.3,5.9	28.4″		
NGC-7026		■	21h 06.3m	+47°51′	11.8p	28″ x 13″		14.2
NGC-7027		□E	21h 07.1m	+42°14′	10.4p	18″ x 11″		
NGC-7048		■	21h 14.2m	+46°16′	11.3p	60″ x 50″		18.0p
NGC-7092	M-39	⊙	21h 32.2m	+48°26′	4.6	30.0′	28	7...
IC 5146		□R	21h 53.3m	+47°16′	9	10′ x 10′		10.0
Collinder 399		⊙	19h 25.0m	+20°00′				

N★ = number of stars, Mag.★ = magnitude range of cluster or magnitude of central star, (...) indicates many fainter.

★²	*Double Star*
LPV	*Long Period Variable*
⊙	*Open Cluster*
□E	*Emission Nebula*
□R	*Reflection Nebula*
SNR	*Supernova Remnant*
■	*Planetary Nebula*

22h
21h
20h
19h
Galactic Equator
CYGNUS
LYRA
VULPECULA
DELPHINUS
SAGITTA
NGC-6826
M-39
IC 5146
NGC-7026
NGC-7048
NGC-7000
IC 5067-70
NGC-6866
NGC-7027
IC 1318
NGC-6910
61
M-29
NGC-6888
NGC-6871
NGC-6979
NGC-6992/5
NGC-6960
M-27
Collinder 399
+50°
+40°
+30°
+20°
+10°

Above: Globular cluster M-71 in Sagitta is so easy to resolve into stars that for years many observers believed it was a rich open cluster. Photo by Bill Iburg.

Far left: Because planetary nebula NGC-6905 in Sagitta appears as a fuzzy ''star'' in small telescopes, a 6- or 8-inch telescope is required for a good view. Photo by Jack B. Marling.

Left: One of the most distant globulars visible in backyard telescopes is Delphinus' NGC-7006, a 3′-diameter ball of gray light. Photo by Jack Newton.

Delphinus

Del
Delphini

Sagitta

Sge
Sagittae

Keith Ward

Delphinus the Dolphin and **Sagitta** the Arrow are both small, bright groups of stars lying near the plane of the Milky Way. Delphinus is a kite-shaped constellation holding five 4th-magnitude stars and a sprinkling of globular clusters and planetary nebulae. Sagitta contains a similar blend of globulars and planetaries and is one of the few constellations that actually looks like what it represents.

The brightest deep-sky object in the area is globular cluster **M-71** (NGC-6838), which is bright enough to be visible in any telescope or pair of binoculars. M-71 is easy to find because it lies midway between the 4th-magnitude stars Gamma (γ) and Delta (δ) Sagittae. To position M-71 in your telescope, move 2° southwest from Gamma Sagittae: you should see a small, fuzzy glow of light from the cluster.

M-71 is thought to have been discovered in 1775 by Johann G. Koehler in Dresden, Germany. However, there is evidence suggesting Phillippe de Cheseaux observed the cluster as early as 1746. In June 1780 the cluster was rediscovered by Charles Messier's colleague Pierre Mechain, which prompted Messier to search for the cluster. Messier's notes describe his seventy-first listing as "very faint . . . it contains no star . . . the least light extinguishes it."

The view through modern backyard telescopes is considerably better. A 3-inch refractor at low power shows a bright knot of light surrounded by a rich field of bright stars. Six-inch telescopes show M-71 as a hazy glow 5′ across, peppered by dozens of 11th- and 12th-magnitude stars. The cluster's total magnitude is 8.3, which makes it appear very bright in large backyard scopes. A 12-inch scope resolves over one hundred stars across a 7′-diameter disk, which demonstrates that M-71 is one of the "loosest" bright globular clusters. The cluster also lacks the typically strong central condensation of normal globular clusters.

BEST VISIBLE DURING
SUMMER

M-71 is so easily resolved that for many years astronomers argued over whether it was a loose globular cluster or an extremely rich open cluster. After years of debate with good evidence supporting each side, most astronomers now believe M-71 is an unusual globular cluster.

Less than one degree southwest of M-71 is the sparse open cluster **Harvard 20**. Measuring 7′ across and glowing at magnitude 7.7, H20 is a fine cluster in binoculars and small scopes at low power. Such instruments show the group as a collection of fifteen stars, the brightest shining at magnitude 10. Large telescopes at low powers improve the view of H20 and show myriad faint field stars in and around the cluster. High-power views of H20 are unimpressive because the sprinkling of stars ceases to appear like a cluster and takes on the look of a typical Milky Way starfield.

Sagitta contains three fine planetary nebulae for small telescopes. Lying 2.5° east of the 6th-magnitude optical double 15 Sagittae is the faint planetary **NGC-6879**. Finding NGC-6879 is fairly easy because NGC-6879 lies along an east-west row of four 7th-magnitude stars, between and slightly north of the two easternmost stars. This object has a photographic magnitude of 13.0 and a diameter of 5″, which mean you'll have to use relatively high power and observe during a period of good seeing. With an 8-inch scope at 200x, NGC-6879 appears like a tiny out-of-focus star with a distinctly bluish cast. NGC-6879's central star glows dimly at magnitude 15, which makes it exceedingly difficult to see.

Three degrees east and slightly south of NGC-6879 is the bright, nearly stellar planetary **IC 4997**. This object shines at photographic magnitude 11.6 but measures only 2″ across. Successfully distinguishing IC 4997 from the stellar backdrop requires a long-focus telescope and very steady seeing. An 8-inch f/10 SCT at 300x reveals the planetary as a disk. The central star in IC 4997 is a variable that usually hovers at around 13th magnitude. The surrounding nebulosity may make this star somewhat difficult to see — especially if seeing is poor — but the star should be faintly visible in a 10-inch telescope.

Now move your telescope 3.5° north and 0.5° east from NGC-6879, and you'll encounter another small planetary, **NGC-6886**. NGC-6886 shines at photographic magnitude 12.2 and spans a mere 4″, so identifying NGC-6886 also requires a large, long-focus telescope and steady seeing. Its central star dimly glows at magnitude 15.7, which makes it very difficult to observe.

Considering its size, Sagitta contains a disproportionately high number of odd stars. **WZ Sagittae**, an unusual recurring

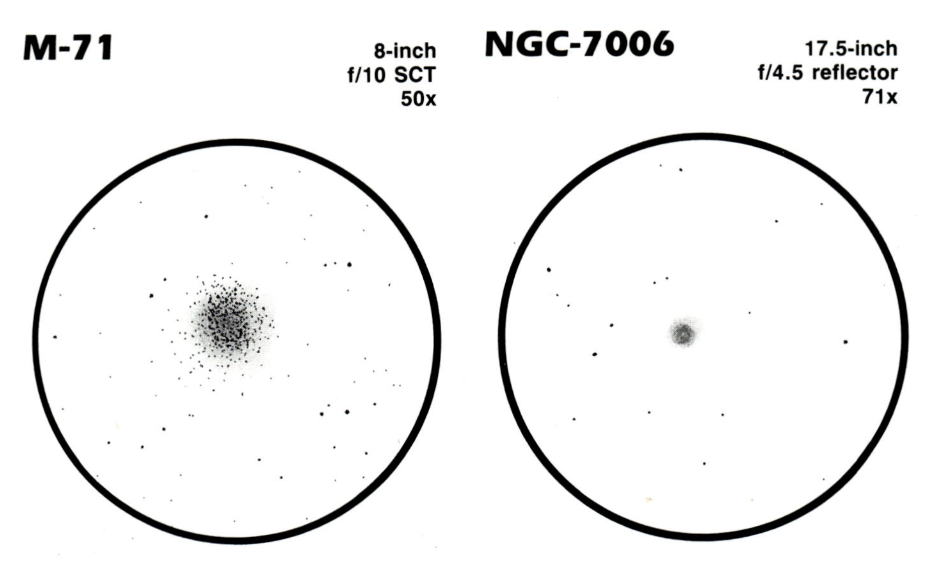

Sketches by David J. Eicher

nova, can be found 1° northwest of NGC-6879, halfway between the row of 7th-magnitude stars around NGC-6879 and an inverted V-shaped asterism of 7th- and 8th-magnitude stars to the northwest. Three major outbursts of WZ Sagittae have occurred: in 1913, 1946, and 1978. This star normally stays around 15th magnitude, but during an outburst it typically rises to 7th magnitude. Careful observations of the previous three outbursts suggest the outbursts may result from interaction between two stars in the WZ Sagittae system, one of which is a white dwarf. Presumably another outburst can be expected sometime near 2010.

Just 25′ northwest of the planetary NGC-6886 is the unusual variable star **FG Sagittae**, which was discovered in 1943 by the German astronomer C. Hoffmeister. Images of FG Sagittae on plates made at Heidelberg Observatory and Harvard Observatory show that the star has gradually increased in brightness over the last seventy-five years. FG Sagittae shone at magnitude 13.7 in 1890, 10.3 in 1959, and 9.5 in 1967. Since the early 1970s the star has stopped its general brightening but began varying by 0.5 magnitude over a period of about two months. A tenuous nebula surrounds this star, but it is too faint to be visible in backyard telescopes.

In the extreme northeastern part of Sagitta is the erratic variable star **V Sagittae**. This star varies irregularly between magnitudes 9.5 and 13.9 with three overlapping periods of variability. Astronomers believe the unusual variability of the V Sagittae system indicates the star may have once been a nova or may soon become one.

The two brightest objects in Delphinus are double stars, each representing one star in the constellation's kite-shaped asterism. **Beta (β) Delphini** is composed of magnitude 4.0 and 4.9 stars separated by 2.5″. Both stars are white and should be visible with a 3-inch scope at high power. **Gamma (γ) Delphini** consists of magnitude 4.5 and 5.5 stars separated by the relatively wide distance of 9.6″. Each of Gamma's stars has a distinctly yellowish cast.

Three degrees north of the distinctive kite-shaped asterism is **HR Delphini**, otherwise known as Nova Delphini 1967. Discovered at magnitude 5.6 by English amateur astronomer George Alcock on July 8, 1967, the star stayed at 5th magnitude until September 1967, when it rose to 4th magnitude. After fading slightly, the star rose to magnitude 3.5 in December of that year and then faded to magnitude 11.5 by 1975. The star has since faded to magnitude 12.4 and is visible as a rather ordinary Milky Way star.

Delphinus contains two planetary nebulae well-suited to small telescopes. **NGC-6891** lies on the western border of Delphinus, some 1.5° southwest of a distinct pair of 6th-magnitude stars. With a photographic magnitude of 11.7 and a diameter of 12″, this nebula is easily visible as a nebular disk with a 4-inch scope. An 8-inch scope shows it as a small bluish green patch of light with a 12th-magnitude central star embedded within. In the northern part of Delphinus lies **NGC-6905**, a photographic magnitude 11.9 object with a diameter of 46″. This object appears as a beautiful ring-shaped nebula with a faint central star in an 8-inch scope. Larger instruments show NGC-6905 as one of the finest planetaries in the summer sky.

NGC-6934 is a 9th-magnitude globular cluster lying in southern Delphinus about 1° northwest of a pair of 6th- and 7th-magnitude stars. Backyard telescopes show this object as a 5′-diameter fuzzy disk of nebulosity with a bright core. Ten degrees away in the northeast part of Delphinus lies the globular cluster **NGC-7006**. Glowing dimly at magnitude 10.6 and spanning only 2.8′, NGC-7006 is certainly not a terrifically impressive globular cluster for small scopes. However, it is the most distant globular visible from the backyard, lying an immense 37 kiloparsecs away.

Object	M#	Type	R.A. (2000)	Dec.	Mag.	Size/Sep./Per.	N★	Mag.★
Harvard 20		⊙	19h 53.1m	+18°20′	7.7	7′	15	9.8
NGC-6838	M-71	●	19h 53.8m	+18°47′	8.3	7.2′		
WZ Sge		N	20h 07.6m	+17°42′	7.0↔15.5	32.6 yr?		
NGC-6879		■	20h 10.5m	+16°55′	13.0_P	5″		15_P
FG Sge		IV	20h 11.9m	+20°20′	9.5↔13.7	irr.		
NGC-6886		■	20h 12.7m	+19°59′	12.2_P	4″		15.7
NGC-6891		■	20h 15.2m	+12°42′	11.7_P	12″		12.4
IC 4997		■	20h 20.2m	+16°45′	11.6_P	2″		13v?
V Sge		N	20h 20.3m	+21°06′	9.5↔13.9	—		
NGC-6905		■	20h 22.4m	+20°07′	11.9_P	46″		13.51_B
NGC-6934		●	20h 34.2m	+7°24′	8.9	5.9′		
Beta (β) Del		★²	20h 37.5m	+14°36′	4.0,4.9	2.5″		
HR Del		N	20h 42.3m	+19°10′	3.7↔12.4	—		
Gamma (γ) Del		★²	20h 46.7m	+16°07′	4.5,5.5	9.6″		
NGC-7006		●	21h 01.5m	+16°11′	10.6	2.8′		

★² *Double Star*
N *Nova*
⊙ *Open Star Cluster*
● *Globular Star Cluster*
■ *Planetary Nebula*
IV *Irregular Variable Star*

N★ = number of stars, Mag.★ = magnitude range of cluster or magnitude of central star, (...) indicates many fainter.

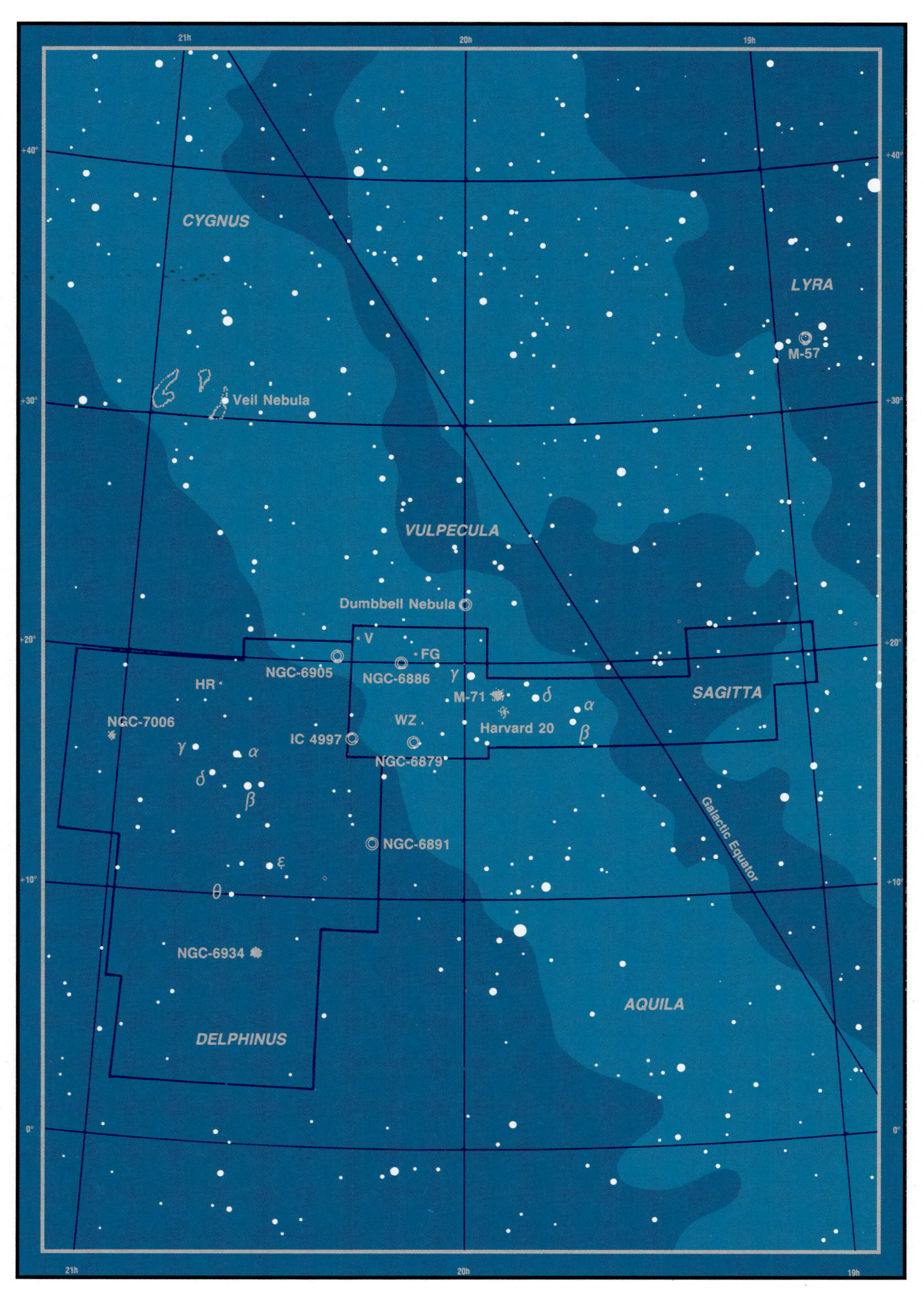
21h
20h
19h
+40°
+30°
+20°
+10°
0°
CYGNUS
LYRA
M-57
Veil Nebula
VULPECULA
Dumbbell Nebula
V
FG
NGC-6905
NGC-6886
HR
γ
M-71
δ
α
SAGITTA
NGC-7006
WZ
Harvard 20
β
IC 4997
NGC-6879
γ
α
δ
β
NGC-6891
ε
θ
Galactic Equator
NGC-6934
AQUILA
DELPHINUS

Above: The slender edge-on galaxy NGC-5907 is one of the finest examples o its class. In a 6-inch telescope NGC-590 is visible as a silvery needle of light with a slight central bulge. Photo by Gene Brings.

Left: Although Charles Messier's confused notes leave its original identity open to question, M-102 is commonly thought to be the bright lenticular galaxy NGC-5866 Photo by Jack Newton.

Below left: The graceful spiral NGC-598 is a superb target for backyard telescopes offering a bright core and subtle spira structure. Photo by K. Alexande Brownlee.

Below: Unimpressive in small telescopes the spiral NGC-4319 nevertheless is fascinating target: near its nucleus is the 15th-magnitude quasar Markarian 205 Both objects can be glimpsed in a single high-power field with a 14-inch telescope Photo by Martin C. Germano.

Draco

Dra
Draconis

Keith Ward

A lengthy and sprawling constellation, **Draco** the Dragon winds its way through 1,083 square degrees of sky around the north celestial pole. This stretched constellation almost completely surrounds Ursa Minor and its Little Dipper asterism. Because it is so far from the plane of the Galaxy, Draco contains few galactic objects, but it is rich in galaxies visible in backyard telescopes.

The identification of one of the brightest deep-sky objects in Draco is embroiled in controversy. When Charles Messier published his description of **M-102** in the *Connaissance des Temps* in 1781, he identified the fuzzy object as "a very faint nebula situated between Omicron Bootis and Iota Draconis; near to it is a 6th-magnitude star." The problem with this description is that Omicron (o) Bootis lies some 40° south of Iota (ι) Draconis, which suggests Messier did not mean to say what he published.

In the *Bedford Catalogue*, published in 1844, William H. Smyth suggested that Omicron Bootis was an obvious misprint for Theta (θ) Bootis, which lies in the right area. Between these stars lies a group of five galaxies including the bright objects NGC-5866 and NGC-5879. Here the matter becomes even more confused: Smyth reported observing NGC-5879 and described this object as the "brightest in the group." However, the brightest in the group is clearly NGC-5866. Confirming this line of reasoning, Harlow Shapley in 1917 identified M-102 with the nebulosity NGC-5866.

In 1947 astronomer Helen Sawyer Hogg revealed that Pierre Mechain, Messier's assistant who discovered M-101, had written to the German astronomer J. Bernoulli in 1786. Mechain claimed that Messier had made a clerical mistake and that M-102 was simply a repeated observation of Ursa Major's M-101. Despite this evidence directly from the pen of the discoverer, most astronomers continue to accept NGC-5866 as M-102, if for no other reason than as a tribute to the human limitations of science.

Its long and controversial history notwithstanding, M-102 is a bright, easy-to-find galaxy for users of small scopes. Located at the southern edge of Draco, 4° southwest of the bright star Iota Draconis, this object is one of the finest examples of a lenticular galaxy in the sky. Shining at magnitude 10.0 and spanning 5.2′ by 2.3′, M-102 is visible as a thin saucer-shaped blotch of nebulosity in large finder scopes. A 6-inch scope shows it as a bright oval nucleus centered in a large, elliptical halo of faint greenish-gray light.

Four more galaxies lie in close proximity. **NGC-5905** and **NGC-5908** form a pair of faint galaxies 2° due east of M-102. NGC-5905 is a barred spiral with a blue-light magnitude of 12.3 and dimensions of 4.2′ by 3.3′; NGC-5908 is a normal spiral dimly glowing at magnitude 11.9 with dimensions of only 3.2′ by 1.3′. Observers with 8- or 10-inch telescopes trained on the field containing these galaxies can expect to see two faint, fuzzy, slightly elongated patches of light separated by 12.5′. NGC-5905 is the one on the northwestern end, while NGC-5908 lies to the southeast of its larger neighbor. The galaxies' redshifts are nearly equal, but they're different enough to suggest the two objects aren't close neighbors in space.

One degree north of the pair is the large, beautiful galaxy **NGC-5907**, surpassed as an edge-on spiral only by NGC-4565 in Coma Berenices. NGC-5907 is an Sb-type spiral inclined almost exactly edge-on to our line of sight; its magnitude of 10.4 and dimensions of 12.3′ by 1.8′ make it easily visible in large binoculars or small telescopes. A 3-inch refractor at moderate power shows this galaxy as a uniformly bright sliver of silvery-white light with a slight central bulge measuring slightly over 1′ across. A 12-inch scope at high power shows a bright streak of grayish light with a 2′-long central bulge and a thin dark lane bisecting the galaxy's long axis. Larger telescopes show NGC-5907 as a brighter streak of light but don't significantly increase the amount of visible detail.

The little Sb-type spiral **NGC-5879**, one of those galaxies temporarily thought to be M-102, lies just 1° north-northwest of NGC-5907. This galaxy has a magnitude of 11.5, measures 4.4′ by 1.7′ across, and features a bright nucleus and faint spiral arms visible as an 11th-magnitude spot surrounded by a faint haze. Some 7′ north of the little galaxy is a 7th-magnitude star.

More bright galaxies lay scattered all across Draco. The fine Sb-type spiral **NGC-5985** is located 4.5° northeast of NGC-5879 in an area devoid of bright stars. This 11th-magnitude object is a mag-

BEST VISIBLE DURING
SPRING

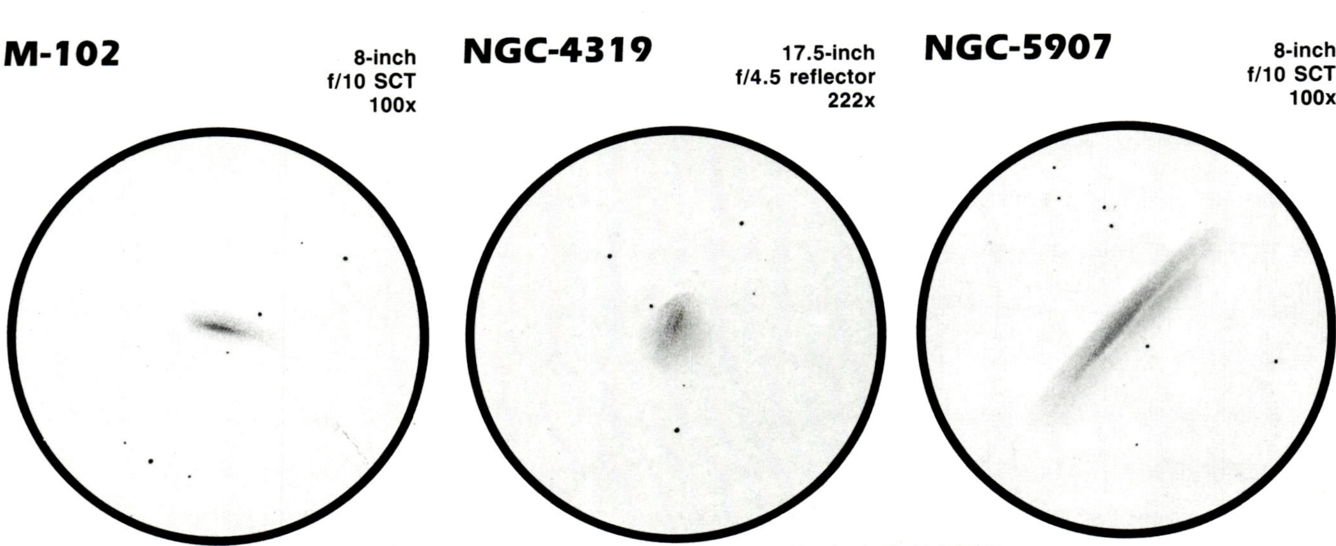

Sketches by David J. Eicher

nificent sight in large backyard telescopes: its small, oval central disk is wrapped in multiple, tightly-wound arms just visible in large scopes. Seven arcminutes west of this grand spiral is the diminutive elliptical galaxy **NGC-5982**, an 11th-magnitude fuzzy spot with a slight central brightening. Three-and-one-half degrees north-northeast of the 5982/5 pair is another worthwhile galaxy, **NGC-6015**. This galaxy shines at magnitude 11.2 and measures 5.4′ by 2.3′ in extent; its nucleus is bright and enveloped in a faint, grayish halo of faint nebulosity. It is a multiple-arm spiral, like NGC-5985, but its detail is somewhat more elusive.

The northeastern part of Draco contains two bright galaxies. **NGC-6503**, a magnitude 10.2 spiral, measures 6.2′ by 2.3′ across and is easily visible in small scopes as an elliptical smear of nebulosity. Ten-inch scopes show a typically bright nuclear region surrounded by a faint oval halo of misty light. Some 4′ east of this galaxy is a 9th-magnitude star. **NGC-6643** is an 11th-magnitude Sc-type spiral spanning only 3.9′ by 2.1′; it appears as a roundish patch of indistinct light in a 6-inch scope.

Lying nearly on top of the north pole of the ecliptic, between Delta (δ) and Zeta (ζ) Draconis, is the bright planetary nebula **NGC-6543**. This little nebula measures only 18″ across but shines at photographic magnitude 8.8, which means its nebulosity has an exceedingly high surface brightness (and can therefore withstand great magnification). This is fortunate because small telescope observers need to use high power on nights of steady seeing to distinguish the planetary's disk from bright stars in the area. NGC-6543 appears as a tiny, bluish-green spot of bright nebulosity in most telescopes; it takes a large scope at very high power to show the 11th-magnitude central star. (Small scopes have difficulty showing the star because it is overwhelmed by the bright nebulous disk.)

Four bright or unusual galaxies lie in the northwestern corner of Draco. **NGC-3735** is a fine Sb-type spiral shining at blue magnitude 12.2 and measuring 4.2′ by 1.0′. **NGC-4236** is an impressive object in small scopes, shining at magnitude 9.7 and measuring 18.6′ by 6.9′ across; it is a barred spiral and offers glimpses of mottled, knotty detail in its spiral arms to owners of large backyard scopes. **NGC-4256** is a blue, magnitude 12.4 Sb-type spiral resembling NGC-3735 and measuring 4.6′ by 1.0′ across.

Our final galaxy, **NGC-4319**, offers an unusual surprise. The galaxy is a barred spiral dimly glowing at blue magnitude 12.2 and measuring 3.1′ by 2.5′ across — hardly unusual. But within its spiral arms is a magnitude 14.5 dot of light visible to owners of 16-inch or larger scopes; this dot is no star but **Markarian 205**, a quasar with a redshift that indicates a distance of 280 megaparsecs. The galaxy itself, measured by its redshift, is a mere 24 megaparsecs away, yet some astronomers claim the two objects are connected by faint nebulosity just visible on long-exposure CCD images. Are the dissident astronomers wrong, or will NGC-4319 and Markarian 205 disprove redshift measurements as indicators of cosmic distances?

Object	M#	Type	R.A. (2000)	Dec.	Mag.	Size/Sep./Per.	H
NGC-3735		§	11h 36.0m	+70°32′	12.2_B	4.2′x 1.0′	Sb
NGC-4236		§B	12h 16.7m	+69°28′	9.7	18.6′x 6.9′	SB+
NGC-4256		§	12h 18.7m	+65°54′	12.4_B	4.6′x 1.0′	Sb:
NGC-4319		§B	12h 21.7m	+75°19′	12.2_B	3.1′x 2.5′	SBb−
Mrk 205		QSO	12h 21.7m	+75°19′	14.5	stellar	
NGC-5866	M-102	§L	15h 06.5m	+55°46′	10.0	5.2′x 2.3′	SA0+ sp
NGC-5879		§	15h 09.8m	+57°00′	11.5	4.4′x 1.7′	Sb
NGC-5905		§B	15h 15.4m	+55°31′	12.3_B	4.2′x 3.3′	S(B)b
NGC-5907		§	15h 15.9m	+56°19′	10.4	12.3′x 1.8′	Sb+
NGC-5908		§	15h 16.7m	+55°25′	11.9	3.2′x 1.3′	Sb−
NGC-5985		§	15h 39.6m	+59°20′	11.0	5.5′x 3.2′	Sb
NGC-6015		§	15h 51.4m	+62°19′	11.2	5.4′x 2.3′	Sc
NGC-6503		§	17h 49.4m	+70°09′	10.2	6.2′x 2.3′	Sb
NGC-6543		■	17h 58.6m	+66°38′	8.8_P	18″	
NGC-6643		§	18h 19.8m	+74°34′	11.1	3.9′x 2.1′	Sc

■	*Planetary Nebula*
§	*Spiral Galaxy*
§B	*Barred Spiral Galaxy*
§L	*Lenticular Galaxy*
QSO	*Quasar*

H = Hubble type for galaxies

Subscript "P" denotes photographic magnitude; subscript "B" denotes blue magnitude.

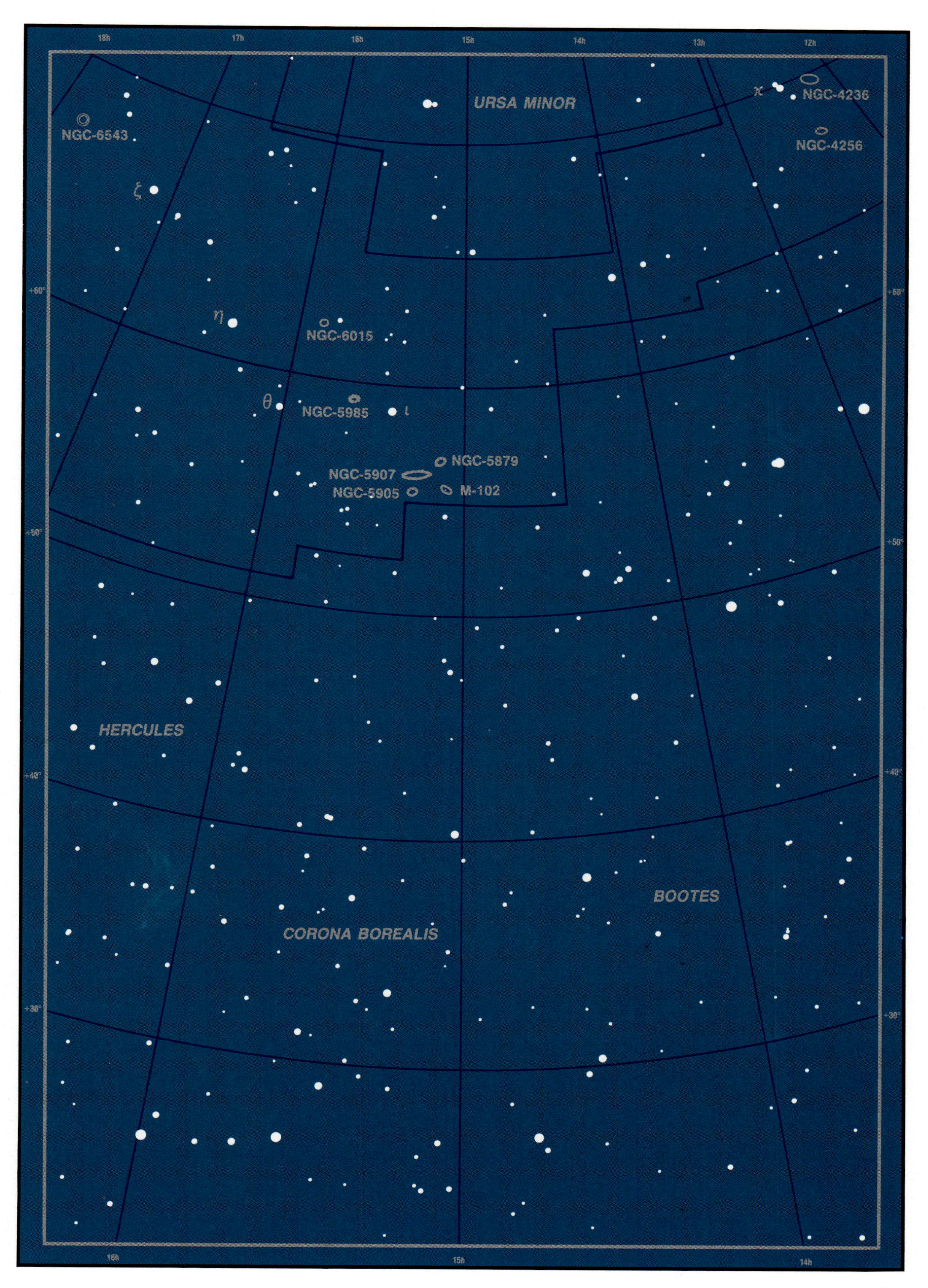

URSA MINOR
NGC-4236
NGC-4256
NGC-6543
NGC-6015
NGC-5985
NGC-5879
NGC-5907
NGC-5905
M-102
HERCULES
CORONA BOREALIS
BOOTES

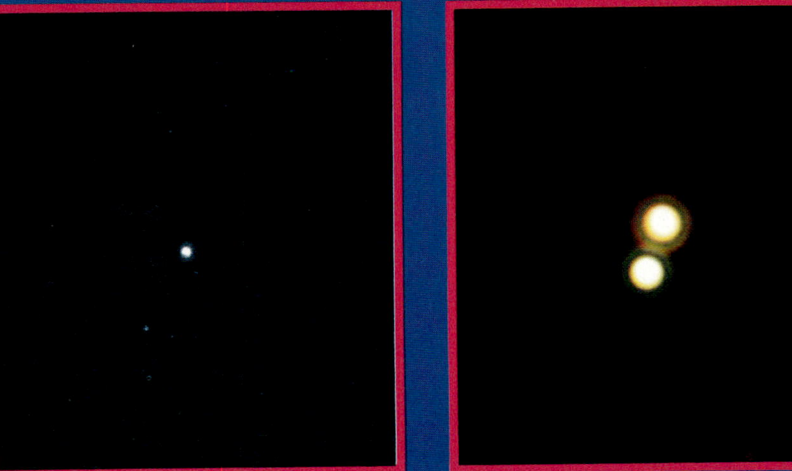

Above: Though it glows at only 11th magnitude, NGC-1300 is one of the finest examples of a face-on barred spiral galaxy. Small telescopes show a bright core surrounded by undefined haze, which large telescopes reveal to be the galaxy's arms. Photo by John Gleason. Far left: The planetary nebula NGC-1535 shows a strong bluish-green color, often the only way to distinguish its starlike disk from ordinary field stars. Photo by Martin C. Germano. Left: One of the finest double stars in Eridanus is 32 Eridani, easily separable with small scopes.

Eridanus

Eri

Eridani

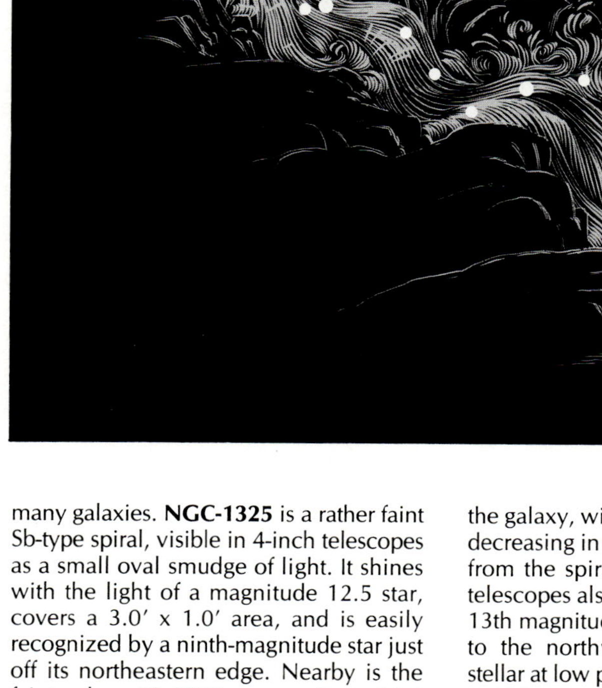

Keith Ward

Eridanus the River is a long, twisting constellation which begins near Orion's bright star Rigel and winds down to Achernar, some 49° to the south. Eridanus is not striking to the naked eye; it lies away from the plane of the Milky Way and contains relatively barren starfields — but it also harbors great numbers of faint galaxies.

The brightest collection of galaxies in Eridanus lies near its center. **NGC-1300** is one of these — a magnitude 11.3 galaxy measuring 6.0′ x 3.2′ across, making it easy to find with small telescopes. It appears as an elongated smudge with a sharply brighter center, surrounded by a large diffuse halo of nebulosity. Dark skies and large optics show NGC-1300 as one of the finest face-on barred spiral galaxies in the sky with two distinct arms emanating from a bright central bar. A nebular filter attached to your scope will help you see subtle details, especially when skies are light-polluted.

Southwest of NGC-1300 is **NGC-1232**, another face-on spiral. A type Sc galaxy, it is one of the "many-armed spirals" which have more than the normal two, three, or four spiral arms. Its magnitude 10.7 light spreads over an area of 7.0′ x 6.0′, making its surface brightness fairly low. This galaxy shows up in 4-inch reflectors as a circular nebulosity without much central condensation. NGC-1232's spiral arms are very difficult to observe, although a 16-inch reflector at moderate power may show hints of bright arms running alongside thin dark lanes. Nearby is another multiple-arm spiral, **NGC-1187**. At magnitude 11.3, this galaxy is slightly fainter than NGC-1232, but it covers only 5.5′ x 4.0′ of sky, and therefore looks equally as bright in the telescope.

Located several degrees east of NGC-1187 are two more of Eridanus' many galaxies. **NGC-1325** is a rather faint Sb-type spiral, visible in 4-inch telescopes as a small oval smudge of light. It shines with the light of a magnitude 12.5 star, covers a 3.0′ x 1.0′ area, and is easily recognized by a ninth-magnitude star just off its northeastern edge. Nearby is the faint galaxy **IC 1953**, a magnitude 12.5 barred spiral, a perfect circle measuring 2.4′ across. It is faintly visible in 6-inch telescopes as a fuzzy spot of white light, showing little or no central condensation.

Several degrees southeast of the NGC-1325 area is the long period variable star **T Eridani**. This star is fascinating to watch as it fluctuates between magnitudes 7.4 and 13 during its 252-day cycle. At maximum brightness, T Eridani is a bright, easily-observed rosy red star in binoculars or small telescopes; at minimum, it glows weakly at less than 13th magnitude and is difficult to identify without a special finder chart.

Toward the eastern fringes of Eridanus lies a pair of little galaxies, **NGC-1531** and **NGC-1532**. The brighter of the two, NGC-1532, is a 12th magnitude Sb-type spiral. It is nearly edge-on to our line of sight — measuring 5.0′ x 1.0′ — and is fairly easy to find because of its high surface brightness. A 6-inch telescope under dark skies shows a silvery sliver of light without detail. A 12-inch scope, however, lets you see that NGC-1532's nucleus is slightly brighter than the rest of the galaxy, with an elongated halo evenly decreasing in brightness as it moves away from the spiral's center. Large backyard telescopes also show NGC-1531 — a tiny 13th magnitude elliptical located just 1.6′ to the northwest. This galaxy appears stellar at low powers (it measures only 30″ x 18″ across) and reveals no detail in backyard telescopes.

Much brighter than NGC-1531 is the fine barred spiral **NGC-1291**. This galaxy covers 5.0′ x 2.0′ and shines at magnitude 10.2. A 2-inch telescope will show this object as an oval patch of light, but for northern observers its southerly declination of −41° may cause problems for those without a clear, dark southern horizon. Under good conditions, a 10-inch telescope shows NGC-1291's very bright core, surrounded by a uniformly illuminated envelope of fainter nebulosity. Owners of large telescopes and observers at low latitudes may see some structure in this galaxy.

Two bright galaxies lie in the northern part of Eridanus, within easy viewing reach of Northern Hemisphere telescopes. The first is **NGC-1337**: a 12th magnitude edge-on Sc-type spiral. Though low in total magnitude, NGC-1337's surface brightness is great because the galaxy is turned sideways to our line of sight. NGC-1337 measures 5.4′ x 0.9′ and is visible in large finder telescopes under dark skies. A 6-inch reflector shows a

BEST VISIBLE DURING
WINTER

NGC-1232 — 8-inch reflector 100x

NGC-1300 — 8-inch reflector 100x

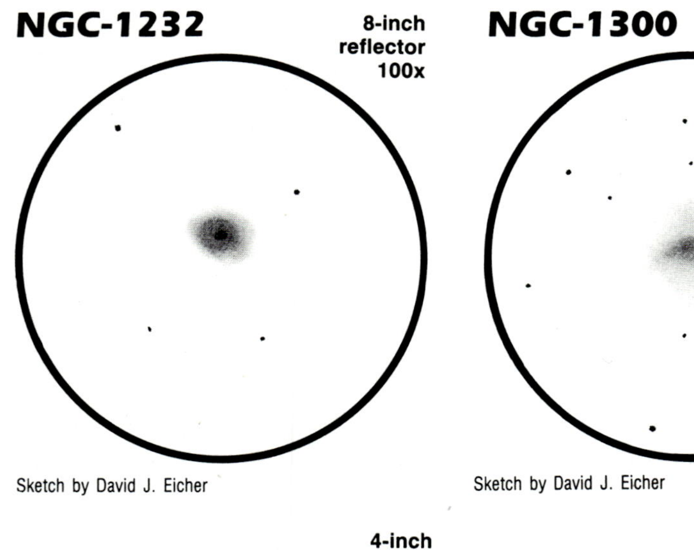

Sketch by David J. Eicher

Sketch by David J. Eicher

NGC-1535 — 8-inch reflector 255x

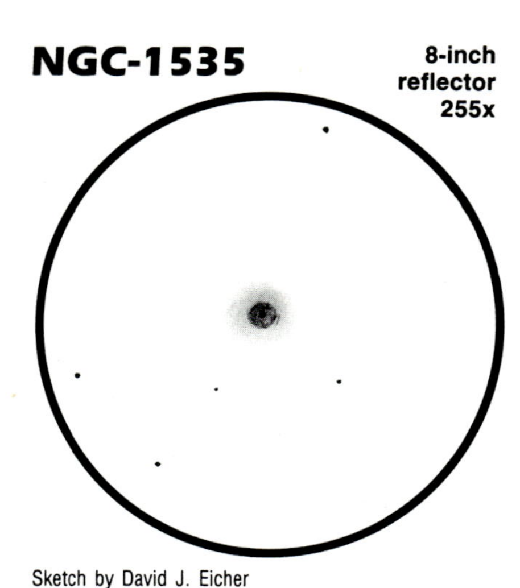

Sketch by David J. Eicher

o^2 Eri — 4-inch reflector 90x

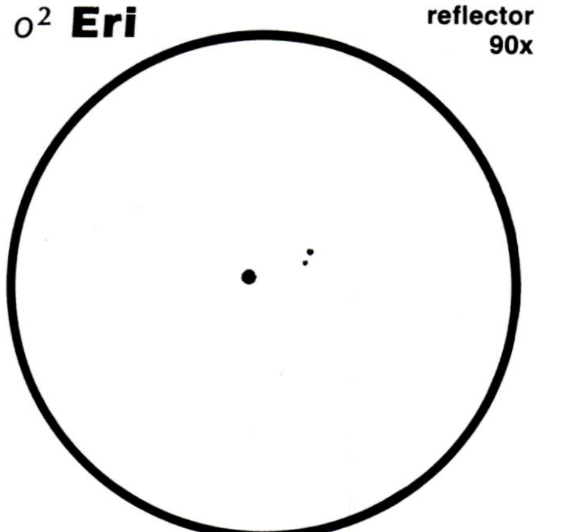

Sketch by Jim Mullaney

32 Eri — 8-inch reflector 100x

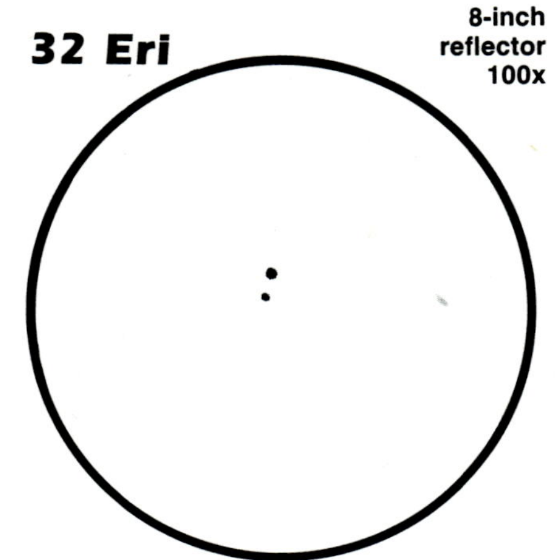

Sketch by Jim Mullaney

bright, sharply-defined core, enclosed in a spindle-shaped nebulosity. Another galaxy to note is NGC-1084, an 11th magnitude Sc-type spiral. Its dimensions measure 2.1′ x 1.0′, and its bright middle shows up well in a 4-inch telescope. Larger telescopes reveal a roundish halo of faint light which is the galaxy's spiral arms.

Tau⁴ Eridani is a bright double star encircled by galaxies NGC-1232, NGC-1300, NGC-1325, and IC 1953. At magnitudes 4 and 10, its components are widely separated in brightness, but the 5.7″ gap between them allows small telescopes at high power to split them. Another double star for small telescope viewing is **32 Eridani**, whose magnitude 5 and 6 members are separated by 6.9″ in p.a. 347°. This star offers excellent color contrasts: the primary shines golden-yellow, while the companion star is sparkling white.

A remarkable triple star system, located in the northeastern part of the constellation, is **Omicron² Eridani**. The system contains two stars, separated by 82.8″ in p.a. 105°, which shine at magnitudes 4.5 and 9.5. The fainter star is one of the most easily observed white dwarf stars. The secondary of this wide pair is itself a double star, harboring an 11th magnitude red dwarf star — again, one of the best examples of such a star for small telescopes. The close pair is separated by 7.6″ in p.a. 347°, and is easily split with high power under good seeing conditions. Omicron² Eridani is only 5 parsecs away — the eighth closest naked-eye star in the sky — and has a large proper motion of some four arc-seconds per year.

Though small, the planetary nebula **NGC-1535** is one of the best for observing the characteristic color found in these objects. At ninth-magnitude, NGC-1535 measures only 20″ x 17″ across, but its bright disk shows a strong and unmistakable bluish-green hue in any telescope. In fact, this distinguishes it as a nebula and not a field star: high-power backyard scopes reveal the nebula's magnitude 11.5 central star and show the disk as clearly nonstellar in appearance.

Object	M#	Type	R.A. (2000)	Dec.	Mag.	Size/Sep./Per.	H
NGC-1084		§	0h 45.9m	− 6°54′	11.3	2.1′x1.0′	Sc(s)
NGC-1187		§	3h 02.6m	−22°52′	10.9	5.5′x4.0′	SBbc(s)
NGC-1232		§	3h 09.8m	−20°35′	10.5	7.0′x6.0′	Sc(rs)
NGC-1291		§B	3h 17.3m	−41°06′	9.4	5.0′x2.0′	SBa
τ^4		★²	3h 19.5m	−21°44′	4.0, 10.0	5.7″	
NGC-1300		§B	3h 19.8m	−19°24′	11.1	6.0′x3.2′	SBb(s)
NGC-1325		§	3h 24.5m	−21°32′	12.3	3.0′x1.0′	Sb
NGC-1337		§	3h 27.9m	− 8°24′	12.3	5.4′x0.9′	Sc(s)
IC 1953		§B	3h 33.6m	−21°29′	12.3	2.4′x2.4′	SBbc(rs)
T		LPV	3h 50.5m	−24°57′	7.4↔13	252d	
32		★²	3h 54.2m	− 2°57′	5.0,6.0	6.9″	
NGC-1531		§	4h 12.0m	−32°51′	12.8	0.5′x0.3′	?
NGC-1532		§	4h 12.1m	−32°52′	11.5	5.0′x1.0′	Sab(s)
NGC-1535		■	4h 14.5m	−12°44′	10.8	20″x17″	
o^2		★³	4h 15.4m	− 6°56′	4.5,9.5,11.0	82.8″,7.6″	

H = Hubble classification type for galaxies

★²	*Double Star*
★³	*Triple Star*
LPV	*Long Period Variable*
■	*Planetary Nebula*
§	*Spiral Galaxy*
§B	*Barred Spiral Galaxy*

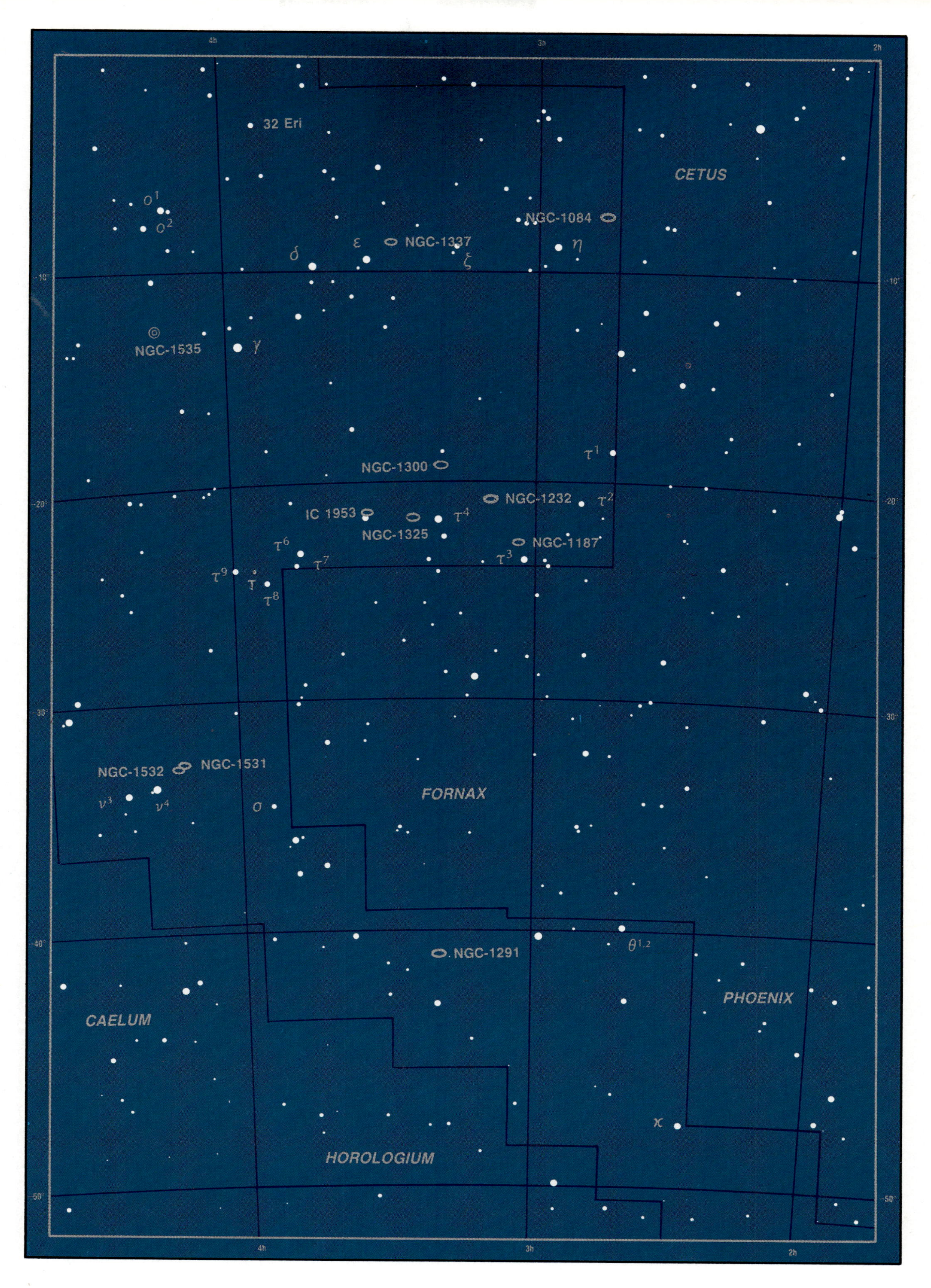

4h
3h
2h
32 Eri
CETUS
NGC-1084
NGC-1337
NGC-1535
NGC-1300
NGC-1232
IC 1953
NGC-1325
NGC-1187
NGC-1532
NGC-1531
FORNAX
NGC-1291
PHOENIX
CAELUM
HOROLOGIUM
-10°
-20°
-30°
-40°
-50°

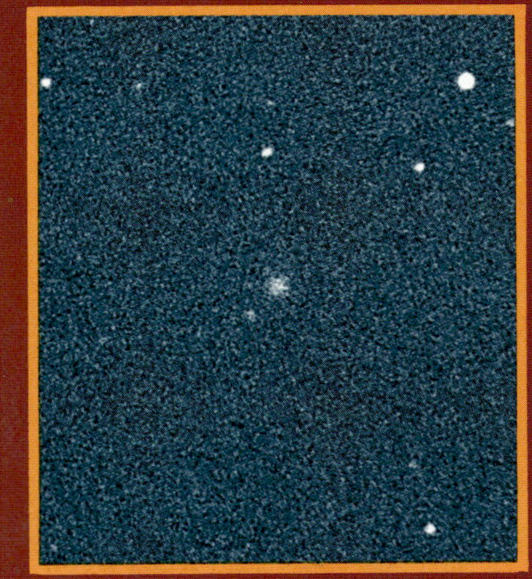

Above: NGC-1097 is the second brightest galaxy in Fornax and one of the best examples of a barred spiral for observers with small telescopes. A 6-inch instrument at medium powers shows the faint arcs of the galaxy's spiral arms.

Far left: One of the brightest planetary nebulae in the southern sky, NGC-1360 is visible in a 2-inch scope as a bluish oval haze surrounding an 11th-magnitude central star.

Left: NGC-1049, barely visible in backyard telescopes as a faint "out-of-focus star," is the brightest globular cluster belonging to the Fornax dwarf galaxy. Ironically the Fornax dwarf itself has such a low surface brightness it is invisible in backyard telescopes. Photos by Martin C. Germano.

Fornax

For Fornacis

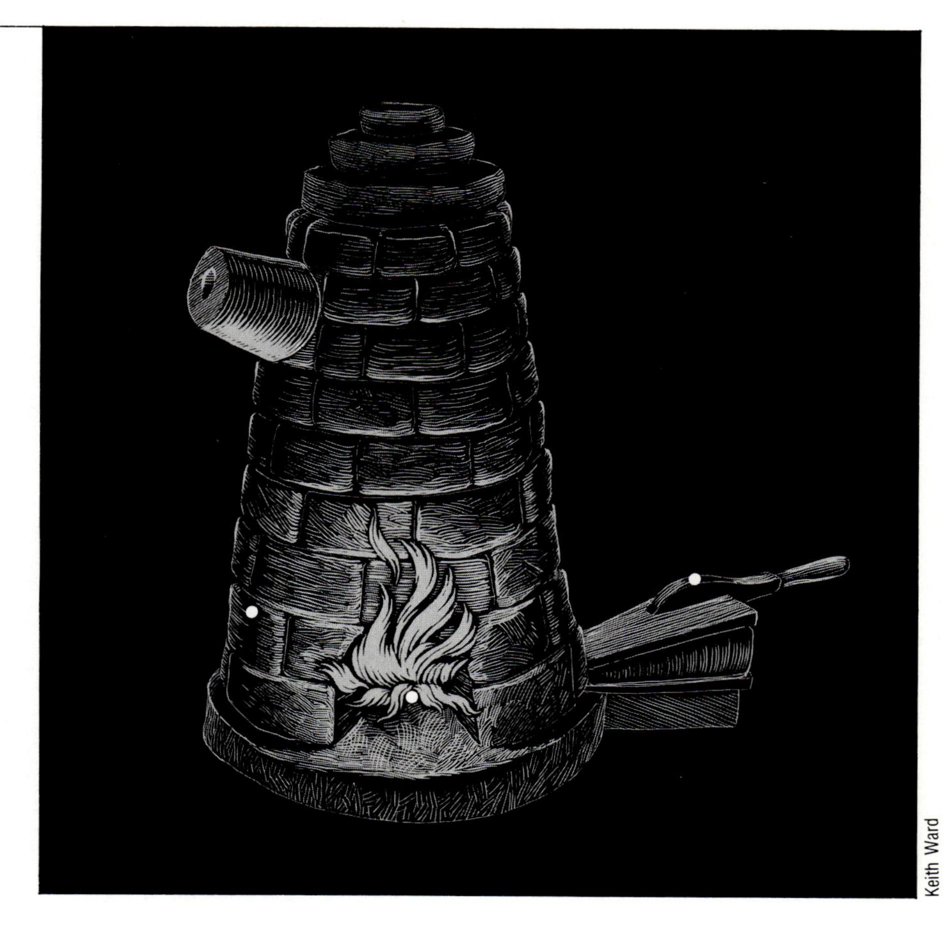
Keith Ward

Fornax the Furnace is a relatively small, squarish constellation lying far enough away from the galactic equator that it contains primarily galaxies. Most of these galaxies are faint objects in small telescopes that even on a dark night will challenge the optical quality of your mirror and the ability of your eye to perceive subtle details. With persistence, however, you'll be able to spot detail in several of the Fornax galaxies.

Perhaps the prettiest galaxy in Fornax is the barred spiral **NGC-1097**, an object glowing at magnitude 9.3 and spanning 9.3′ by 6.6′. To find this galaxy, use a low-power eyepiece and aim your finder scope at the bright double star Alpha (α) Fornacis. Move 5° southwest to equally bright Beta (β) Fornacis, and then continue to a point just over 2° north-northwest of Beta. You should see a bright nebulous glow in the telescope's field of view. (If you don't, try slowly sweeping around this area and see if you don't run across the galaxy.)

Once you've found NGC-1097, increase your telescope's magnification to something like 12x per inch of telescope aperture and study the galaxy in detail. NGC-1097 is one of the best barred spirals in the sky, because it is bright and oriented nearly face-on. With a small telescope you should see a bright, round nucleus surrounded by an oval-shaped halo of slightly fainter nebulosity. Through this halo runs a long, thin rectangular nebulosity, visible in 6-inch scopes on nights of good transparency. This feature is NGC-1097's "bar," the characteristic feature of barred spirals. If you have an 8-inch or larger scope, you may see faint semicircular extensions on the bar's ends. These extensions represent light from the galaxy's spiral arms. Keen-eyed observers with large telescopes will see a tiny detached patch of light 2′ northwest of the galaxy's northern spiral arm. This is NGC-1097A, a tiny, 13th-magnitude elliptical galaxy.

Some 4° south-southwest of NGC-1079 on the other side of Beta Fornacis lies a faint and extremely unusual globular star cluster. Because its magnitude is 13.0 and its disk only 24″ across, **NGC-1049** appears as a pale spot of gray light in a 6-inch scope. When the seeing isn't terrifically steady, this cluster appears like an out-of-focus star or a small, colorless planetary nebula, though actually NGC-1049 is an enormous globular cluster located some 170 kiloparsecs away in a Local Group galaxy called the **Fornax dwarf galaxy**. Discovered at Harvard College Observatory in 1938, the Fornax dwarf is one of several tiny elliptical galaxies scattered throughout our Local Group. It is one of the smallest galaxies known, contains about fifty times as much mass as a typical Milky Way globular cluster, and measures only 2 kiloparsecs across. The Fornax dwarf has such a low surface brightness that it is well beyond the reach of backyard telescopes. Yet, ironically, its brightest globular cluster, NGC-1049, is readily visible in moderate-sized instruments.

In addition to an elusive dwarf Local Group galaxy, Fornax contains a full-fledged galaxy cluster. The so-called Fornax Galaxy Cluster contains eighteen bright galaxies and at least ten faint galaxies compressed into a 6°-diameter circle; most members lie between 20 and 25 megaparsecs away. Located near the heart of the cluster is **NGC-1380**, a lenticular galaxy with a blue light magnitude of 11.1 and dimensions of 4.9′ by 1.9′. To find this galaxy, aim your finder scope toward the bright triangle of stars f, g, and h Eridani. From g Eridani move 3° due west and very slightly north. This should place you on NGC-1380 and the heart of the Fornax cluster.

With a low-power eyepiece centered on NGC-1380, you may notice several other smudges of light, especially if you have an 8-inch or larger telescope. It is possible to see nine galaxies in a single 1°-field: NGC-1380, NGC-1374 (a blue magnitude 12.4 elliptical), NGC-1379 (a blue magnitude 12.3 elliptical), NGC-1381 (a blue magnitude 12.5 lenticular), NGC-1386 (a blue magnitude 12.1 spiral), NGC-1387 (a blue magnitude 12.0 lenticular), NGC-1389 (a blue magnitude 12.6 elliptical), NGC-1399 (a magnitude 9.9 elliptical), and NGC-1404 (a magnitude 10.3 elliptical).

Lying about 1° southwest of the main group is another member of the Fornax cluster, the large face-on barred spiral **NGC-1365**. NGC-1365 shines at magnitude 9.5 and measures 9.8′ by 5.5′, which make it one of the brightest galaxies in the area and easily visible as a luminous haze in small telescopes. The bar structure and spiral arms of NGC-1365 are

BEST VISIBLE DURING
AUTUMN

NGC-1049

17.5-inch
f/4.5 reflector
142x

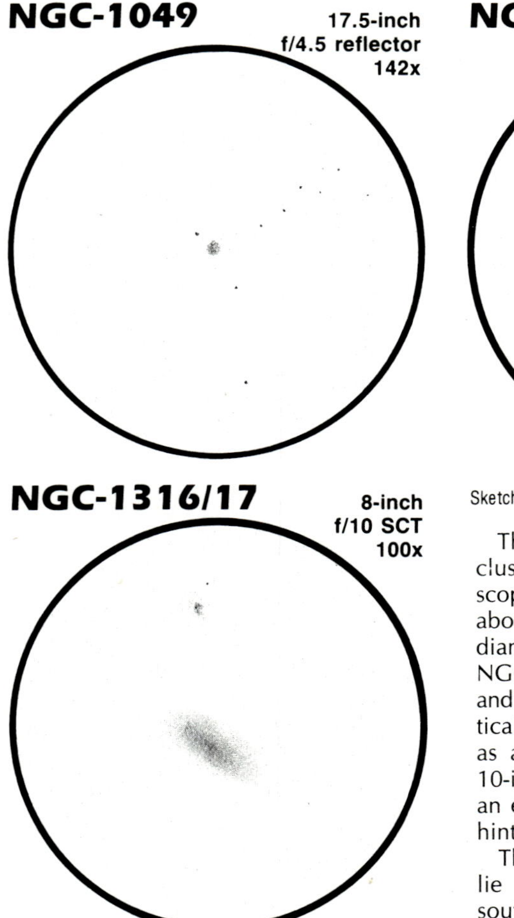

NGC-1316/17

8-inch
f/10 SCT
100x

visible in backyard telescopes, although they can't be seen as easily as those of NGC-1097. A 6-inch telescope shows NGC-1365 as a small bright knot of nebulosity (its core), surrounded by a uniformly illuminated halo of faint nebulosity. A 10-inch telescope shows the bar structure easily and on good nights reveals the galaxy's subtle spiral structure. A faint star is visible in front of the galaxy's face about 2′ northwest of the nucleus.

NGC-1097

8-inch
f/10 SCT
100x

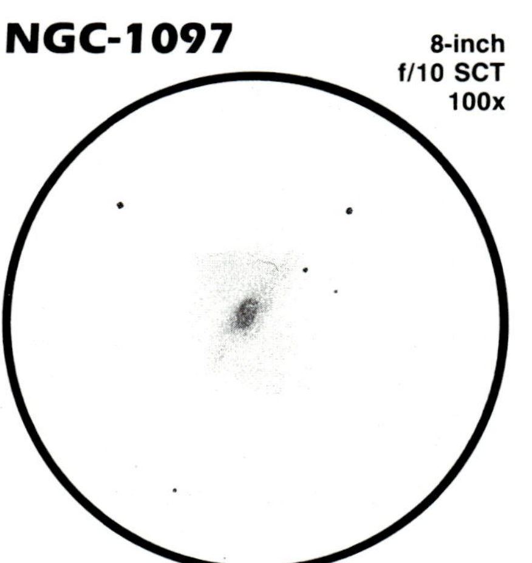

Sketches by David J. Eicher

Three more members of the Fornax cluster are impressive objects in small scopes. The first of these, **NGC-1427**, lies about 1° northeast of the cluster's center, diametrically across from NGC-1365. NGC-1427 has a blue magnitude of 12.2 and covers 2.8′ by 2.0′ of sky. It is an elliptical galaxy, visible in a 4-inch telescope as a faint oval patch of nebulosity. A 10-inch telescope reveals NGC-1427 as an elongated smear of gray light with a hint of central condensation.

The remaining Fornax cluster members lie in close proximity some 3° west-southwest from the cluster center. **NGC-1316** is a bright peculiar barred spiral of uncertain classification. Somewhat resembling the galaxies NGC-5128 in Centaurus and NGC-1275 in Perseus, this object is an unusually dusty galaxy with a disproportionately bright nucleus. NGC-1316 is also an intense radio energy emitter and is designated radio source Fornax A. Telescopically this galaxy appears as a bright round ball of light with a very faint oval halo superimposed upon it. Large telescopes fail to reveal significant detail in this object. About 6′ north of NGC-1316 is the tiny galaxy **NGC-1317**, which appears as a small knot of faint gray light.

More bright galaxies lie to the north. **NGC-1406**, a blue magnitude 12.6 spiral, lies 1° northwest of the 5th-magnitude star Delta (δ) Fornacis. This object is highly inclined to our line of sight and consequently appears as a dim streak of fuzzy light some 4′ long. Large backyard scopes show a bright disk of nebulosity surrounding its center but fail to show detail in the galaxy's arms. The bright spiral **NGC-1255** can be found 3.5° north of Alpha Fornacis. This galaxy shines at magnitude 11.1 and measures 4.1′ by 2.8′ across, so it is easily identifiable in 3-inch telescopes under a dark sky. With an 8-inch scope NGC-1255 looks like a mottled patch of light surrounding a bright, condensed nucleus. A 12th-magnitude star lies 2′ southwest of the galaxy's center.

In the northeastern corner of Fornax are three more galaxies. **NGC-1371** is a blue magnitude 11.5 barred spiral with a bright center and a subtle dust band visible with large backyard scopes. A bright, unequal double star lies about 5′ east of this object. One degree northeast is **NGC-1385**, a magnitude 11.2 spiral easily visible in small scopes. Slightly over 2° south-southeast is the bright barred spiral **NGC-1398**, visible in 4-inch scopes as a large bright circle of nebulosity.

About 1.5° southwest of NGC-1371, **NGC-1360** forms an isosceles triangle with 6th- and 7th-magnitude stars. One of the southern sky's best planetary nebulae, this object has a remarkably high surface brightness that makes finding it in small telescopes relatively easy. There is no modern published magnitude for this object, but estimates place it at around 8th magnitude. The nebula measures 390″ — or 6.5 arcminutes — across and can be seen as a large disk of bluish green nebulosity in any telescope. An 8-inch scope reveals the 13th-magnitude central star.

Object	M#	Type	R.A. (2000)	Dec.	Mag.	Size/Sep./Per.	H
Fornax dwarf		0	2h 39.9m	−34°32′	9.0_B	20.0′x 13.8′	dE3
NGC-1049		●	2h 39.9m	−34°16′	13.0	0.4′	
NGC-1097		§B	2h 46.3m	−30°17′	9.3	9.3′x 6.6′	S(B)b
NGC-1255		§	3h 13.5m	−25°44′	11.1	4.1′x 2.8′	Sc
NGC-1316		§B	3h 22.7m	−37°12′	8.9	7.1′x 5.5′	S(B)0pec
NGC-1317		§B	3h 22.8m	−37°06′	11.0	3.2′x 2.8′	SBa
NGC-1360		■	3h 33.3m	−25°51′	—	390″	
NGC-1365		§B	3h 33.6m	−36°08′	9.5	9.8′x 5.5′	SBb
NGC-1371		§B	3h 35.0m	−24°56′	11.5_B	5.4′x 4.0′	S(B)a
NGC-1380		§L	3h 36.5m	−34°59′	11.1_B	4.9′x 1.9′	S0
NGC-1385		§	3h 37.5m	−24°30′	11.2	3.0′x 2.0′	Sc
NGC-1398		§B	3h 38.9m	−26°20′	10.3	2.5′x 2.3′	$S(B)b^-$
NGC-1406		§	3h 39.4m	−31°19′	12.6_B	3.9′x 1.0′	Sb^+
NGC-1425		§	3h 42.2m	−29°54′	11.7_B	5.4′x 2.7′	Sb
NGC-1427		0	3h 42.3m	−35°25′	12.2_B	2.8′x 2.0′	E3

H = Hubble type for galaxies
Subscript "P" denotes photographic magnitude; subscript "B" denotes blue magnitude.

- ● *Globular Star Cluster*
- ■ *Planetary Nebula*
- § *Spiral Galaxy*
- §B *Barred Spiral Galaxy*
- §L *Lenticular Galaxy*
- 0 *Elliptical Galaxy*
- # *Irregular or Peculiar Galaxy*

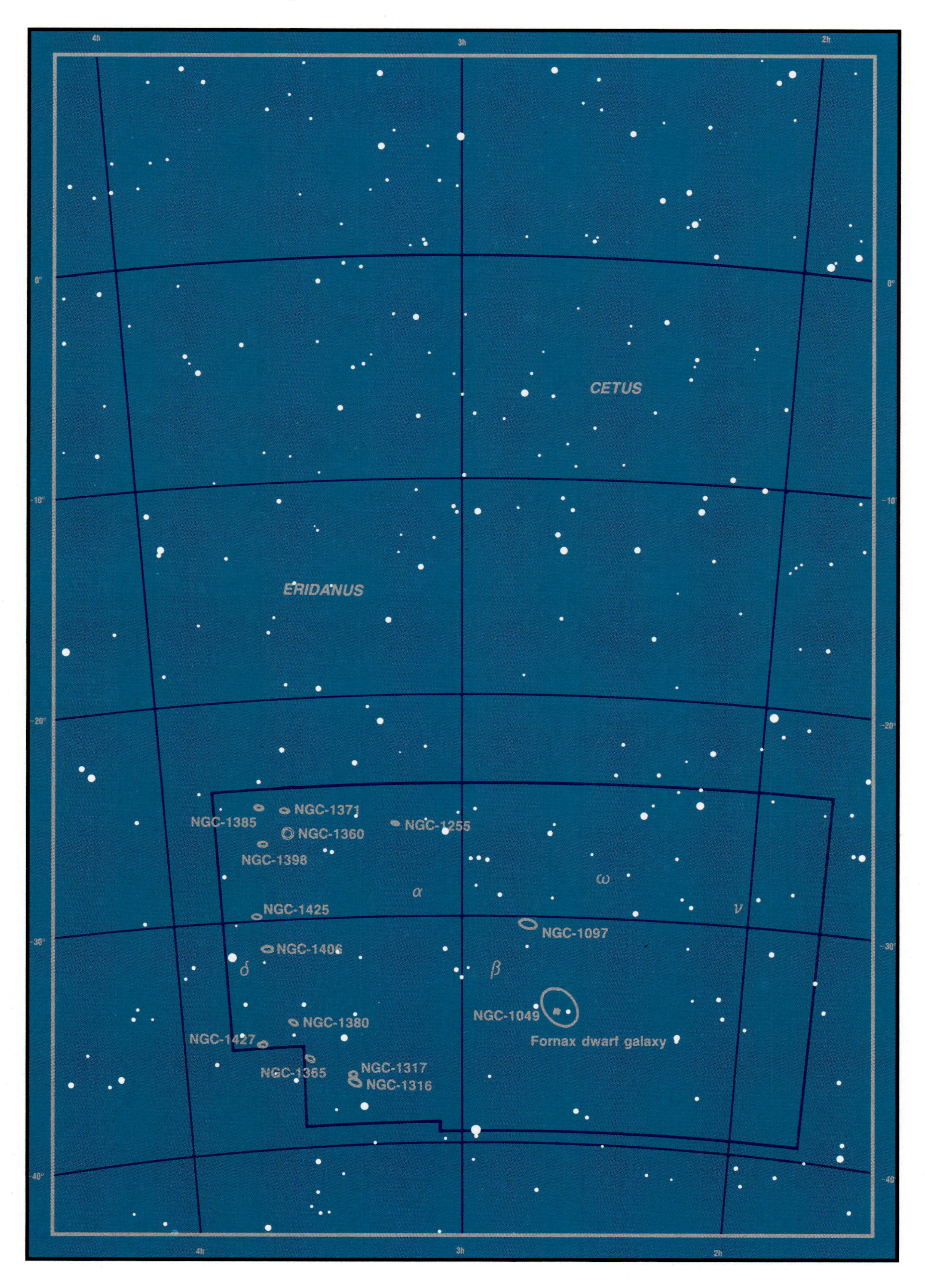

4h
3h
2h
0°
-10°
-20°
-30°
-40°
CETUS
ERIDANUS
NGC-1385
NGC-1371
NGC-1360
NGC-1255
NGC-1398
α
ω
ν
NGC-1425
NGC-1097
NGC-1406
δ
β
NGC-1049
NGC-1380
Fornax dwarf galaxy
NGC-1427
NGC-1365
NGC-1317
NGC-1316

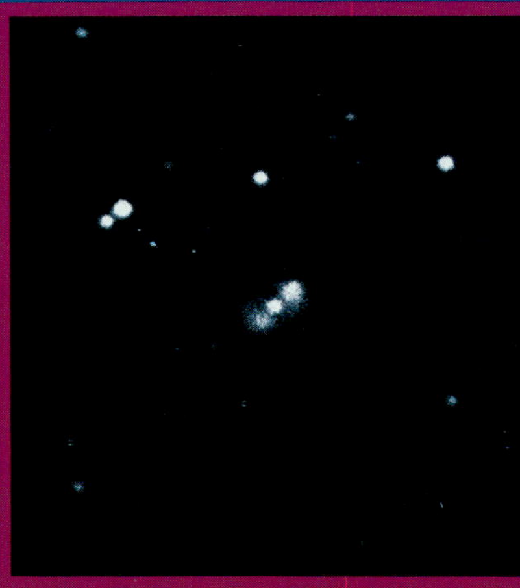

Above: One of the brightest open star clusterrs, M-35, is visible to the naked eye on dark nights. In this photo by Evered Kreimer, the more distant cluster NGC-2158 lies at bottom right. Below: Supernova remnant IC 443 is an extremely difficult object to observe. The bright star Eta (η) Geminorum lies at bottom right. Photograph by Rick Dilsizian. Bottom left: NGC-2392, the Eskimo Nebula, appears as a bluish-green, out-of-focus star in small telescopes. Photograph by Jack Newton. Middle left: Backyard telescopes show the faint planetary nebula NGC-2371/2 as a tiny double knot of nebulosity surrounding a moderately bright star.

Gemini

Gem
Geminorum

Canis Minor

CMi
Canis Minoris

The region of sky incorporating the constellations **Gemini** the Twins and **Canis Minor** the Little Dog represents one of the richest areas of the winter Milky Way for hunting deep-sky objects. Here, looking away from the Galaxy's nucleus and toward the thick of its Perseus arm, we find nice examples of open star clusters, variable and double stars, and planetary and gaseous nebulae.

The most striking object in this area is the brilliant fifth-magnitude open cluster **M-35** (NGC-2168). This group of 120 stars measures one-half degree across, and its brightest members — B-type main sequence stars and G- and K-type giants — shine with 400 times the luminosity of the Sun, making their combined glow visible to the unaided eye on good nights. A pair of 7 x 50 binoculars shows a rich field surrounding a large misty spot, partially resolved into stars. The cluster's members are evenly spread about its diameter, and there are many bright stars among the faint ones; a 6-inch or 8-inch telescope at low power shows a field full of stars, many arranged in curves and rows, with a bright orange star near the center.

While you're looking through your finderscope at M-35, you may notice a small hazy spot about 30′ to the southwest — this is **NGC-2158**, another open cluster. NGC-2158 appears much smaller, fainter, and less resolved than M-35. M-35 lies only 2,200 light-years away, while NGC-2158 lies at a distance of 16,000 light-years. Here is a chance to observe — side by side — two physically similar objects at dramatically different distances from us.

NGC-2158's 150 stars combine for a total magnitude of 11, and the group's soft glow spans only 4′. It is not an easy target for small telescopes, but is visible nevertheless as a circular nebulous smudge.

BEST VISIBLE DURING
WINTER

An 8-inch scope at high power shows a dazzling field of stars around a granular patch, giving a strong impression of great distance. The richness and relative old age of NGC-2158 suggest that it may be a transition cluster between galactic and globular types, but most astronomers class it as a galactic type.

Another nearby open cluster is **NGC-2129**, a small group of 50 stars of 8th to 15th magnitude. Although it is about the same size as NGC-2158, this group shines at seventh magnitude, largely due to a triangle of eighth and ninth-magnitude stars at the group's center. Since most of this cluster's stars are faint, small telescopes show NGC-2129 as a nebulous patch with several bright stars. An 8-inch reflector reveals an irregular mass of faint stars with one apparent double.

About 3° southeast of the M-35 area is the bright double star **Eta** (η) **Geminorum** (also known as Propus), one of the finest in the area for medium-size telescopes. The magnitude 3.3 primary is a type M3 red giant and lies 200 light-years away, making it 160 times as luminous as the Sun. The secondary is a 6.5 magnitude G8-type subgiant star, lying only 1.5″ from the primary at position angle (p.a.) 266°. Thus the star is difficult to split with anything less than a 10-inch telescope. The primary is also a semi-regular variable star with a period of 233 days and a range of almost one magnitude. The entire η Gem system is enshrouded in a cloud of cold gas about 300 astronomical units in diameter.

With your telescope centered on η Gem, plop in a low-power eyepiece and swing the telescope 1° northeast. If you are observing in dark skies, you may just pick up the supernova remnant IC 443, a tenuous arc of gas 25′ long. This is a much fainter version of the Veil Nebula in Cygnus, and is probably not visible with anything less than a 12-inch telescope on a night of good transparency.

Three notable planetary nebulae lie within the boundaries of Gemini, each offering a different shape and structure. The brightest is **NGC-2392**, a ninth-magnitude object located midway between Kappa (κ) and Lambda (λ) Geminorum. This object became known as the "Eskimo Nebula" because of its appearance — mottled disk and faint outer ring — on long-exposure photographs. With small telescopes, the Eskimo's diminutive size of 40″ sometimes poses problems to observers hunting with low power; it appears as a tiny, out-of-focus star with a noticeable bluish-green tint. Once you locate it, switch to higher powers which reveal the 10th magnitude central star and fuzzy encircling disk. NGC-2392 is one of the youngest planetaries known, lies at a distance of 3,000 light-years, and measures half a light-year across.

NGC-2371/2 is somewhat larger than the Eskimo and shows a curious double-lobed structure reminiscent of M-76 in

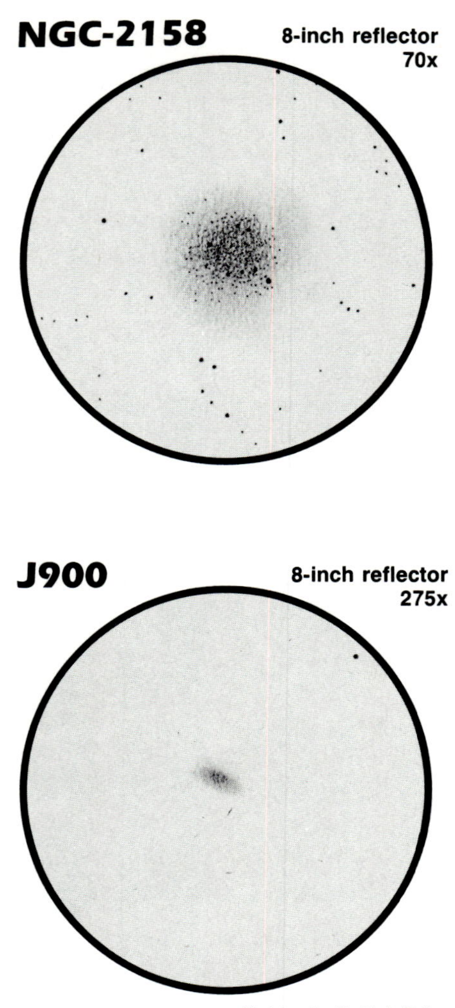

Sketches by David J. Eicher

Perseus and M-27 in Vulpecula (hence its two NGC designations). It glows at only 13th magnitude, however, and has a low surface brightness — an 8-inch scope shows two fuzzy half circles almost in contact and the magnitude 12.5 central star. The third planetary in Gemini is the 12th magnitude object **J900**, a tiny dim oval that is not impressive even in medium-size backyard instruments.

Near the region of the Eskimo nebula is the fine double star **Delta** (δ) **Geminorum**, a pair of magnitude 3.5 and 8 stars separated by 6.3″ in p.a. 218°. Also named Wasat, this star is relatively easy to split in small telescopes; its separation of 95 astronomical units is slowly widening; this will eventually make δ Gem even easier to resolve. In February 1930, when Clyde Tombaugh discovered Pluto, the planet lay near this star. (Incidentally, Uranus lay beside the previous double, η Gem, when William Herschel discovered it in 1781.)

East of δ Gem and the Eskimo are two more objects. The first is open cluster **NGC-2420**, a ninth-magnitude group of 30 stars packed into a 7′ diameter. All of the stars in this cluster are faint — 11th to 18th magnitude — so a small scope shows a mottled hazy arrangement of stars and nebulosity. Nearby is the irregular variable star **U Geminorum**, an SS Cygni-type peculiar star that seems to mimic a recurring nova. About once every 100 days, this type of cataclysmic variable can rise from its normal 14th magnitude to 9th magnitude in as little as 24 hours! It then stays at maximum for either 9 or 14 days before fading back to magnitude 14. Astronomers hypothesize that the system contains a red giant star coupled with a tiny blue dwarf and accompanying ring of gas.

One of the brightest stars in this region is **Castor** (Alpha [α] Geminorum), also one of the finest double stars in the sky for backyard observers. The system's total magnitude is 1.6, its components shining at 2.0 (Castor A) and 2.9 (Castor B) and separated by 2.3″ in p.a. 98°. Currently somewhat difficult for small instruments, the separation will widen to a comfortable 4.0″ by the end of the century. A third component, Castor C, shines at ninth magnitude and lies 72.5″ away in p.a. 164°. Yet Castor is not simply a triple star system: Each of the three stars is also a spectroscopic binary, making Castor a system with six suns! Castor A and Castor B are each composed of two A-type main sequence stars; Castor C is a system with two red dwarf stars. All three of the spectroscopic companions are invisible in backyard telescopes.

The final bright open cluster in Gemini is **NGC-2266**, a fairly rich ninth-magnitude group of stars whose individual magnitudes range from 11 to 15. The 35 members are packed into an area only 4′ across, affording a nice view in 4-inch to 8-inch telescopes at medium power.

The plane of the Milky Way is a strange place to go galaxy hunting, but one bright specimen, **NGC-2339**, is located in Gemini southwest of the δ Gem/Eskimo nebula area. NGC-2339 is a type Sc spiral glowing at magnitude 12.5 and measuring 2.0′ x 1.4′ across. It is visible as a faint oval smudge of grey nebulosity in an 8-inch telescope at high power, suspended in back of a rich field of our Galaxy's stars.

Unfortunately, Gemini's neighbor, Canis Minor, does not contain any nebular deep-sky objects or clusters of stars. Its sole attraction is the bright star **Procyon** (Alpha [α] Canis Minoris), a magnitude 0.4 type F5 subgiant which is the eighth brightest star in the sky. Procyon is a difficult if not impossible double star for amateur observers, not because of its separation — a comfortable 3.9″ in p.a. 113° — but because of the great magnitude difference between the components. Procyon B is a 10th magnitude white dwarf star whose feeble light is simply overpowered by the bright primary.

Gemini and Canis Minor lie in the midst of a rich region. Nearby lie the star cluster M-37 in Auriga, a bright binocular object, and the faint gaseous nebulae NGC-2237/9 (the Rosette Nebula), NGC-2264 (the Cone Nebula), and NGC-2261 (Hubble's Variable Nebula). The Rosette and Cone nebulae surround large bright star clusters, while NGC-2261 is a small triangular patch of misty light.

Object	M#	Type	R.A. (2000)	Dec.	Mag.	Size/Sep./Per.	N★	Mag.★
NGC-2129		⊙	6h 01m	+23°18′	6.7	5′	50	8-15
NGC-2158		⊙	6h 07m	+24°06′	11.6	4′	150	13...
NGC-2168	M-35	⊙	6h 09m	+24°19′	5.5	30′	120	8...
Eta (η) Gem		SRV	6h 15m	+22°30′	3.1↔3.9	233d		
IC 443		□SNR	6h 17m	+22°47′	14?	25′x5′		
J900		■	6h 26m	+17°47′	12.0	10′		
NGC-2266		⊙	6h 44m	+26°59′	9.1	4.5′	35	11-15
NGC-2339		∮	7h 08m	+18°47′	12.5	2.0′x1.4′		
Delta (δ) Gem		★^{2}	7h 21m	+21°59′	3.5+8.0	6.3″		
NGC-2371/2		■	7h 26m	+29°29′	12.5	50″x30″		12.5
NGC-2392		■	7h 29m	+20°55′	8.0	40″		10.0
Castor (α Gem)		★^{2}	7h 35m	+31°53′	2.5+3.5	1.8″		
NGC-2420		⊙	7h 38m	+21°34′	9.0	7′	30	10-18
Procyon (α CMi)		★^{2}	7h 39m	+50°14′	0.4+11.0	3.9″		
U Gem		IV	7h 56m	+22°00′	8.9↔14.0	Irr		

★^{2}	*Double Star*
SRV	*Semi-Regular Variable*
IV	*Irregular Variable*
⊙	*Open Cluster*
■	*Planetary Nebula*
□SNR	*Supernova Remnant*
∮	*Spiral Galaxy*

N★ = number of stars, Mag.★ = magnitude range of cluster or magnitude of central star, (...) indicates many fainter.

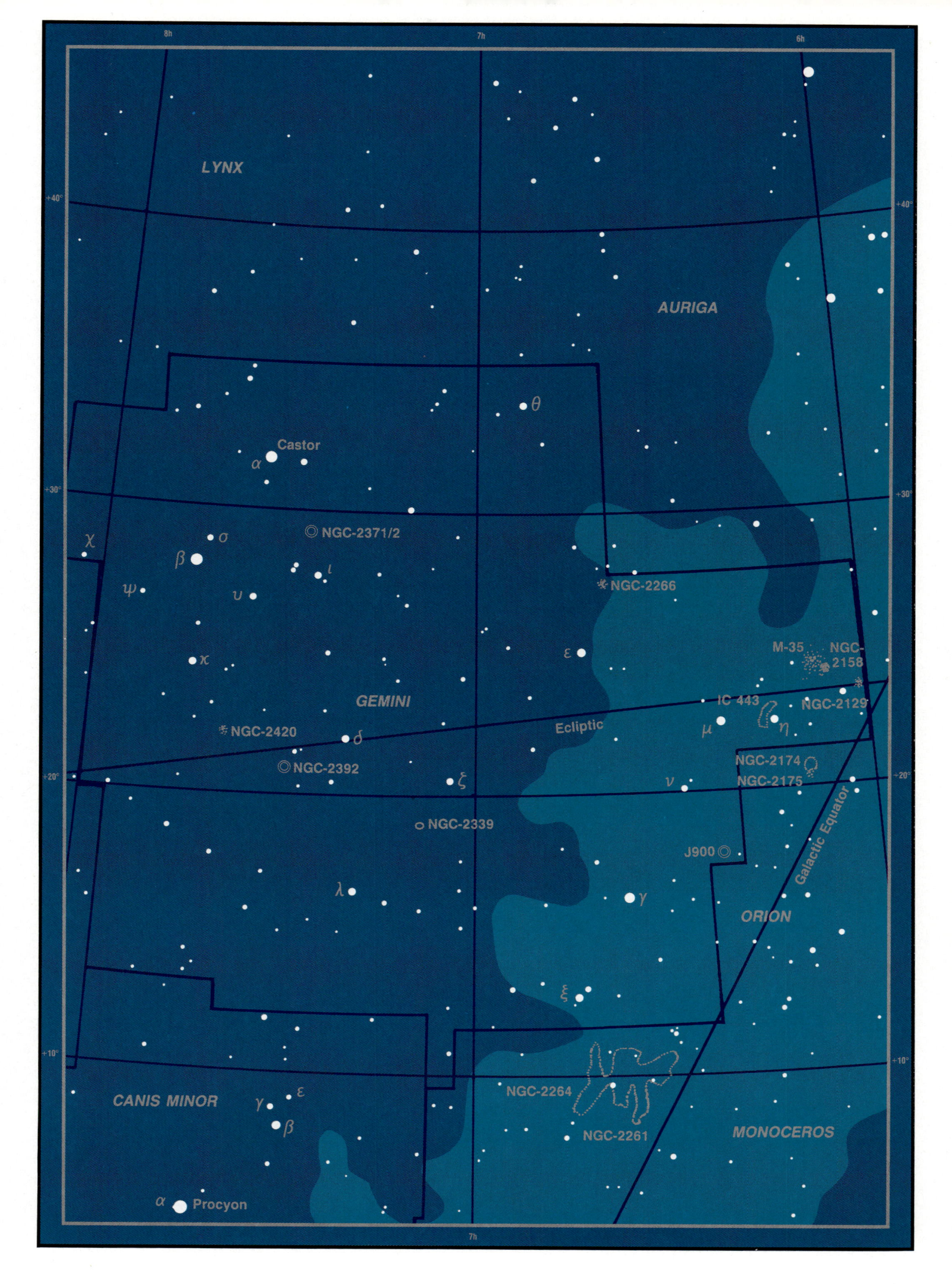

8h
7h
6h
LYNX
AURIGA
Castor
NGC-2371/2
NGC-2266
M-35
NGC-2158
NGC-2129
IC 443
GEMINI
Ecliptic
NGC-2420
NGC-2392
NGC-2174
NGC-2175
NGC-2339
J900
Galactic Equator
ORION
NGC-2264
NGC-2261
CANIS MINOR
MONOCEROS
Procyon
+40°
+30°
+20°
+10°

Above: M-13, the Hercules cluster, is one of the largest and brightest globular clusters in the sky. It is faintly visible to the naked eye under a dark sky, but its outer edges can be resolved into stars with telescopes as small as 4 inches in aperture. Photo by Bill Iburg. Far left: Overlooked because of its famous neighbor is M-92, also a fine globular cluster for small telescopes. Photo by Jack Newton. Left: NGC-6210, a tiny blue-green planetary nebula, requires high power for identification. Photo by Jack Newton.

Hercules

Her
Herculis

Keith Ward

Lying between the rich plane of the Galaxy in Cygnus and Lyra and the realm of faint galaxies in Bootes and Corona Borealis, **Hercules** the Strongman holds a mixture of easy globulars and faint galactic objects.

As constellations go, Hercules is large but rather faint. Its brightest stars shine at 3rd magnitude, making it traceable under dark skies but elusive where light pollution invades. The constellation's central asterism is the "Keystone," a trapezoid formed by Eta (η), Zeta (ζ), Epsilon (ε), and Pi (π) Herculis. It is recognizable even under typical suburban backyard conditions.

The constellation's best offerings are: M-13 and M-92, two of the best globulars in the sky; NGC-6210, a fine planetary nebula; and 95 Herculis, a double star showing one of the most striking examples of color contrast.

The appropriate starting point for deep-sky viewing in Hercules is **Messier 13** (NGC-6205), popularly known as the Hercules cluster. M-13 lies 2.5° due south of Eta Herculis, one of the Keystone stars, and is bright enough at magnitude 5.9 to spot with the naked eye under a perfectly black sky. Typically, however, it is visible only as a fuzzy "star"with finderscopes or binoculars.

You can find it most easily by starting at Eta Herculis (the northernmost star in the Keystone) and moving the telescope south about one-third of the way toward Zeta Herculis, the star at the small end of the Keystone. There you'll see the fuzzy disk of M-13, which forms a right triangle with two 7th-magnitude stars.

A 3-inch refractor shows M-13 as a uniformly lit disk of milky light some 8′ across; the view with a 4-inch refractor or 6-inch reflector is much better since the cluster's edges are resolved into faint stars.

BEST VISIBLE DURING
SUMMER

M-13 is a breathtaking sight in large backyard telescopes. A 10-inch or larger instrument and good seeing and transparency resolves stars across the entire face of M-13, showing hundreds of pinpoints against a velvety black backdrop of sky. On steady nights, don't be reluctant to use high power on the Hercules cluster. Its high surface brightness and remarkable number of component stars will reward you with a telescopic field jam-packed with tiny suns.

High powers may also produce two curious effects with M-13. First, notice that many of the cluster's outer stars seem to be arranged in long arcs, giving the appearance of appendages extending out from the cluster's center. Next, look at the distribution of stars across the cluster's face: there appear to be dark voids intermingled with patches of numerous stars, three of which meet in a Y-shape on the core's southeastern side. These are not "holes" in the cluster but rather perceptual effects of seeing more bright stars at one place and less at another.

M-13 is regarded by many Northern Hemisphere observers as the finest globular cluster, largely because it lies high in the sky from their vantage point. Actually, ten globular clusters are larger and six are brighter than M-13, although all of them are either invisible or low on the horizon for North American and European skywatchers. M-13's true diameter is some 37 parsecs and its distance is 7.7 kiloparsecs. It appears big and bright due to its close proximity, not because it is intrinsically large or luminous.

If you observe M-13 on a dark night and use a wide-field eyepiece, you may notice a tiny fuzzy patch of light in the same field, 1.5° north and slightly east of the great cluster's center. This is **NGC-6207**, a diminutive Sc-type spiral galaxy tilted about 45° to our line of sight. This little galaxy isn't terribly impressive in most backyard instruments (the only detail visible is a starlike nucleus surrounded by a misty envelope of light), but it does offer one of the sky's best examples of extreme depth of field. The galaxy lies 14.2 megaparsecs away — nearly 2,000 times more distant than its neighboring globular cluster.

Nearly 10° northeast of the M-13/NGC-6207 pair, in a relatively blank section of sky, is the fine globular cluster **M-92** (NGC-6341). This object is overshadowed by the popular Hercules cluster and would rate far higher on observing lists if it were located in another part of the sky. Shining at magnitude 6.5 and spanning some 11.2′ across, M-92 is nearly as large and bright as M-13, but since it is more compact it is harder to resolve into kindred stars. A 6-inch scope at high power will show a smattering of suns lining M-92's edges; a 12-inch instrument under ideally dark skies will essentially duplicate the view of M-13 in a smaller scope. M-92 is physically smaller and fainter than M-13 and lies slightly farther away.

A third bright globular in Hercules is **NGC-6229**, a magnitude 9.4 cluster located 7° northwest of M-92. A finderscope under a dark sky will show NGC-6229 as a tiny smudge of greenish gray light. Spotting it is easy because the cluster is located several arcminutes east of a bright, wide pair composed of 7th- and 8th-magnitude stars. NGC-6229 has a physical diameter about the same as M-92's; it is relatively faint and unimpressive only because it lies more than twice as far away as its brighter neighbors to the south.

Hercules contains three bright planetary nebulae observable in backyard telescopes. The best of the lot is **NGC-6210**, a bluish disk 14″ across shining at photographic magnitude 9.3. (Since photographically derived magnitudes tend to run faint, NGC-6210 will appear as an 8th-magnitude object in your eyepiece.) To find this little planetary, start at Beta (*β*) Herculis, a 3rd-magnitude star located 10° south of the Keystone. Then move 4° northeast and you'll see a wide pair of 7th-magnitude stars, one of which is a double. NGC-6210 lies several arcminutes northwest of the single 7th-magnitude star.

At first you may have trouble identifying this object due to its tiny size. Once you've found the right spot, switch to medium power and look for a bluish "star" that appears slightly out of focus — that will be the nebula.

The other two planetaries in Hercules are considerably fainter than NGC-6210 but still bright enough to see in small scopes. **NGC-6058** lies in a barren patch of sky near the Hercules-Corona Borealis border, some 8° northwest of M-13. It measures 23″ across and glows at photographic magnitude 13.3, causing it to appear as a pale ghostly patch of light. **IC 4593** shines at photographic magnitude 10.9 and is about the same size as NGC-6210. It lies in the constellation's southern end near the border with Serpens Caput.

Three unusual stars lie within Hercules' boundaries. Alpha (*α*) Herculis is a semi-regular variable with an approximate period of 90 days that witnesses a change in brightness of nearly one magnitude, between 3.1 and 3.9. Observe it throughout the summer and compare its brightness with that of the other 3rd- magnitude stars in Hercules. **DQ Herculis**, now a 15th-magnitude speck of light barely perceivable with amateur instruments, went nova in 1934 and shone at magnitude 1.3 for nearly two months. **95 Herculis**, a pretty double star consisting of twin 5th-magnitude suns, offers one of the most stunning color contrasts of any pair in the sky. The famed 19th-century observer Admiral Smyth described the colors as "apple green and cherry red." Do you agree?

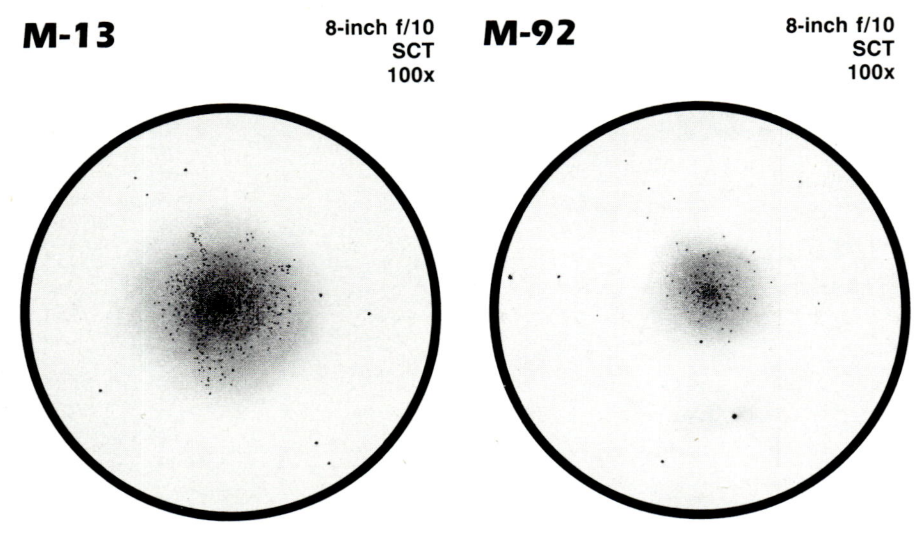

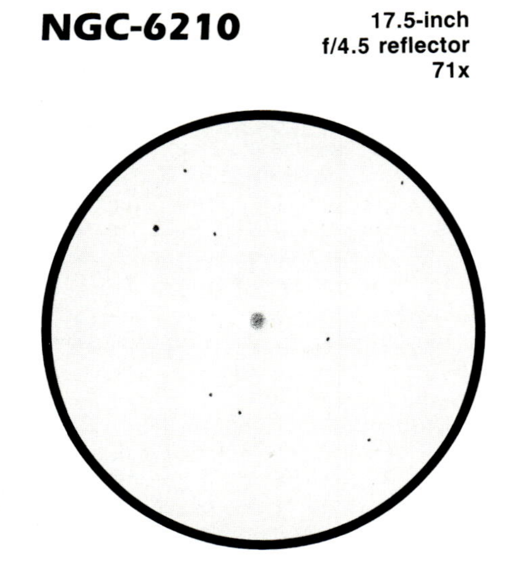

Sketches by David J. Eicher

Object	M#	Type	R.A. (2000)	Dec.	Mag.	Size/Sep./Per.	H
NGC-6058		■	16h 04.4m	+40°41′	13.3p	23″	
NGC-6052		§	16h 05.2m	+20°32′	13.0	1.0′x0.7′	Sc
IC 4593		■	16h 12.2m	+12°04′	10.9p	12″	
NGC-6106		§	16h 18.8m	+ 7°25′	12.2	2.6′x1.5′	Sb+
NGC-6181		§	16h 32.3m	+19°50′	11.9	2.6′x1.3′	Sc
NGC-6205	M-13	●	16h 41.7m	+36°28′	5.9	16.6′	
NGC-6207		§	16h 43.1m	+36°50′	11.6	3.0′x1.4′	Sc
NGC-6210		■	16h 44.5m	+23°49′	9.3p	14″	
NGC-6229		●	16h 47.0m	+47°32′	9.4	4.5′	
NGC-6239		§B	16h 50.1m	+42°44′	12.3	2.8′x1.3′	Sb
Alpha (α)		SRV	17h 14.6m	+14°23′	3.1↔3.9	90:d	
NGC-6341	M-92	●	17h 17.1m	+43°08′	6.5	11.2′	
NGC-6482		()	17h 51.8m	+23°04′	11.3	2.3′x2.0′	E3pec
95		★²	18h 01.5m	+21°36′	5.0,5.1	6.3″	
DQ		N	18h 07.5m	+45°51′	1.3↔15.6	—	
NGC-6574		§	18h 11.9m	+14°59′	12.0	1.4′x1.1′	S:

H = Hubble classification type for galaxies

★²	*Double Star*
SRV	*Semi-regular Variable*
N	*Nova*
●	*Globular Star Cluster*
■	*Planetary Nebula*
§	*Spiral Galaxy*
§B	*Barred Spiral Galaxy*
()	*Elliptical Galaxy*

19h
18h
17h
16h
+50°
+40°
+30°
+20°
+10°
LYRA
DQ
NGC-6229
M-92
NGC-6239
NGC-6058
NGC-6207
M-13
NGC-6210
95
NGC-6181
NGC-6574
OPHIUCHUS
18h
17h

Top left: NGC-3628 in Leo is a spiral galaxy turned almost edge-on. The broad dust lane in its plane cuts across the nuclear bulge. Photo by Jack Newton. Top right: NGC-2903 (also in Leo) is seen close to face-on; its spiral arms delineated by lanes of dust stand out clearly in this photo by Martin Germano. Above left: M-95 is a barred spiral galaxy in Leo resembling a Greek theta; photo by Martin Germano. Above right: NGC-3184 (in Leo) is a faint spiral galaxy tilted so that we are looking down onto its disk. Photo by Martin Germano. Left: M-65 and M-66 in Leo form a pretty pair in Jack Newton's photo. Both galaxies have bright nuclei surrounded by a faint haze difficult to see in small instruments.

Leo
Leo
Leonis

Leo Minor
LMi
Leonis Minoris

Sextans
Sex
Sextantis

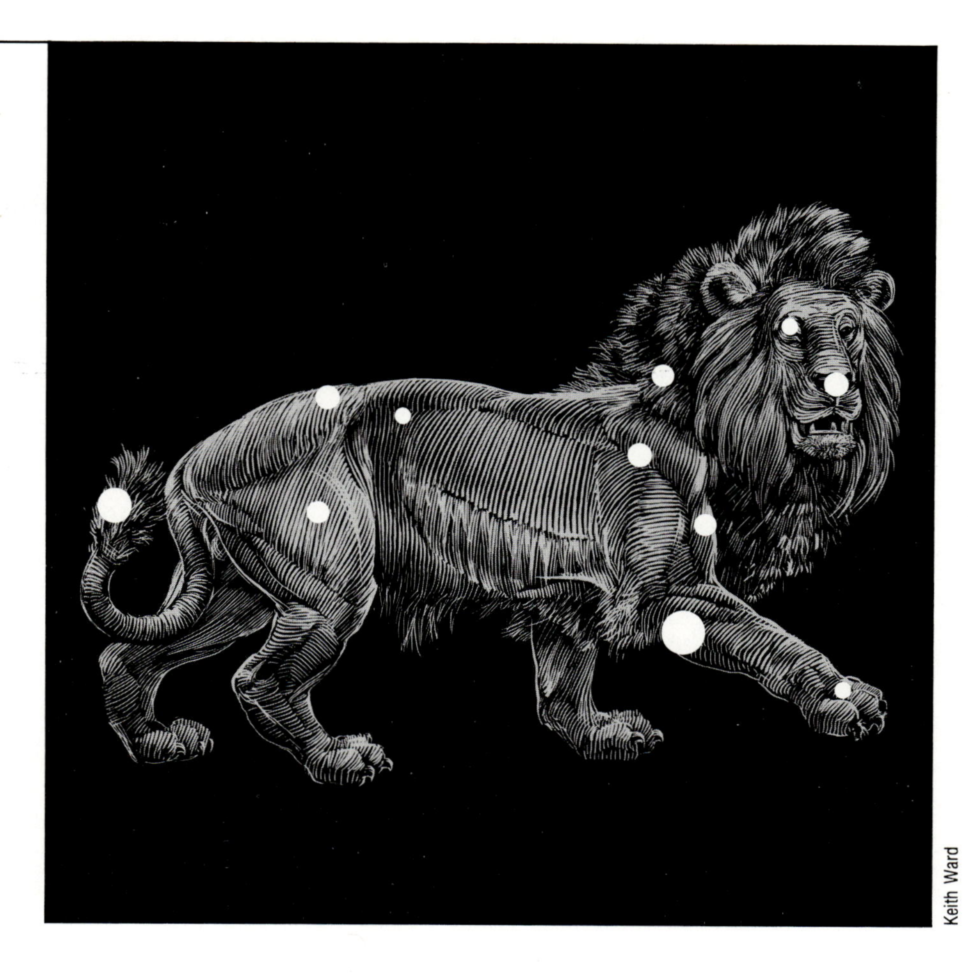
Keith Ward

The constellations **Leo** the Lion, **Leo Minor** the Little Lion, and **Sextans** the Sextant encompass a region of sky with lots of a single type of deep-sky object — galaxies. Though many 9th and 10th magnitude spirals dominate the scene, there is also room left for the occasional double star to test your optics and variable star to test your eye's keenness.

The brightest galaxy in Leo — located about 1.5° south of Lambda (λ) Leonis — is **NGC-2903**, a large, loosely wound spiral, which shines at magnitude 9.5. Its dimensions of 11.0′ x 4.7′ make it visible as a tiny smudge even in a 2-inch telescope; a 4-inch at high power shows a mottled oval core surrounded by a tenuous haze. Eight-inch telescopes on good nights show this galaxy's condensed nucleus encapsulated in a vast halo of faint nebulosity, including mottled dark areas indicating dust in the spiral arms. The distance to this great spiral is 7 megaparsecs (Mpc); its mass is on the order of 43 billion solar masses.

Almost as luminous as NGC-2903 is the Sb-type spiral **M-66** (NGC-3627), which glows at magnitude 9.7 and is the kingpin of a group of three bright galaxies clustered in one telescopic field. Midway between the stars Theta (θ) and Iota (ι) Leonis — about 2.5° south-southeast of θ — lie M-66, **M-65** (NGC-3623), and **NGC-3628**. M-66 measures 8.0′ x 2.5′ across and is visible in finder telescopes on dark nights, appearing as a tiny patch of fuzz. Larger telescopes show a bright elongated region of grey nebulosity wrapped in a fainter halo; 10-inch scopes hint at spiral structure. The galaxy M-65 is a type Sa/Sb and appears much flatter, measuring 7.8′ x 1.6′. Its magnitude of 10.2 keeps its surface brightness fairly high, so small telescopes reveal it as a silvery disk of light viewed close to edge-on — something like seeing an illuminated quarter-dollar from far away. The thin bright disk portion of M-65 shows up in all telescopes, but you'll need a good 8-incher to see much of the galaxy's perimeter, composed of a lightly mottled haze. M-65 and M-66 lie 21′ apart in the sky, corresponding to about 55 kpc; they measure 18.3 kpc and 15.4 kpc across, respectively, and their distance from Earth is around 9.7 Mpc.

The third member of the trio of galaxies that form the core of the M-65 group is NGC-3628, a large, faint galaxy aligned edge-on to our line of sight. Lying just 35′ north of M-66, it is one of the sky's best edge-on galaxies, though its surface brightness is considerably lower than NGC-4565 in Coma Berenices. Nevertheless, a good 6-inch reflector at high power shows the magnitude 10.3 disk intertwined with a thin but obvious dust band running along the galaxy's edge.

The second bright trio of galaxies in Leo is the group consisting of **M-95** (NGC-3351), **M-96** (NGC-3368), and **M-105** (NGC-3379). The brightest of the three is M-96, which glows at magnitude 10.1 and measures 6.0′ x 4.0′ across; its tight structure is elusive in telescopes, but an oval core is visible even in a good 3-incher. An 8-inch scope at high power begins to show some outer low-surface-brightness nebulosity. About 42′ west of M-96 is the small barred spiral system M-95; it shines at magnitude 10.5 and spans 4.0′ x 3.0′. This system is composed of a bright small core surrounded by a smooth bar that extends out to an encircling ring of matter. The galaxy's core and even its bar structure are relatively easy marks for medium-sized backyard telescopes on nights of good transparency; the outer ring structure, however, has a dismally low surface brightness. The distance to these two galaxies is about 8.9 Mpc.

About 48′ north-northwest of M-96 is the elliptical galaxy M-105. This little ball of fuzz measures only 2.1′ x 2.0′ across and shines at magnitude 10.3; it is visually unimpressive compared with the spirals in Leo. Being an elliptical, it has a high surface brightness, but doesn't show any detail, regardless of the telescope used. This galaxy lies 130 kpc from M-95 and M-96. The faint galaxies NGC-3384 and NGC-3389 (magnitudes 11.0 and 12.2, respectively) form a little triangle with M-105, which measures about 8′ on a side. See if you can pick up all three in a low-power eyepiece field.

A large, impressive spiral galaxy is

BEST VISIBLE DURING
SPRING

M-65 & M-66

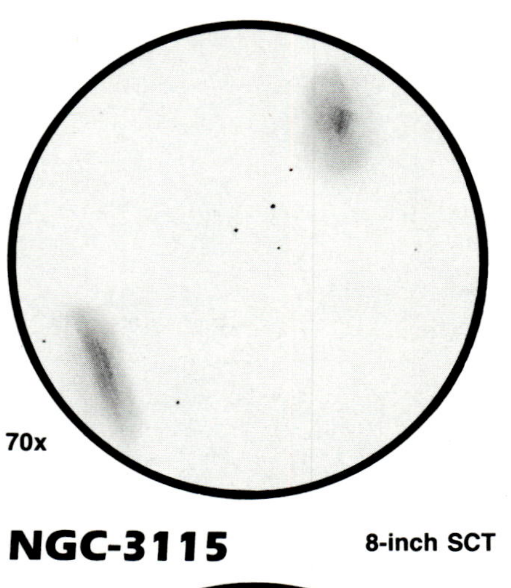
70x

NGC-3115

8-inch SCT

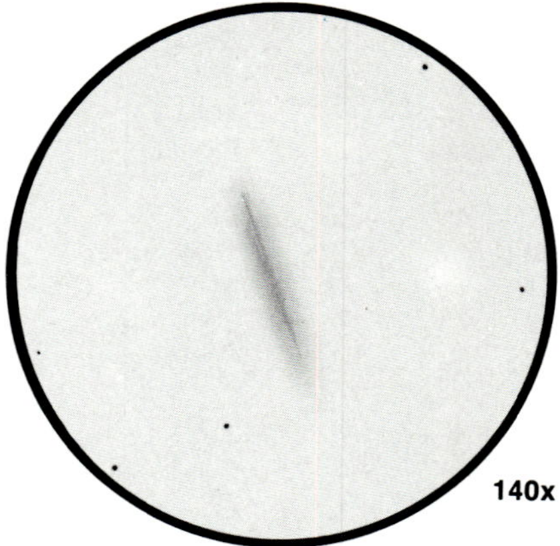
140x

Sketches by David J. Eicher

NGC-3521, located in the far southern reaches of Leo. This is a good example of a tightly wound multi-arm spiral of type Sb; its magnitude is 9.6, and its dimensions 6.0′ x 4.0′. Affix a low-power ocular when sweeping to find this galaxy, and you'll note a bright, obvious streak of grey misty light. A 4-inch telescope using medium powers clearly reveals an elongated shape around a condensed core of bright nebulosity. An 8-incher shows this, as well as some of the hazy envelope of low surface brightness nebulosity composing the galaxy's spiral arms.

Regulus

8-inch SCT
222x

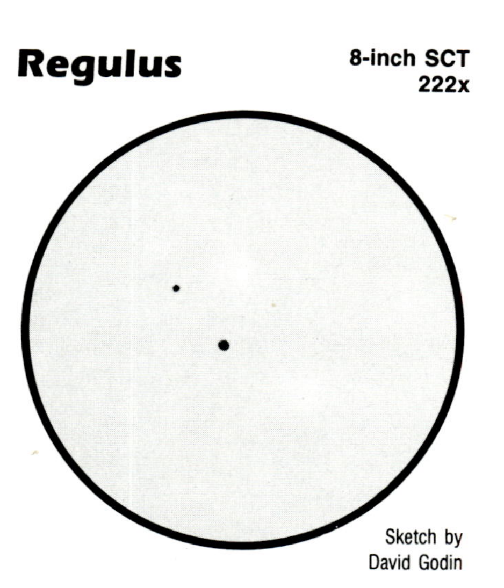

Sketch by David Godin

The Lion contains some good double and variable stars as well as innumerable galaxies. One of the most popular doubles is **Regulus** (Alpha [α] Leonis), the brightest star in this swath of sky. Regulus means "the little king," and it is a fitting title for the 1.4 magnitude star whose secondary — shining at magnitude 7.9 and located 177″ away in p.a. 307° (1924) — is an easy-to-find, attractive sight in small telescopes. Regulus is about 26 pc distant, and shines with the luminosity of 160 Suns. A curious object only 20′ north of Regulus is the dwarf elliptical galaxy known as **Leo I**, one of the smallest and faintest known galaxies and — at a distance of 230 kpc — a member of the Local Group. On photographic plates it appears as a smudge about 10′ across; Regulus' glare and the galaxy's low surface brightness make it extremely difficult to observe visually.

A fine double star is **Algeiba** (Gamma [γ] Leonis), a magnitude 2.0 system including 2.1 and 3.4 magnitude stars separated by 4.4″ in p.a. 123° (1966). First find this star, the brightest in the "Sickle" of Leo, by moving 8° north and slightly east from Regulus. You can then increase your telescopic magnification to the point where the double is cleanly split.

Nearly due west of Regulus is the Mira-type star **R Leonis**, a long-period variable with a cycle of 312 days. The fourth known LPV, discovered by J.A. Koch in 1782, this star ranges from magnitude 5.2 to 10.5. Fortunately, R Leonis is bracketed by some very useful comparison stars — one shining around 5th magnitude and, closer yet, a pair of 11th magnitude stars forming a triangle with the variable. When you train your scope on this star, do you see a pair of bright stars surrounded by fainter ones or a solitary bright star accompanied by a triangle of fainter ones? The former indicates R Leo near its maximum, the latter near minimum.

The area of sky immediately surrounding Leo incorporates some fine galaxies, one of which is **NGC-3115** in Sextans. This is a 10th magnitude edge-on spindle galaxy, which shows as a slender streak of grey light in most backyard telescopes. Although it measures 4.0′ x 1.0′ in extent, this galaxy does not show much detail in backyard telescopes. **NGC-3344**, a galaxy in Leo Minor, is an magnitude 10.4 face-on late-type spiral, measuring 6.0′ x 5.1′ across and offering fine views to medium-sized telescopes. An 8-incher shows a soft glow surrounding a stellar nucleus. **NGC-3486**, also a resident of Leo Minor, is a multi-arm spiral measuring 5.5′ x 4.2′ and glowing at magnitude 10.8. Small telescopes show it as a circular haze with little detail of its inner workings.

Object	M#	Type	R.A. (2000)	Dec.	Mag.	Size/Sep./Per.	H
NGC-2903		∮	9h 32m	+21°27′	9.5	11.0′x4.7′	Sb/Sc
R Leo		LPV	9h 48m	+11°26′	5.2↔10.5	312d	
NGC-3115		()L	10h 05m	−07°43′	10.0	4.0′x1.0′	E7/S0
Regulus (α Leo)		★²	10h 08m	+11°58′	1.5,8.0	177″	
Leo I		()	10h 09m	+12°17′	11.1	11′x8′	E4
Algeiba (γ Leo)		★²	10h 20m	+19°51′	2.5,3.5	4.4″	
NGC-3344		∮	10h 43m	+24°54′	10.5	6.0′x5.1′	Sc
NGC-3351	M-95	∮B	10h 44m	+11°42′	10.5	4.0′x3.0′	SBb
NGC-3368	M-96	∮	10h 47m	+11°49′	10.1	6.0′x4.0′	Sb
NGC-3379	M-105	()	10h 48m	+12°35′	10.3	2.1′x2.0′	E1
NGC-3486		∮	11h 01m	+28°59′	10.9	5.5′x4.2′	Sc
NGC-3521		∮	11h 06m	−00°02′	9.6	6.0′x4.0′	Sb
NGC-3623	M-65	∮	11h 19m	+13°07′	10.2	7.8′x1.6′	Sa/Sb
NGC-3627	M-66	∮	11h 20m	+13°01′	9.7	8.0′x2.5′	Sb
NGC-3628		∮	11h 20m	+13°36′	10.3	12.0′x2.0′	Sb

H = Hubble classification type for galaxies

★²	*Double Star*
LPV	*Long Period Variable*
∮	*Spiral Galaxy*
()	*Elliptical Galaxy*
∮B	*Barred Spiral Galaxy*
()L	*Lenticular Galaxy*

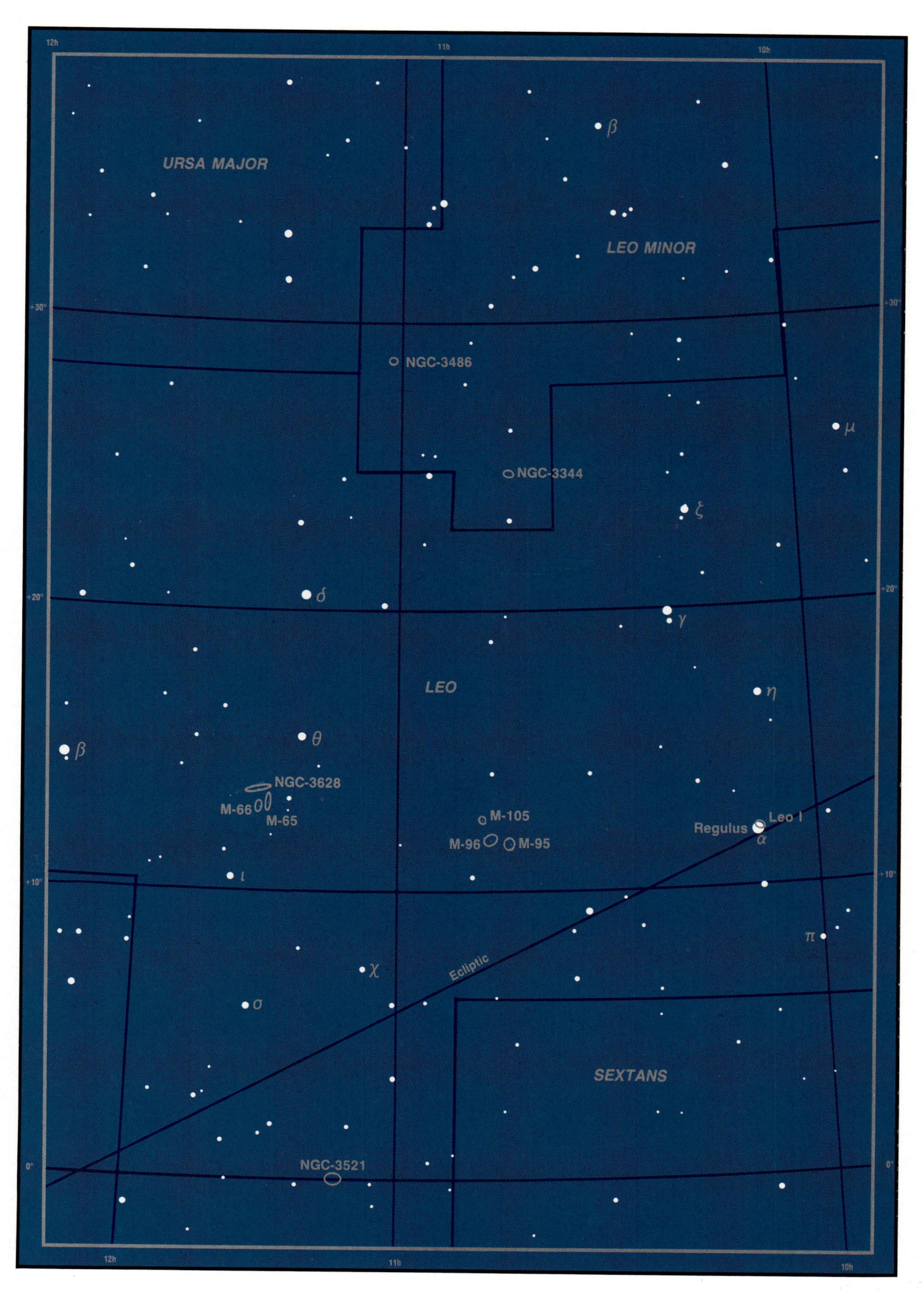

12h
11h
10h
+30°
+20°
+10°
0°
URSA MAJOR
LEO MINOR
LEO
SEXTANS
NGC-3486
NGC-3344
NGC-3628
M-66
M-65
M-105
M-96
M-95
Regulus
Leo I
NGC-3521
Ecliptic

Above: As one of the sky's best globular clusters, M-5 offers a bright core and resolved edges to most backyard telescopes. Photo by Ben Mayer. Far left: The galaxy NGC-5921 appears as a bright core surrounded by faint nebulosity. Photo by Ron Potter. Left: The globular cluster NGC-5897 has a low surface brightness and is difficult to spot on less-than-dark nights. Photo by Martin C. Germano.

Libra

Lib
Librae

Serpens Caput

Ser
Serpentis

Keith Ward

The region of sky containing Libra and Serpens Caput is overshadowed by the Milky Way's riches in nearby Scorpius and Sagittarius. Nevertheless, **Libra**, the Scales, and **Serpens Caput**, the Serpent's Head, both hold many small galaxies and each features a unique globular cluster.

Libra's most unusual object, **NGC-5897**, is difficult to observe owing to low surface brightness. It is a large, loosely-structured globular cluster lying in the south-central part of the constellation. Although NGC-5897's total magnitude is 8.6, its light is spread out over an area nearly half the Moon's diameter, making it appear extremely faint in small telescopes: only under very dark skies does a 4-inch telescope reveal NGC-5897, and a 6-inch scope doesn't show much more than a dim mottled patch of greyish-green light. A 12-inch telescope, however, distinctly shows the globular's large nebulous circle.

Because of NGC-5897's distance of 11.7 kiloparsecs, its brightest stars — red giants — glow weakly at fainter than 17th magnitude. It is impossible to clearly see individual stars in this cluster with a small backyard telescope. Even though NGC-5897 is not one of the most impressive globulars, the challenge here lies in simply finding the object.

Several degrees southeast of NGC-5897 is a diminutive planetary nebula known as **Merrill 2-1**. Skipped by the compilers of both the *New General Catalogue* and the two *Index Catalogues*, this object was discovered in the 1940s by astronomer Paul Merrill when he noticed its strong emission spectrum. Also studied by the Soviet astronomer Boris Vorontsov-Velyaminov, Me 2-1 is additionally known as VV 72. A problem with observing Me 2-1 is its tiny diameter: it spans a mere 7″, appearing as a star with anything but high power. It shines at about 12th magnitude and its elusive central star is estimated at magnitude 15.4. To find this small planetary, wait until a very dark, steady night. With a low-power eyepiece, find the nebula's approximate position. Now carefully study all objects of 12th magnitude — if one looks slightly fuzzy, switch to high magnification and see if your suspicions will be confirmed.

BEST VISIBLE DURING
SPRING

Libra also contains many small groups of galaxies. Lying between Zubenelgenubi (Alpha [α] Librae) and Libra's western border is the small Sc-type spiral **NGC-5595**. This petite galaxy shines at magnitude 12.4 and measures 2.0′ x 1.2′, making it visible as a fuzzy patch in telescopes as small as 6 inches in aperture. A 12-inch scope trained on NGC-5595 shows a small bright core enveloped in a smooth cottony halo. Just 4′ southeast of NGC-5595 is a companion galaxy catalogued as NGC-5597. This object is slightly fainter than NGC-5595, but — at 2.0′ x 1.8′ — it is also slightly larger. Its nucleus is quite bright and condensed into a nearly stellar ball of light. Both of these galaxies lie together in a group some 35 megaparsecs distant.

Midway between the NGC-5595/NGC-5597 pair and Zubenelgenubi is the fine barred spiral **NGC-5728**. In long-exposure photos, this galaxy shows a "theta" shape similar to that of the much brighter galaxy M-109 in Ursa Major. NGC-5728 shines at magnitude 11.3 and covers 2.8′ x 1.6′ of sky; it is an easy target for a good 6-inch telescope. The galaxy doesn't show much detail, however. All it reveals when viewed with a backyard telescope is a bright core surrounded by faint mottled haze.

Three more galaxies lie in the central part of Libra. **NGC-5878** is a magnitude 11.5 Sb-type spiral measuring 3.5′ x 1.7′ across. Although easily visible in small telescopes, this galaxy generally appears as nothing more than a misty blur set against a barren starfield; its nucleus is diffuse and unconcentrated. **NGC-5861**, a 12th magnitude spiral, is somewhat more enticing: it measures 3.0′ x 1.8′, contains an unusually weak and motley nucleus, and features peculiar distorted spiral arms. These arms are not easily visible, but on a good night, a large backyard telescope — one in the 16-inch plus category — hints that strange disruptions are taking place inside this otherwise innocent-looking spiral. **NGC-5885**, a magnitude 11.7 spiral galaxy, is more typical. It's viewed nearly face-on, measures 3.5′ x 3.2′ across, and shows a bright stellar nucleus wrapped in a faint mottled circle of nebulosity.

The best galaxy in Libra lies in its northwestern corner. **NGC-5792** is a highly-inclined 12th magnitude spiral measuring 7.2′ x 2.1′. Easily visible in a 6-inch telescope on dark nights, it appears as an oval smudge of light with a bright stellar nucleus. Larger telescopes don't reveal much more structure in NGC-5792,

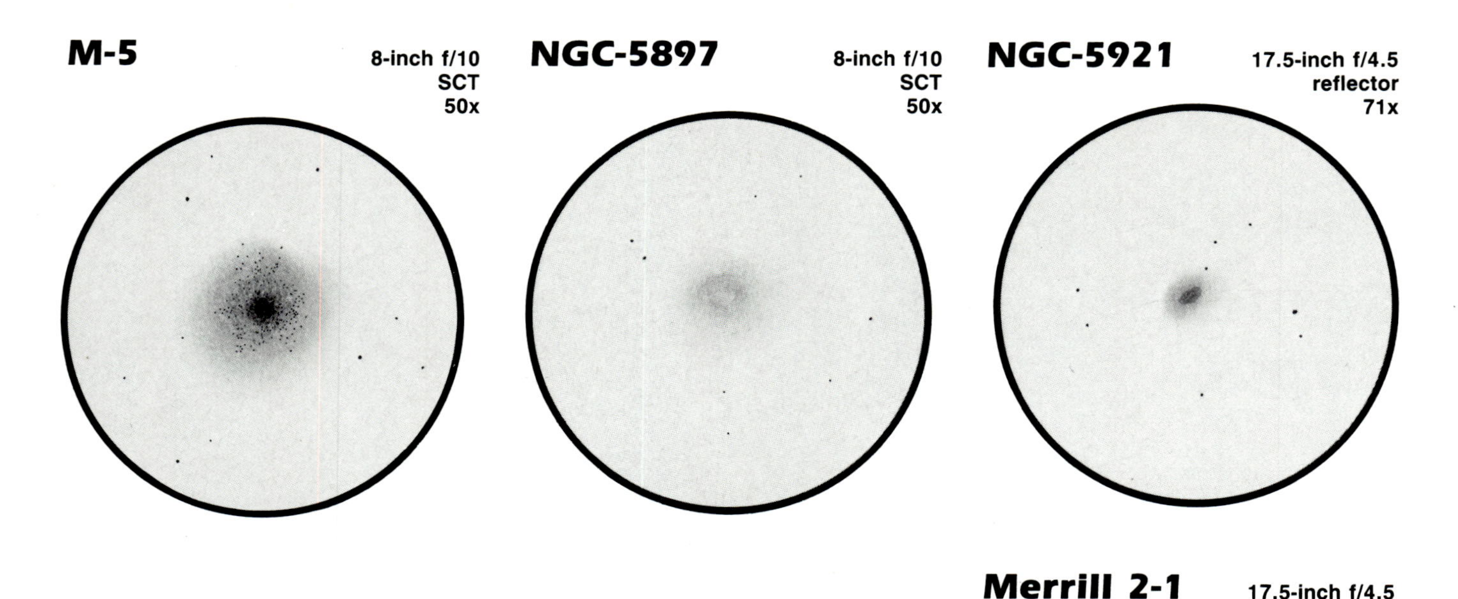

Merrill 2-1

17.5-inch f/4.5
reflector
142x

Sketches by David J. Eicher

but show its halo of nebulosity — the spiral arms — as knottier as telescope aperture increases.

An unusual star in Libra is **48 Librae**, the classic example of a so-called shell star, a late B or early A-type star which is slightly variable and shows peculiar emission lines in its spectrum. These features occur because the star is surrounded by a turbulent shell of gas whirling to and fro at velocities of 60 miles per second. This strange star is easy to observe, as it normally shines at magnitude 4.9. Its range of variance is slight: from magnitude 4.8 to 5.0.

Serpens Caput's brightest galaxy is **NGC-5921**, a magnitude 10.8 barred spiral measuring some 4.9′ x 4.2′ across. This galaxy has a bright compact core surrounded by faint barred spiral arms, which are separated by dusty patches. Viewed in a backyard telescope of 5 inches' aperture, NGC-5921 looks like a small, roundish blotch of light with a bright center. A 10-inch telescope shows the nucleus as an intensely bright ball encircled by faint wisps of nebulosity suggesting spiral arms. A 17.5-inch scope at medium power reveals a mottled, indefinite spiral structure. **NGC-6118**, another bright galaxy in Serpens Caput, is a 12th magnitude Sb-type spiral. Its 4.7′ x 2.3′ dimensions make NGC-6118 visible in a 6-inch telescope on a dark night.

One of the finest globular clusters in the sky is **Messier 5** (NGC-5904), a large globe of stars shining at magnitude 5.8. M-5 measures 17.4′ across, corresponding to an actual diameter of 39 parsecs. It is fairly close as globulars go, lying high above the galactic plane 7.7 kiloparsecs away. Because it is so bright, M-5 is visible to the naked eye as a "star" under very dark skies. In binoculars, this unresolved patch of light appears as a fuzzy disk resembling an out-of-focus star. A good 3-inch telescope shows a bright glowing core wrapped inside a much fainter halo of nebulosity. A 6-inch scope at high power resolves the outer edges of M-5, revealing its stellar nature and transforming it into a fascinating object. Large backyard telescopes under dark skies resolve stars fully across M-5's face, providing unforgettable views.

Object	M#	Type	R.A. (2000)	Dec.	Mag.	Size/Sep./Per.	H
NGC-5595		§	14h 24.2m	−16°43′	12.5	2.0′x1.2′	Sc
NGC-5728		§B	14h 42.4m	−17°15′	11.3	2.8′x1.6′	SBb
NGC-5792		§	14h 58.4m	− 1°05′	12.4	7.2′x2.1′	Sbpec
NGC-5861		§	15h 09.3m	−11°19′	12.2	3.0′x1.8′	Sbpec
NGC-5878		§	15h 13.8m	−14°16′	11.0	3.5′x1.7′	Sb
NGC-5885		§	15h 15.1m	−10°05′	11.7	3.5′x3.2′	S
NGC-5897		●	15h 17.4m	−21°01′	8.9	12.6′	
NGC-5904	M-5	●	15h 18.6m	+ 2°05′	5.8	17.4′	
NGC-5921		§B	15h 21.9m	+ 5°04′	10.8	4.9′x4.2′	SBb
Merrill 2-1		■	15h 22.3m	−23°38′	12?	7″	
48 Lib		★S	15h 58.2m	−14°17′	4.8-5.0	Irr	
NGC-6118		§B	16h 21.8m	− 2°17′	12.3	4.7′x2.3′	SBb

H = Hubble classification type for galaxies

★S *Shell Star*
● *Globular Cluster*
■ *Planetary Nebula*
§ *Spiral Galaxy*
§B *Barred Spiral Galaxy*

16h
15h
VIRGO
SERPENS CAPUT
NGC-5921
M-5
NGC-5792
NGC-6118
LIBRA
NGC-5885
NGC-5861
NGC-5878
48
NGC-5595
NGC-5728
NGC-5897
SCORPIUS
Merrill 2-1
M-4
CENTAURUS
LUPUS
0°
10°
20°
30°
40°

Above: The oddly shaped planetary nebula IC 4406 in Lupus shines at photographic magnitude 10.6 and covers nearly 30″. It is visible in 6-inch or larger scopes as a box-shaped smudge of light.

Left: Far more elegant but not as bright is the ring-shaped planetary Shapley 1 in Norma. Although it is larger than Lyra's famous Ring Nebula, Shapley 1 glows dimly at photographic magnitude 13.6 and therefore has a very low surface brightness. Photos by Jack B. Marling.

Lupus

Lup
Lupi

Norma

Nor
Normae

Keith Ward

The constellations **Lupus** the Wolf and **Norma** the Carpenter's Square lie between the bright, star-filled Milky Way star groups Centaurus and Scorpius. Although neither Lupus nor Norma contains nearly as many deep-sky objects as its larger neighbors, both constellations do offer bright examples of open and globular star clusters and several unusual planetary nebulae. Most of these objects are visible in binoculars or small telescopes when Lupus and Norma are relatively high in the sky. For many North American observers, however, this never happens.

Of Lupus' varied objects, probably the most intriguing is the oddly shaped planetary nebula **IC 4406**. This object shines at photographic magnitude 10.6 and has a long dimension of only 28″, so its average surface brightness is relatively high. IC 4406's central star, embedded in a dense cloud of nebulosity, glows dimly at photographic magnitude 14.7. It is exceedingly difficult to spot with backyard telescopes.

Locating IC 4406 is easy: start by centering your telescope's finder on the triangle of 4th- and 5th-magnitude stars formed in part by Iota (ι) and Tau (τ) Lupi. (This triangle lies 10° east and 2° north of the great globular cluster Omega [ω] Centauri.) The planetary lies about 2° north-northeast of the 5th-magnitude star that marks the triangle's apex.

Once you find IC 4406, don't be afraid to use high power to inspect its minute details. A 4-inch scope at 100x shows IC 4406 as merely a tiny, faint cloud of gray light. A 6-incher at high power, however, reveals that IC 4406 is distinctly box-shaped. A 12-inch scope shows this nebula as a bright, bluish-gray patch of light that is clearly rectilinear.

BEST VISIBLE DURING
SPRING

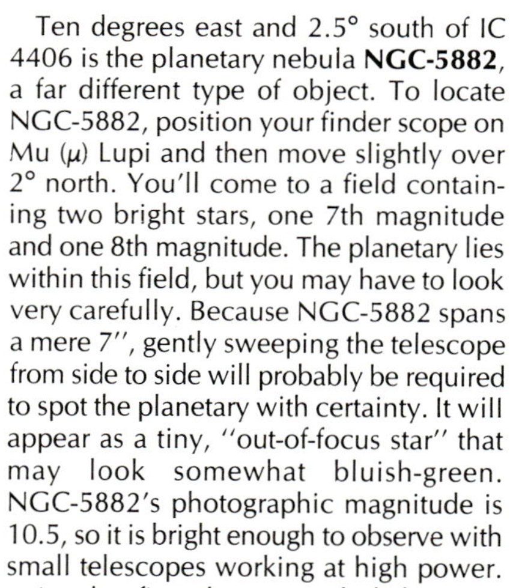

Ten degrees east and 2.5° south of IC 4406 is the planetary nebula **NGC-5882**, a far different type of object. To locate NGC-5882, position your finder scope on Mu (μ) Lupi and then move slightly over 2° north. You'll come to a field containing two bright stars, one 7th magnitude and one 8th magnitude. The planetary lies within this field, but you may have to look very carefully. Because NGC-5882 spans a mere 7″, gently sweeping the telescope from side to side will probably be required to spot the planetary with certainty. It will appear as a tiny, "out-of-focus star" that may look somewhat bluish-green. NGC-5882's photographic magnitude is 10.5, so it is bright enough to observe with small telescopes working at high power.

Another fine planetary nebula lies near the eastern border of Lupus roughly midway between the bright stars Chi (χ) and Theta (θ) Lupi. **NGC-6026** is a 45″-diameter planetary, thought to be a galaxy until 1955 when the University of Texas astronomer Gerard de Vaucouleurs identified it as a member of our Galaxy. This object lies 2° southeast of Chi Lupi and can be located easily by following an imaginary line from Chi through the 5th-magnitude double star Xi (ξ) Lupi and then extending the line for an equal distance.

NGC-6026 is a rather large planetary nebula, and with a magnitude of 12.5 it has a very low surface brightness. A 10-inch scope shows this object as a uniformly illuminated, faint haze with a condensed, nearly stellar nucleus. Successively larger apertures show little more than an interesting field full of stars.

The remaining bright deep-sky objects in Lupus are star clusters. Northwestern Lupus holds the fine globular cluster **NGC-5824**, found less than 1° southeast of a 5th-magnitude star. Shining at 9th magnitude and measuring 6.2′ across, NGC-5824 is recognizable as a cluster when viewed through just about any telescope. This cluster is strongly condensed toward its center and is therefore difficult to resolve with backyard telescopes. But on a dark, transparent night partial resolution is provided by a 12.5-inch scope at high power.

Bigger and brighter than NGC-5824 is the globular **NGC-5986**, a cluster sandwiched between the 5th-magnitude star h Lupi and a row of three 7th- and 8th-magnitude stars on the constellation's eastern side. NGC-5986 measures 9.6′ across and glows at magnitude 7.1, which make it easily visible as a large, fuzzy disk of light. It is relatively simple to resolve NGC-5986 into constituent stars, and a 6-inch scope at high power shows many individual points of light ringing the cluster's edges.

Still more impressive is the globular cluster **NGC-5927**, lying in the southeast part of Lupus about 3° northeast of Zeta (ζ) Lupi. Shining at magnitude 8.3 and covering 12′, NGC-5927 appears in small telescopes as a nebulous haze peppered

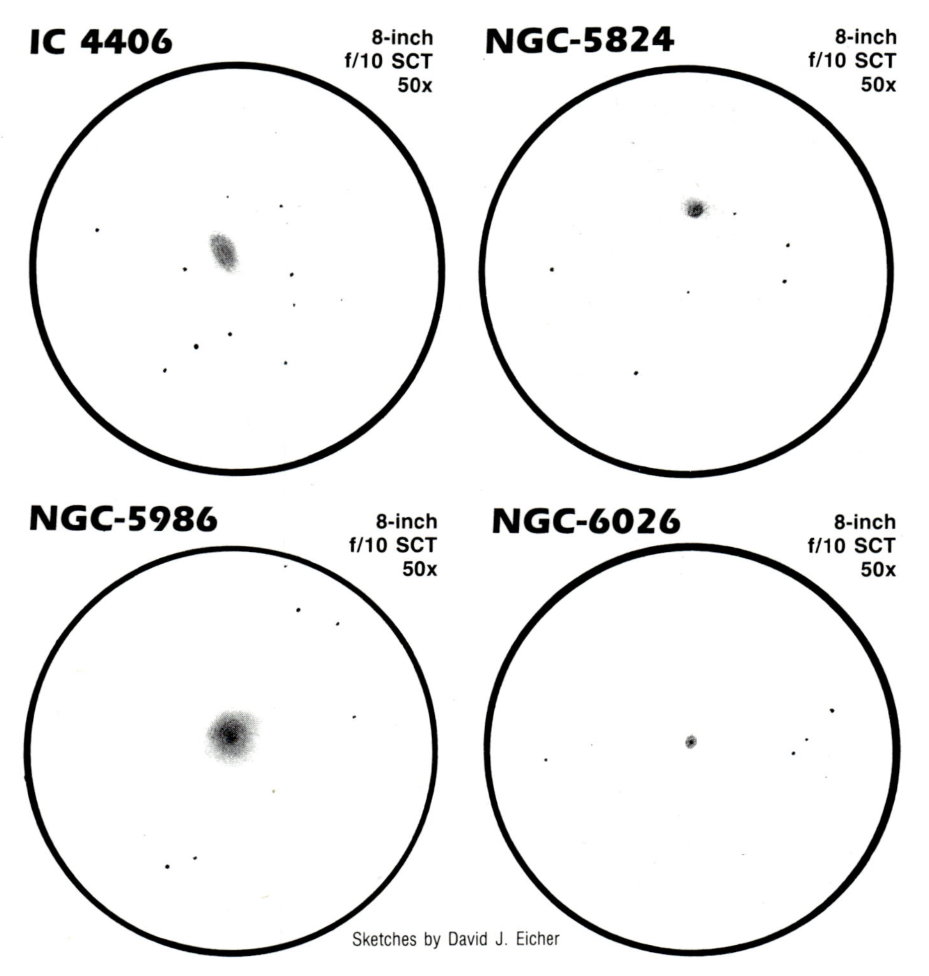

Sketches by David J. Eicher

with tiny stellar points; its central core is bright white, unresolvable, and surrounded by a field full of sparkling stars. It is a magnificent sight.

Two bright open clusters can be found in south-central Lupus. **NGC-5749** is a 9th-magnitude group of thirty stars compressed into an area measuring only 8′ across. This cluster appears like a small clump of light in binoculars and is transformed into a respectable and captivating collection of stars in small telescopes. **NGC-5822** is a much brighter and larger group; consequently, it requires a large field of view. This cluster contains 150 stars in an area spanning 40′ and has a total photographic magnitude of 6.5. The cluster's brightest stars shine at 10th magnitude, which makes NGC-5822 visible in small telescopes as a large, scattered field of bright and faint stars loosely comprising a cluster.

As with Lupus, the most captivating object in Norma is a planetary nebula. **Shapley 1**, named for American astronomer Harlow Shapley, is one of the most elegant planetaries in the sky, a perfectly round ring of nebulosity 76″ across. Although it is as large as the famous Ring Nebula in Lyra, Shapley 1 is nowhere near as bright. With a photographic magnitude of 13.6, the light from Shapley 1 is spread out over a large enough area that it has a low surface brightness. Consequently, this planetary is difficult to spot with anything less than an 8-inch telescope under a very dark sky. If you see a faint, ring-shaped nebulosity, look for the 14th-magnitude central star from which the nebula formed.

Another planetary in Norma is the well-observed **Vorontsov-Velyaminov 78**. Named for the Russian astronomer who identified the nebula, VV 78 lies less than 1° southwest of the 5th-magnitude star Kappa (κ) Normae. Although it glows as dimly as Shapley 1, VV 78 is much smaller (27″ in diameter) and therefore has a far higher surface brightness. VV 78 appears like a bright, round, gray patch of light with no central star and is surrounded by a pretty star field.

Four open star clusters are strewn across Norma, and each offers a different combination of qualities for small scope observers. About 1.5° east of VV 78 is the fine cluster **NGC-6067**, a group of one hundred stars of 8th magnitude and fainter packed into an area some 13′ across. Among the stars in this group are a number of bright pairs that form an attractive sight in nearly any size telescope. Three degrees west of NGC-6067 is the splendid star group **NGC-6087**, a cluster of forty stars of 8th magnitude and fainter in an area 12′ across. This group's total magnitude is 5.4.

Lying near the eastern edge of Norma are the bright open clusters **NGC-6134** and **NGC-6152**. NGC-6134 is a bright group of 9th-magnitude and fainter stars some 7′ across; its total magnitude is 7.2. NGC-6152 is a group of seventy stars shining at 8th magnitude and is a wonderful sight in binoculars. Its stars, 11th magnitude and fainter, cover as much sky as the Full Moon.

Object	M#	Type	R.A. (2000)	Dec.	Mag.	Size/Sep./Per.	N★	Mag.★
IC 4406		■	14h 22.4m	−44°09′	10.6_P	28″		14.7_B
NGC-5749		⊙	14h 48.9m	−54°31′	8.8_P	8′	30	
NGC-5824		●	15h 04.0m	−33°04′	9.0	6.2′		
NGC-5822		⊙	15h 05.2m	−54°21′	6.5_P	40′	150	10_P
NGC-5882		■	15h 16.8m	−45°39′	10.5_P	7″		12.0_B
NGC-5927		●	15h 28.0m	−50°40′	8.3	12.0′		
NGC-5986		●	15h 46.1m	−37°47′	7.1	9.6′		
Shapley 1		■	15h 51.7m	−51°31′	13.6_P	76″		13.9
NGC-6026		■	16h 01.4m	−34°32′	~12.5	45″		
NGC-6067		⊙	16h 13.2m	−54°13′	5.6	13′	100	8.3
VV 78		■	16h 14.5m	−54°57′	12.6_P	27″		
NGC-6087		⊙	16h 18.9m	−57°54′	5.4	12′	40	7.9
NGC-6134		⊙	16h 27.7m	−49°09′	7.2	7′		9.3
NGC-6152		⊙	16h 32.7m	−52°37′	8.1_P	30′	70	11_P

⊙ *Open Cluster*
● *Globular Cluster*
■ *Planetary Nebula*

N★ = number of stars; Mag.★ = magnitude range of cluster or magnitude of central star

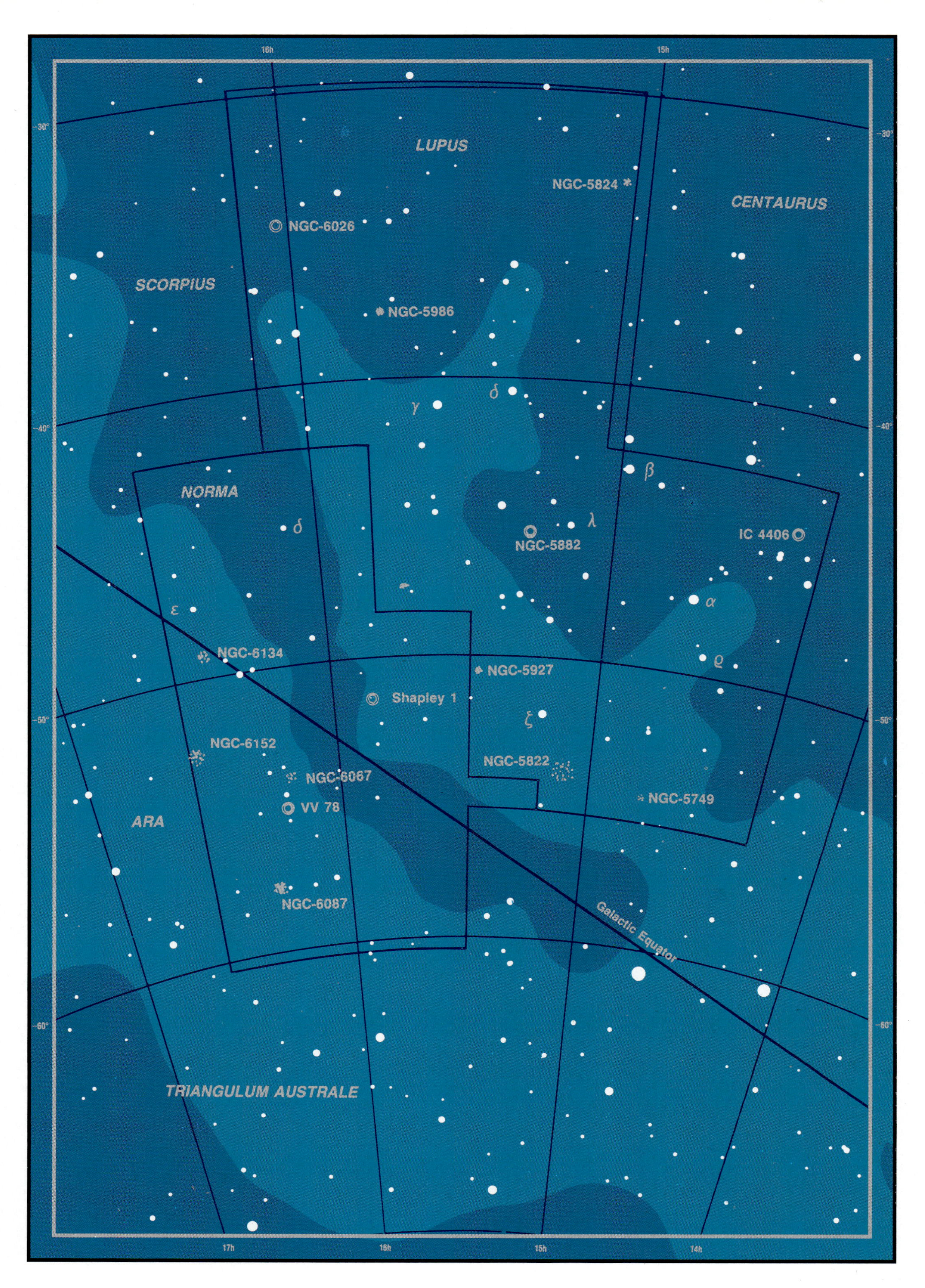
16h
15h
LUPUS
NGC-5824
CENTAURUS
NGC-6026
SCORPIUS
NGC-5986
γ
δ
NORMA
β
δ
λ
NGC-5882
IC 4406
ε
α
NGC-6134
NGC-5927
ϱ
Shapley 1
ζ
NGC-6152
NGC-5822
NGC-6067
NGC-5749
VV 78
ARA
NGC-6087
Galactic Equator
TRIANGULUM AUSTRALE
-30°
-40°
-50°
-60°
17h
16h
15h
14h

Above: The Sb-type spiral NGC-2683 is one of the brighter springtime galaxies; i reveals subtle spiral structure to owners of medium-sized telescopes. Photo by K Alexander Brownlee. Left: Located 90 kiloparsecs distant, NGC-2419 is one o the remotest globular clusters visible ir backyard telescopes. Photo by Martin C Germano.

Lynx

Lyn
Lyncis

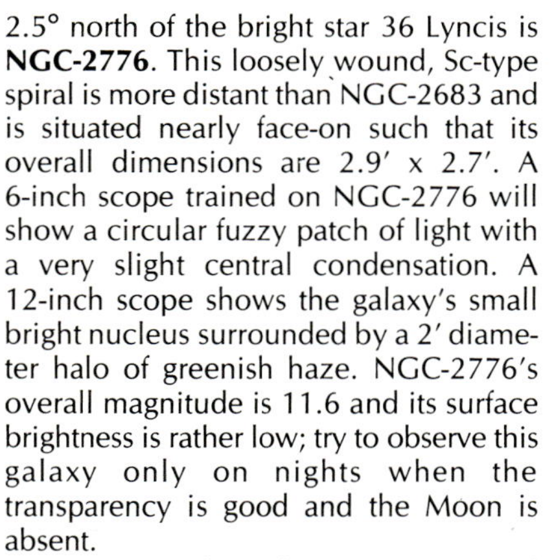

Because it is almost invisible to the naked eye, **Lynx** the Lynx is often thought to contain virtually no interesting deep-sky objects. But this faint collection of stars lies on the extreme western edge of a rich cloud of galaxies that stretches from Ursa Major in the north through Virgo in the south. Lynx is home to a number of galaxies that appear quite impressive in small telescopes. The constellation also contains two unusual deep-sky objects: an extremely distant globular cluster and a very large, faint, and challenging planetary nebula.

The finest galaxy in this region of sky is the large Sb-type spiral **NGC-2683**, a galaxy that lies just inside Lynx's border with Cancer. To find this 10th-magnitude misty oval of light, find the star Sigma² (σ^2) Cancri in your finderscope and move the telescope about 2° northwest. A low-power eyepiece should provide a good overall view of NGC-2683, which is inclined at a nearly edge-on angle to our line of sight. With a telescopic aperture of 4-inches or more, you should see a small, thin, bright oval nucleus measuring some 2′ x 0.5′ across. This is enshrouded in a far larger envelope of hazy light spanning 6′ x 1.5′. (The photographic extent of NGC-2683 is 9.3′ x 2.5′.)

Experiment with different eyepieces; since NGC-2683's surface brightness is rather high, it can take fairly high magnification without becoming a featureless blob. With a 10-inch or larger telescope you may detect some of the galaxy's subtle spiral structure, which is fairly easy to see in the largest backyard telescopes (those over 16 inches in aperture). If you have one, try using a blue-sensitive nebula filter on this galaxy. You may get impressive results.

Several smaller galaxies lie in the constellation's northeastern corner. About 2.5° north of the bright star 36 Lyncis is **NGC-2776**. This loosely wound, Sc-type spiral is more distant than NGC-2683 and is situated nearly face-on such that its overall dimensions are 2.9′ x 2.7′. A 6-inch scope trained on NGC-2776 will show a circular fuzzy patch of light with a very slight central condensation. A 12-inch scope shows the galaxy's small bright nucleus surrounded by a 2′ diameter halo of greenish haze. NGC-2776's overall magnitude is 11.6 and its surface brightness is rather low; try to observe this galaxy only on nights when the transparency is good and the Moon is absent.

BEST VISIBLE DURING
SPRING

Moving southward past 36 Lyn and continuing another 3.5° will bring you to the fine Sb-type spiral **NGC-2782**. This galaxy is slightly larger and brighter than NGC-2776, appearing as a dish-shaped oval of light in a 4-inch telescope. A 10-incher reveals a strongly condensed nuclear area wrapped in a uniformly bright halo of nebulosity. Midway between 36 Lyn and NGC-2782 and slightly to the east is the faint, barred spiral **NGC-2798**. This galaxy shines at magnitude 12.3 and spans 2.8′ x 1.1′ across, appearing as a faint oval spot of light in small telescopes. Large backyard telescopes don't show much more detail but do reveal a bright nucleus and show the haze as asymmetrical.

A curious collection of galaxies lies in central Lynx, bracketed by the bright star 31 Lyncis and the double star 27 Lyncis. The most unusual of these, **NGC-2537**, is about 4° northwest of 31 Lyn. NGC-2537 is probably a face-on spiral (its classification is a bit uncertain) measuring 1.7′ x 1.5′ and glowing softly at magnitude 11.7. A 4-inch scope shows this galaxy as a uniformly illuminated, pale roundish patch. Larger telescopes, however, make NGC-2537 more interesting. An 8-incher under a clear, dark sky shows the galaxy evenly lit save for a bright, incomplete ring of nebulosity around its perimeter. A 16-inch scope reveals some knotty structure in this ring and several bright clumps over the galaxy's face. This weird feature gives NGC-2537 its nickname — the "Bear-Paw Galaxy."

If you observe NGC-2537 at low power with a scope of 8 inches or more aperture, you'll probably get an added treat. In the same field lies **IC 2233**, one of the thinnest galaxies visible in backyard telescopes. IC 2233 is a 13th-magnitude barred spiral oriented almost exactly edge-on. Its dimensions are 4.7′ by a slender 0.6′. An 8-inch scope shows it as a hair-thin, silvery needle of milky light. Larger scopes show only a slight central bulge around the nucleus.

About 3° south and a little east of 27 Lyn is the large spiral **NGC-2541**, a galaxy whose relatively high surface brightness makes it visible as an oval-shaped smudge in 3-inch scopes. NGC-2541's total magnitude is 11.8 and its dimensions are 6.6′ x 3.5′, causing it to appear as a large, nebulous halo centered on a bright, compact nucleus in 6 inch and larger telescopes. Nearby is **NGC-2500**, a barred spiral that is 1.5° southwest of 27 Lyn. NGC-2500 is relatively small — 2.9′ x 2.7′ in diameter — but its total magnitude of 11.6 gives it a high surface brightness.

NGC-2549, our last galaxy in Lynx, lies in the constellation's northeastern corner less than a degree west of the bright star 30 Lyncis. This elliptical galaxy measures 4.2′ x 1.5′ and shines at magnitude 11.1. Small telescopes show it as a bright, featureless oval haze in the same field as 30 Lyn. On a dark clear night, try observing NGC-2549 at high power to get 30 Lyn out of the field. This object is a fine example of a bright E6-type elliptical. The galaxy lies at a high enough declination that it remains visible to many Northern Hemisphere viewers for long periods of time.

One of the largest planetary nebulae is located in a blank field of sky some 6° southwest of NGC-2549. This is **Perek-Kohoutek 164-31.1**, a large amorphous mass of greenish white light measuring a whopping 400″ — nearly 7′ — across. This nearby, intrinsically faint planetary has a total magnitude of roughly 14 and a very low surface brightness, making it a good challenge for owners of large backyard scopes and nebula filters. The nebula is easily photographed with long exposures under a moonless sky.

Lynx also contains a variety of double and variable stars. **V Lyncis** is an irregular variable haphazardly flitting between magnitudes 9.5 and 12. **R Lyncis** is a long-period variable that regularly fluctuates between magnitudes 7.2 and 15 over a period of 379 days. When you look at either of these objects, make a sketch showing the relative brightnesses of field stars around them. By making regular observations every few weeks, you'll be able to compile a log detailing each star's behavior.

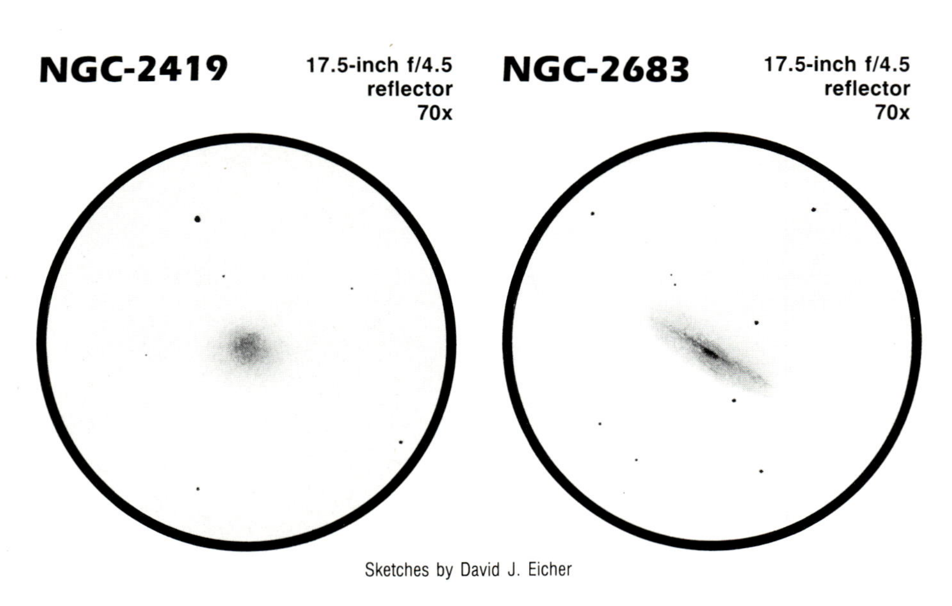

Sketches by David J. Eicher

Easy to find and observe is **38 Lyncis**, a bright triple star just 3° north of Alpha (*α*) Lyncis in the constellation's southeastern corner. When you find this star, you'll see that it consists primarily of a rather close pair of bright stars with magnitudes of 3.9 and 6.6 (separation 2.7″, p.a. 229°). In the same field of view is an 11th-magnitude star 1½ arc-minutes away at p.a. 212°. All three stars have common proper motions and are, therefore, traveling together through the Galaxy.

In the northern part of Lynx lies **19 Lyncis**, a fine quadruple star. The main pair is wide and easy in all scopes. It consists of magnitude 5.6 and 6.5 stars separated by 14.8″ at p.a. 315°. About 4′ away at p.a. 3° is a magnitude 8.9 star, and 74.2″ distant at p.a. 287° is an 11th-magnitude member of the same system. This star is a beautiful sight in telescopes both small and large.

One of Lynx's best deep-sky treats is **NGC-2419**, a small globular cluster located in the same field as a 7th-magnitude star at Lynx's southwestern edge. NGC-2419 isn't a tremendous sight as globulars go: it's 4.1′ in diameter and glows feebly at magnitude 10.4. The key word with NGC-2419 is distance — it lies about 90 kiloparsecs away, making it one of the most distant globulars known in the Milky Way. (The entire disk portion of the Galaxy is a mere 35 kiloparsecs across!) Because of its great distance, the eminent astronomer Harlow Shapley dubbed NGC-2419 the "Intergalactic Tramp," believing that it may have broken the gravitational chains of the Milky Way and headed off into deep space. More recent observations, however, indicate that NGC-2419 is moving in a highly eccentric galactic orbit but is destined to remain a member of our local territory in the universe.

Object	M#	Type	R.A. (2000)	Dec.	Mag.	Size/Sep./Per.	H
V		IV	6h 29.7m	+61°33′	9.5↔12.0	irr.	
R		LPV	7h 01.3m	+55°20′	7.2↔15	379d	
19		★4	7h 22.9m	+55°17′	5.6,6.5 8.9,10.9	14.8″,214.9″, 74.2″	
NGC-2419		●	7h 38.1m	+38°53′	10.4	4.1′	
P-K 164+31.1		■	7h 57.8m	+53°25′	14p	400″	
NGC-2500		§B	8h 01.9m	+50°44′	11.6	2.9′x2.7′	S+
NGC-2537		§	8h 13.2m	+46°00′	11.7	1.7′x1.5′	S:
IC 2233		§B	8h 14.0m	+45°44′	13.0	4.7′x0.6′	SBd:
NGC-2541		§	8h 14.7m	+49°04′	11.8	6.6′x3.5′	S+
NGC-2549		()	8h 19.0m	+57°48′	11.1	4.2′x1.5′	E6
NGC-2683		§	8h 52.7m	+33°25′	9.7	9.3′x2.5′	Sb−
NGC-2776		§	9h 12.2m	+44°57′	11.6	2.9′x2.7′	Sc
NGC-2782		§	9h 14.1m	+40°07′	11.5	3.8′x2.9′	Sb
NGC-2798		§B	9h 17.4m	+42°00′	12.3	2.8′x1.1′	SBap
38		★3	9h 18.8m	+36°48′	3.9,6.6,10.8	2.7″,87.7″	

H = Hubble classification type for galaxies

★3	*Triple Star*
★4	*Quadruple Star*
LPV	*Long Period Variable*
IV	*Irregular Variable*
●	*Globular Cluster*
■	*Planetary Nebula*
§	*Spiral Galaxy*
§B	*Barred Spiral Galaxy*
()	*Elliptical Galaxy*

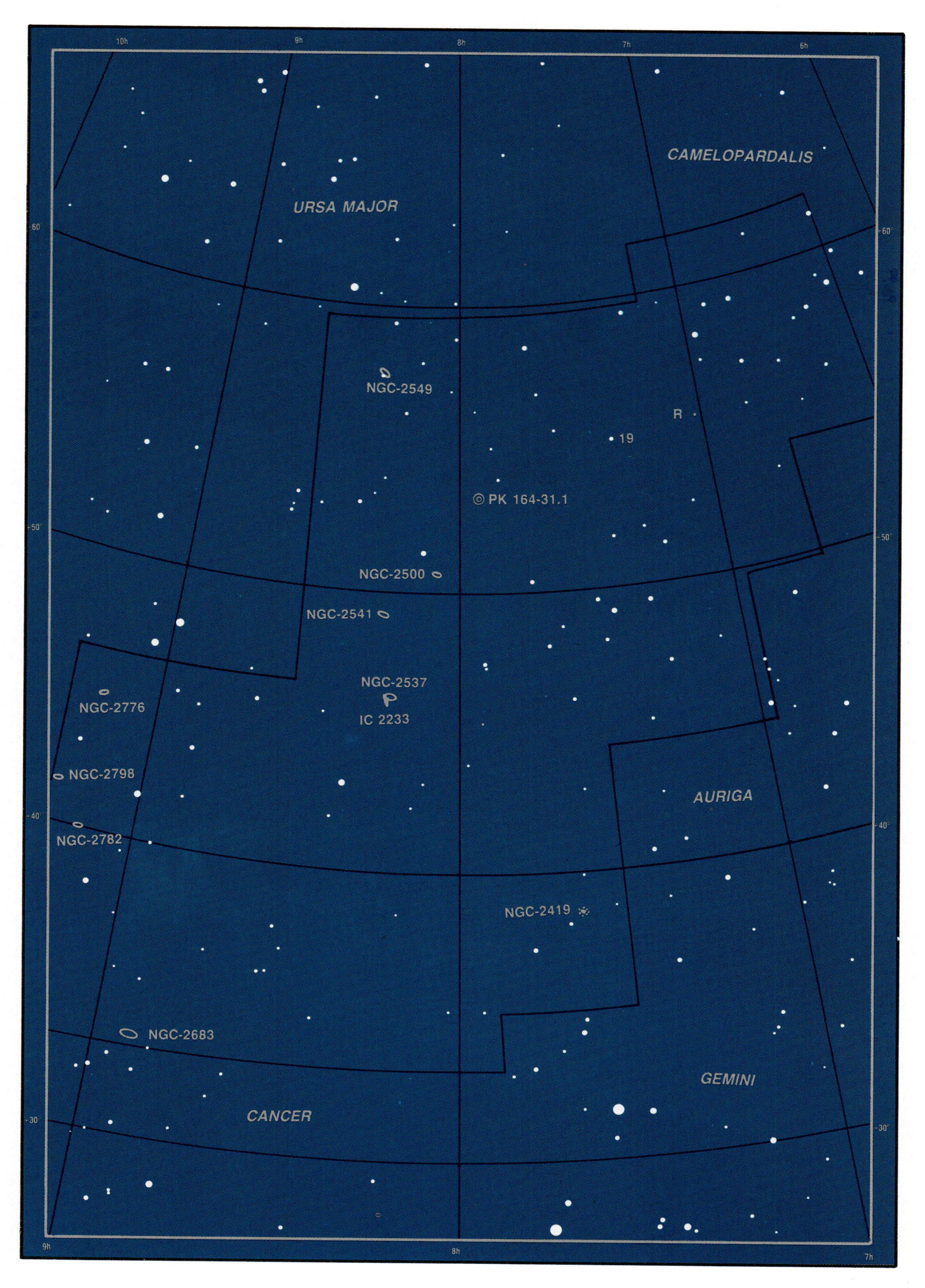

10h
9h
8h
7h
6h
CAMELOPARDALIS
URSA MAJOR
60°
NGC-2549
R
19
PK 164-31.1
50°
NGC-2500
NGC-2541
NGC-2537
IC 2233
NGC-2776
NGC-2798
AURIGA
40°
NGC-2782
NGC-2419
NGC-2683
GEMINI
CANCER
30°
9h
8h
7h

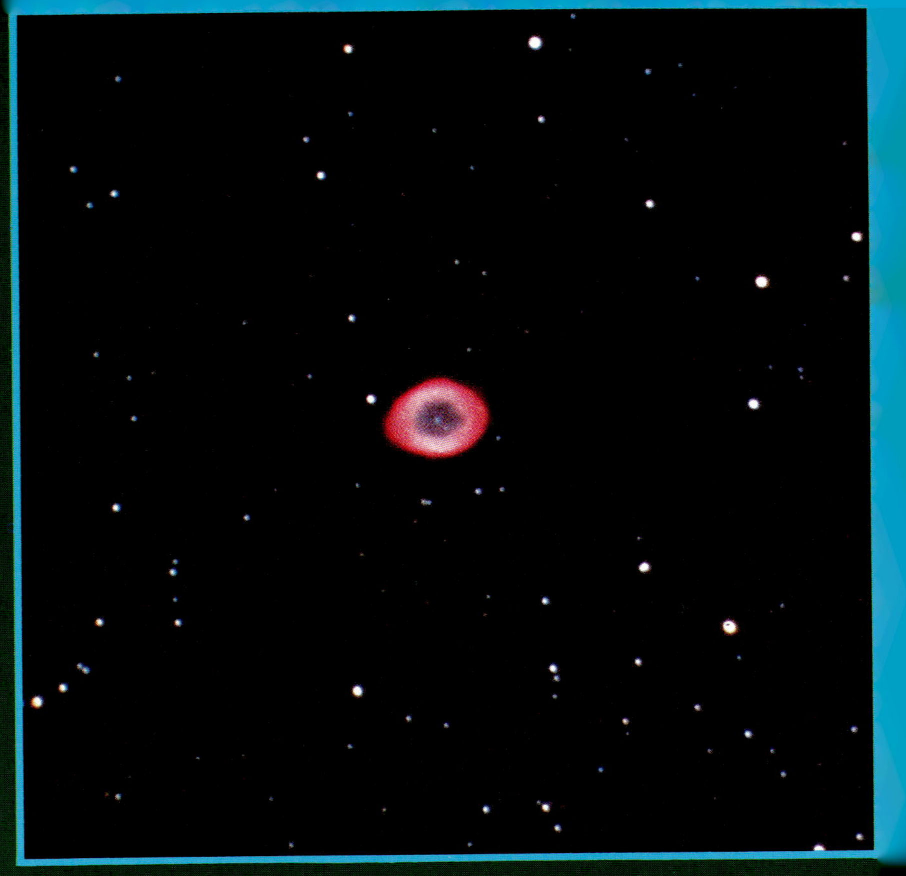

Left: NGC-6791, an exceedingly rich open cluster, is visible in binoculars on dark nights. Photo by Martin C. Germano Above left: Often overlooked, globular M-56 shines at eighth magnitude and spans over 7′ across. Photo by Al Ernst Above: The Ring Nebula, M-57, is the mos observed planetary nebula in the sky Photo by John P. Gleason.

Lyra

Lyr
Lyrae

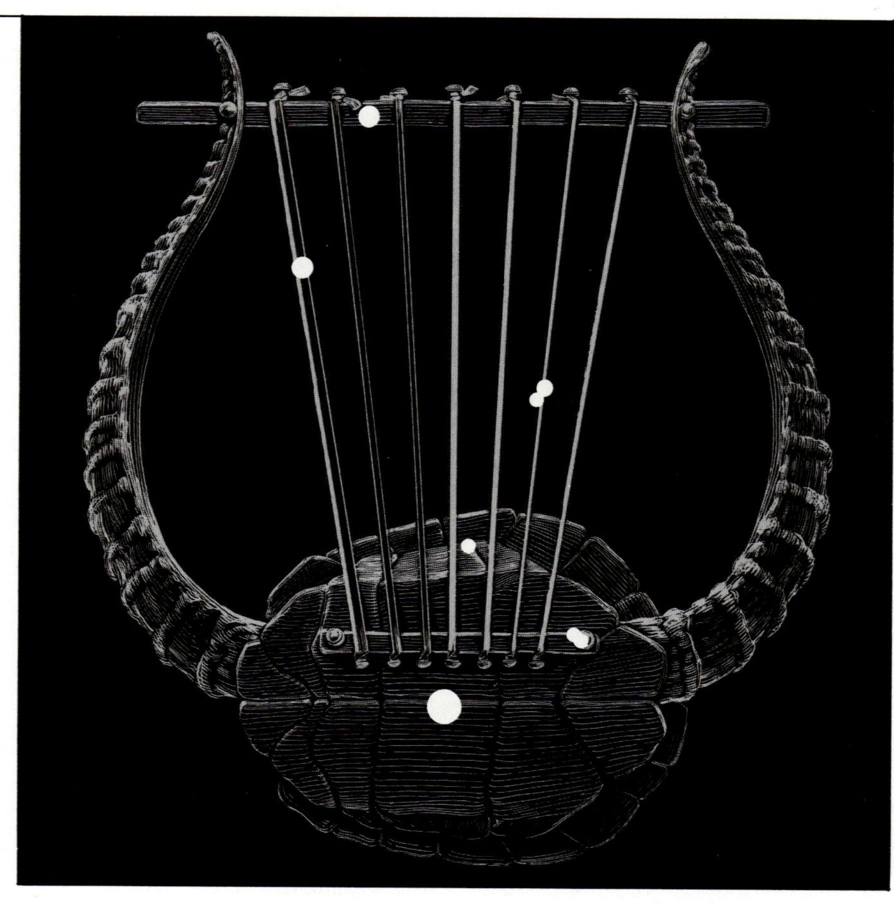
Keith Ward

Lying between the constellations Cygnus and Hercules in the northern summer Milky Way is the petite but bright grouping of stars **Lyra** the Lyre. Although one of the smallest constellations in this part of sky, this very distinctive group lies in a direction where many unusual stars, nebulae, and even a few galaxies reside.

The most observed object in Lyra is the planetary nebula **Messier 57**, popularly known as the Ring Nebula. Measuring some 70″ across and glowing at magnitude 8.8, M-57 is a small but easily recognizable "smoke ring" of hazy light in even a 3-inch refractor. A 6-inch telescope operating at high power clearly shows the central dark hole, which is actually the thinnest part of a spheroidal shell of gas, but M-57 appears as a ring because we see only the thickest areas of nebulosity. A good 8-inch scope shows M-57 as distinctly elliptical with a faint star just off its eastern edge.

The visibility of M-57's central star, a tiny bluish dwarf weakly glowing at about 15th magnitude, is the subject of much debate. Curiously, several observers have reported seeing this faint speck inside the Ring using telescopes as small as 5 inches aperture. Then again, others have failed to spot it with telescopes in the 24 to 40-inch range! What is the explanation for this apparent discrepancy? The star may be variable, or — since it is enveloped in nebulosity — extremely sensitive to observing conditions.

Unknown to many observers is a second planetary nebula in Lyra, **NGC-6765**. This object is much fainter than the Ring and measures some 38″ across — about half as much as M-57. This planetary's central star is a 16th magnitude object, so don't expect to have an easy time spotting it. Try looking for NGC-6765 with a large, low-power telescope; if you find it, you'll be among a select group of observers to do so.

One of many fascinating stars in Lyra is the constellation's brightest, **Vega** (Alpha [α] Lyrae). At magnitude 0.04, Vega is the fifth most luminous star in the sky; along with Deneb in Cygnus and Altair in Aquila, it is recognizable as a member of the so-called Summer Triangle. Lying at a distance of 8.3 parsecs, Vega is a type A0 main sequence star with a surface temperature of 9200 Kelvins — nearly twice as hot as the Sun. Vega has a 10th magnitude bluish companion located some 1′ away in position angle 173°; the two stars are not physically connected but happen to lie along the same line of sight.

Another notable star in Lyra is **Sheliak** (Beta [β] Lyrae), the prototypical eclipsing binary star. At maximum, Sheliak is a magnitude 3.4 brilliant blue-white star with a spectral type of B8; it lies just under 300 parsecs away. Sheliak regularly varies in brightness from about magnitude 3.4 down to 3.8 or 4.1 at alternate minima. This variation is due to mutual eclipses of two stars in the system; the period is 12.91 days.

You can observe Sheliak's regular dimming and brightening by comparing its magnitude — over a period of around two weeks — to equally bright surrounding stars. Make a sketch of the field using binoculars or a small wide-field telescope; given enough time, you'll spot a pattern of changing brightness.

One of the finest double stars for small scopes is **Delta** (δ) **Lyrae** whose magnitude 4.5 and 5.5 components are separated by a whopping 10½ arc-minutes. Despite their great separation and a distance from Earth of about 250 parsecs, the two stars seem to be traveling through space together. They are best observed with binoculars, and form a beautiful color contrast of bluish-white and ruddy orange. A number of other bright stars lie in this same field, and they probably form a physical cluster which includes the Delta Lyr system. If so, this is one of the closest open star clusters to Earth.

One of Lyra's stately variable stars is **R Lyrae**, a so-called semiregular variable in the northern part of the constellation. Over a period of about 46 days, this star slowly and fairly regularly varies between magnitudes 4.1 and 5.0. As are most stars of this type, R Lyr is a reddish star of spectral type M6; this poses a problem in observing it because of the Purkinje effect — the retina's oversensitivity to red light. So when viewing this and other red stars, take quick glances rather than lengthy stares.

Not to be confused with R Lyr is the short-period star **RR Lyrae**, located a few degrees eastward. This is the prototype "cluster variable," so named because many of these stars are found in globular clusters. RR Lyrae itself has a period of

BEST VISIBLE DURING
SUMMER

$\varepsilon^{1,2}$ 3-inch f/10 reflector 120x	**M-56** 8-inch f/10 reflector 50x	**M-57** 8-inch f/10 reflector 100x

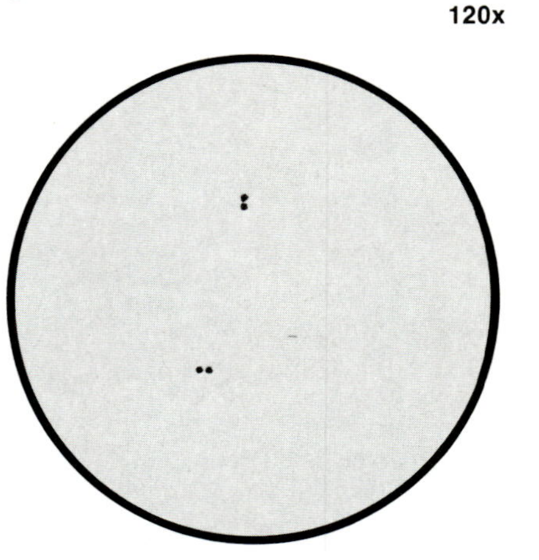

Sketch by Glenn F. Chaple

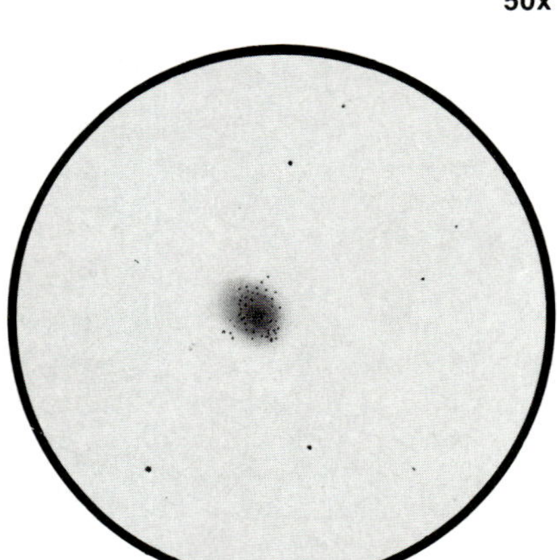

Sketch by David J. Eicher

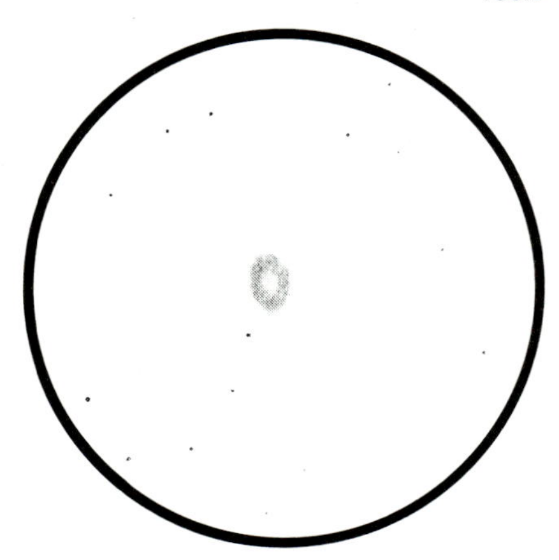

Sketch by David J. Eicher

0.567 day, and a magnitude range between 7.1 and 8.0.

The finest multiple star in the area is **Epsilon^{1,2}** (ε) **Lyrae**, better known as the "Double-Double." Lying about 1½° east of Vega, Epsilon Lyr appears as a double star in binoculars or to those with very sharp eyesight; when viewed with a medium-power telescope, each star transforms into a double in its own right, making the system quadruple. The two main systems of stars are separated by some 208". Epsilon^1 is composed of magnitude 5.5 and 6.5 stars, separated by 2.8" in position angle 359°, while Epsilon^2 is made up of magnitude 5 and 5.5 stars separated by 2.2" in p.a. 98°. The ability to split these stars using a small telescope serves as a good indicator of steady seeing on any given night.

The southern part of Lyra holds the unusual star **HR Lyrae**, dimly glowing at about 15th magnitude. In 1919, it brightened suddenly to magnitude 6.5 before slowly fading back to anonymity. This star is unspectacular to observe now, but is verifiably the burned-out remnant of a nova explosion that occurred nearly 70 years ago.

Toward the eastern edge of Lyra — several degrees southeast of the bright star Theta — is the unusually rich star cluster **NGC-6791**. This object measures about 16′ across and contains some 300 stars — resulting in a total magnitude of 9.5. The cluster appears as a mass of fuzz in small telescopes since the brightest members are 13th magnitude objects. Large telescopes at medium powers show a rich blend of tiny stars strewn across the entire field. This cluster lies at a great distance of 5.1 kiloparsecs.

The second Messier object in Lyra is **M-56** (NGC-6779), a small but bright globular cluster with a characteristically tight core. Overshadowed by M-13 in Hercules, M-56 is a fine object deserving more attention. It measures 7.1′ across and shines at magnitude 8.3. Visible as a fuzzy "star" in finderscopes, M-56 appears as a glob of nebulosity in a 3-inch refractor. An 8-inch telescope doesn't resolve it, but shows a mottled edge surrounding an intensely bright core.

Despite Lyra's proximity to the galactic plane, it contains three notable galaxies. The first, **NGC-6703**, lies in the north-central part of the constellation: shines at magnitude 11.5, spans 2.6′ x 2.5′, and is a lenticular galaxy viewed nearly face on. Lying adjacent to NGC-6703 is **NGC-6702**, a slightly dimmer and smaller elliptical galaxy. Along the southern boundary of Lyra is the galaxy **NGC-6710**, a 13th magnitude lenticular measuring 2.0′ x 1.2′. All three of these galaxies appear as circular or elliptical smudges with little detail in most backyard telescopes.

Object	M#	Type	R.A. (2000)	Dec.	Mag.	Size/Sep./Per.	H
Vega (α)		$\star^2$	18h 36.9m	+38°47′	0.0,10	62.8″	
Epsilon^{1,2} ($\varepsilon^{1,2}$)		$\star^4$	18h 44.3m	+39°40′	5.5,6.5 5.0,5.5	2.8″x2.2″	
NGC-6702		()	18h 47.0m	+45°42′	12.2	2.1′x1.6′	E3:
NGC-6703		§L	18h 47.3m	+45°33′	11.4	2.6′x2.5′	S0
Sheliak (β)		EV	18h 50.1m	+33°22′	3.4-4.3	12.9d	
NGC-6710		§L	18h 50.6m	+26°50′	12.8	2.0′x1.2′	S0:
NGC-6720	M-57	■	18h 53.6m	+33°02′	8.8	80″x60″	
Delta (δ)		$\star^2$	18h 53.7m	+36°58′	5.5,4.5	10.5′	
HR		N	18h 53.7m	+29°14′	6.5-15	Irr.	
R		SRV	18h 55.3m	+43°57′	4.1-5.0	46d	
NGC-6765		■	19h 11.1m	+30°33′	—	38″	
NGC-6779	M-56	●	19h 16.6m	+30°11′	8.3	7.1′	
NGC-6791		⊙	19h 20.7m	+37°51′	9.5	16′	
RR		CV	19h 25.5m	+42°47′	7.1-8.0	0.57d	

H = Hubble classification type for galaxies

$\star^2$	*Double Star*
$\star^4$	*Quadruple Star*
EV	*Eclipsing Variable*
SRV	*Semi-Regular Variable*
CV	*Cluster Variable*
⊙	*Open Star Cluster*
●	*Globular Star Cluster*
■	*Planetary Nebula*
()	*Elliptical Galaxy*
§L	*Lenticular Galaxy*

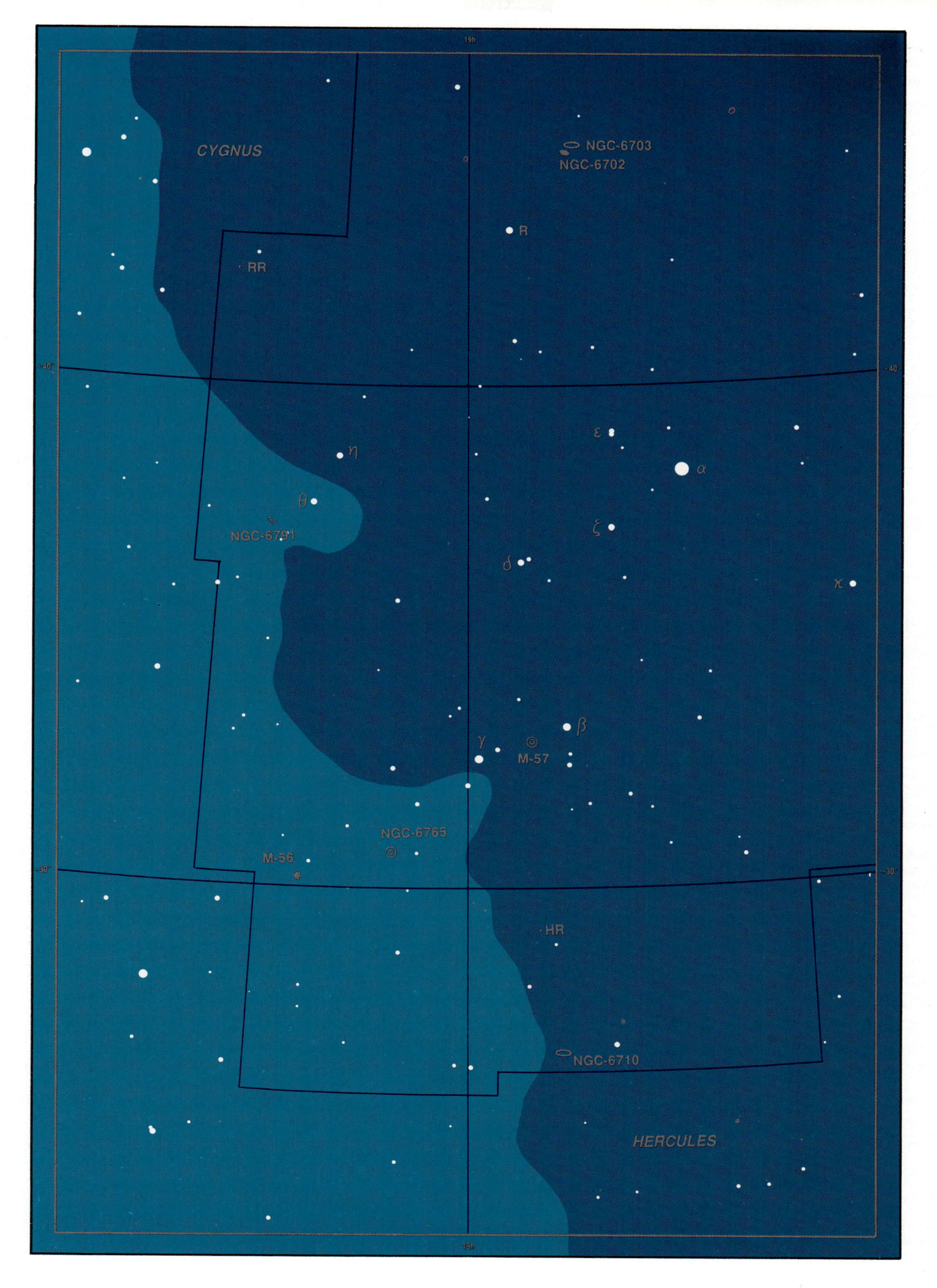
19h
CYGNUS
NGC-6703
NGC-6702
R
RR
40°
40°
η
ε
α
θ
NGC-6791
ζ
δ
κ
β
γ
M-57
NGC-6765
M-56
30°
30°
HR
NGC-6710
HERCULES
19h

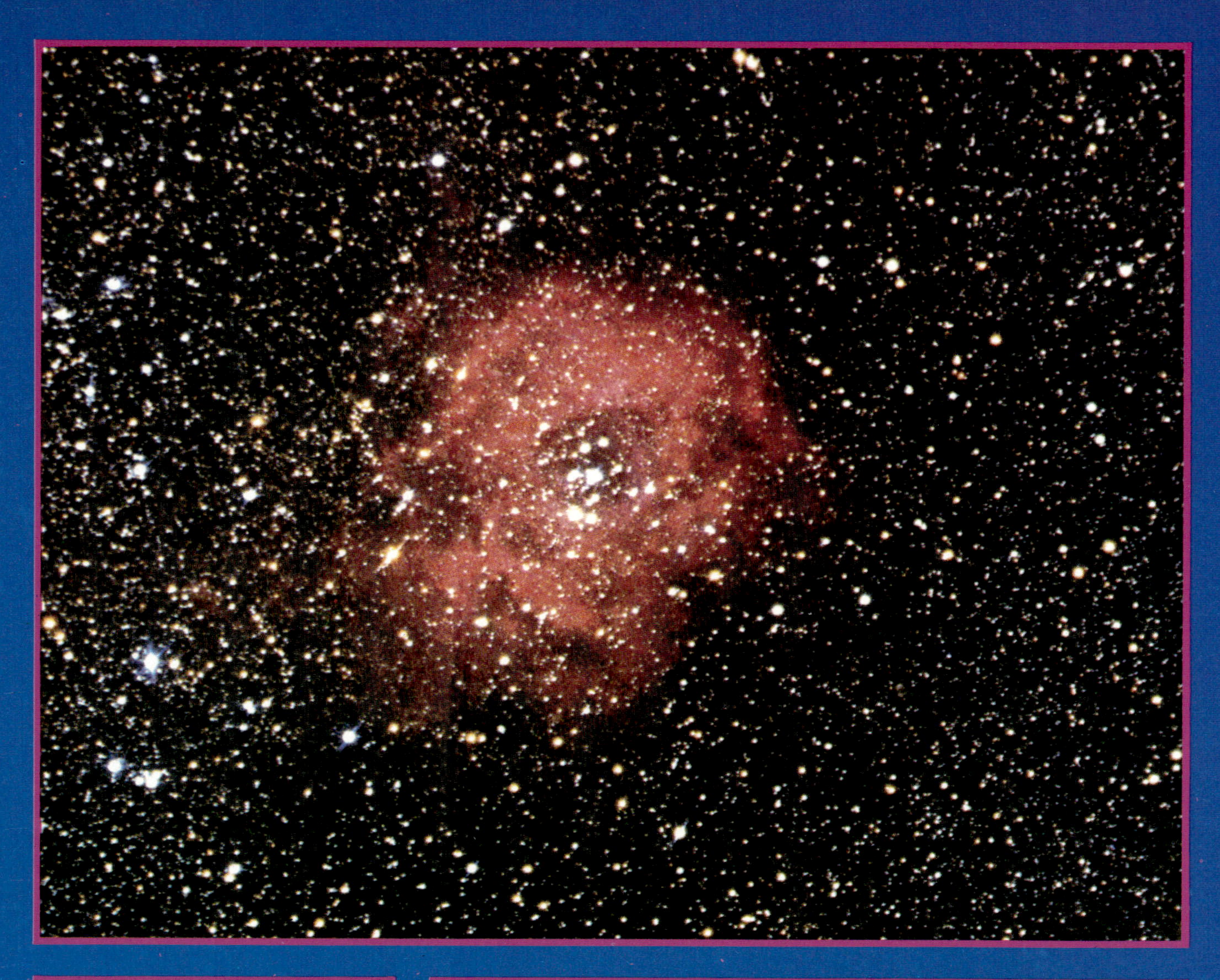

Top: The Rosette Nebula in Monoceros is one of the largest emission nebulae in the sky. The brightest sections of its faint, wreath-shaped outline may be glimpsed in a 6-inch telescope at low power. Photo by Mace Hooley.

Above: The Cone Nebula is a 5′ by 3′ cloud of dust particles backlit by a piece of the emission complex called NGC-2264. It is a difficult object to view because of the low surface brightness of the NGC-2264 nebulosity. Photo by Jack Newton.

Left: Variable in size and luminosity along with the irregular variable R Monocerotis, the tiny triangular nebula NGC-2261 is visible, even at its "minimum," in small backyard scopes. Photo by John P. Gleason.

Monoceros

Mon
Monocerotis

Canis Minor

CMi
Canis Minoris

Keith Ward

Lying amidst the winter Milky Way **Monoceros** the Unicorn and **Canis Minor** the Little Dog each contain bright objects for backyard deep-sky observers. Monoceros, nearly centered on the galactic equator, is rich in open clusters and complex regions of bright and dark nebulosity. Canis Minor, covering a mere 183 square degrees (as opposed to Monoceros' 482 square degrees) and lying a little farther from the galactic equator, is devoid of bright nebulous objects. But it contains a bright and unusual star, a joy to observe for devotees of optical double stars.

The most unusual star cluster in Monoceros is **NGC-2244**, a grouping of some 100 stars found 2.5° east of the bright double star Epsilon (ε) Monocerotis. Since it has a total magnitude of 4.8, you can see NGC-2244 as a fuzzy "star" with your unaided eye under a very dark sky, if you know exactly where to look. Finder scopes show this cluster as a sprinkling of tiny stars some 24′ in diameter superimposed on a milky nebulous glow. NGC-2244's brightest star shines at magnitude 5.8 and is a type O5 bluish white giant. Nearly as bright is 12 Monocerotis, a yellowish K0 giant. Can you spot its distinct color in your scope? A 6-inch scope at low power shows dozens of stars in the cluster, the most prominent being the cluster's six brightest stars, arranged in pairs that make up a rectangle some 20′ long. Scattered inside this rectangle, and to a lesser degree outside of it, lie the remaining ninety-plus stars.

If you view NGC-2244 with a larger telescope under a dark sky, you may be in for a surprise. Surrounding the cluster is a faint emission nebula that is catalogued as NGC-2237, NGC-2238, NGC-2239, and NGC-2246, but popularly known as the **Rosette Nebula**. This wreath-shaped cloud of hydrogen gas is the material from which NGC-2244's constituent stars formed. The cluster is exceedingly young, probably not more than 3 million years old, and as the cluster continues to grow the Rosette's gas will continue to either gravitationally contract into the cluster or be driven off by stellar winds from the already-formed suns. In several tens of millions of years — a blink on the astronomical time scale — the Rosette Nebula will disappear.

The Rosette Nebula and its attendant cluster lie at a distance of about 1.7 kiloparsecs, giving the 80′ by 60′ wreath an actual diameter of some 40 parsecs and making the Rosette one of the largest and most massive nebulae known. The star cluster itself has a physical diameter of about 12 parsecs.

The Rosette Nebula is a difficult, low-contrast object to observe. With a 10- or 12-inch telescope at low power, try spotting it on a night of good transparency when the Moon is out of the sky. You may see it as a large, circular shell of faint gray nebulosity about 1° across. You'll probably see a dark central "hole" — although some observers claim that the nebula appears to be uniformly lit in small telescopes — and dimly glowing, knotty patches of light along the brightest part of the nebulosity.

Using high power, scan along the brightest parts of the nebula and you may well spot tiny, dark patches that appear generally circular, silhouetted against the bright nebulosity. These are Bok globules, small dark nebulae physically involved with the Rosette (and other nebulae) and thought to be regions where dust is collapsing to form stars.

Five degrees north-northeast of the Rosette Nebula is a giant complex of bright and dark nebulae, the brightest part of which is designated **NGC-2264**. NGC-2264 is a magnitude 3.9 open cluster that lies much closer than NGC-2244 — some 750 parsecs — and contains only forty stars. NGC-2264's stars are arranged in a pattern that seems so familiar that the well-known amateur astronomer Leland S. Copeland dubbed NGC-2264 the "Christmas Tree Cluster." The cluster's brightest star, a 5th-magnitude O7 giant called S Monocerotis, lies at the base of the tree; nine other bright stars are nearly perfectly placed to outline the tree. Other observers simply call NGC-2264 diamond-shaped.

S Monocerotis is itself an object for backyard telescopes. Irregularly variable between magnitudes 4.2 and 4.6, the star has a computed absolute magnitude of −5.0, meaning that it shines with the light of 8,500 Suns, and is super-hot; it is also a visual double star with a magnitude 7.5 companion some 2.8″ away in p.a. 213°.

A great cloud of nebulosity covers the star cluster NGC-2264 and its environs, a nebula that is much fainter than the Rosette. The brightest section of this nebulosity carries the same NGC number as the

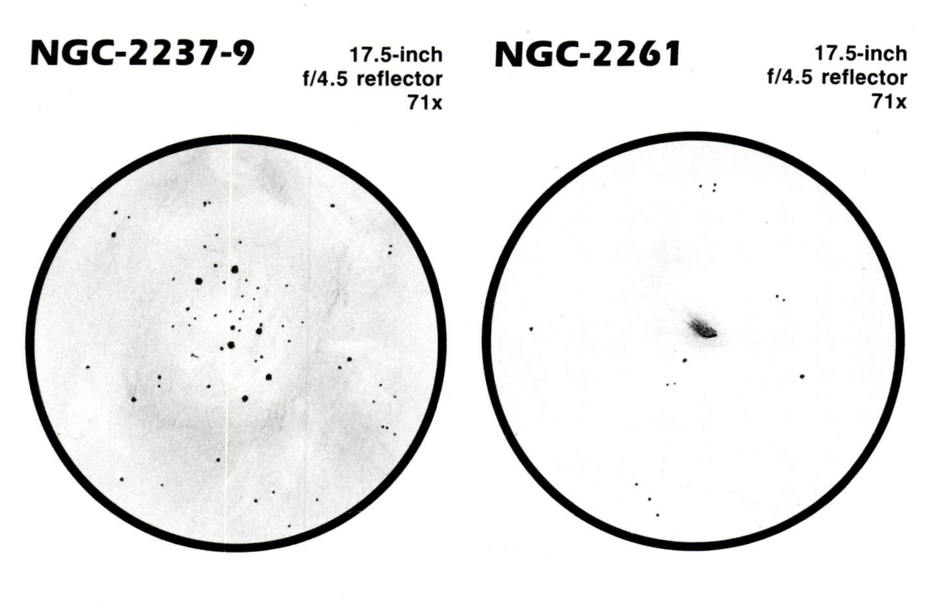

Sketches by David J. Eicher

cluster, 2264, and measures 60′ by 30′ in extent. Floating in front of the NGC-2264 nebulosity, some 40′ south of the star cluster, is one of the sky's finest examples of a contrasty, opaque dark nebula — the **Cone Nebula**. This cone-shaped dark patch is about 5′ long and 40″ to 3′ wide, and is visible as a black smear against gray nebulosity on very dark nights with the largest backyard telescopes.

Also a part of the NGC-2264 nebular complex, and lying some 2° southwest of the star cluster, is the curious reflection nebula **NGC-2261**, also called Hubble's Variable Nebula. Discovered by William Herschel in 1783, this peculiar triangular (2′ by 1′) nebula envelops the bright variable star **R Monocerotis**, and therefore varies in size and brightness along with the star.

The variability of R Mon was discovered at the Athens Observatory in 1861, but it wasn't until 1916 that Edwin P. Hubble discovered that the nebula itself varies, even to the point where photographs made several weeks apart — even amateur photos — show significant structural changes in the nebulosity.

The most intense study of NGC-2261 was done by Carl O. Lampland at Lowell Observatory, who made over 900 photographs of the object over a 30-year period. Lampland found no periodicity in the variance, but to his surprise he found that details in the nebula changed by as much as 1.0″ displacement in just four days! Evidently the changes are produced by shadows cast on the nebulosity by dark material orbiting the star R Mon.

Although reflection nebulae are notoriously faint, Hubble's Variable Nebula is unusual in that it has a very high surface brightness. This allows you to observe it using high powers on nights of very steady seeing and crisp transparency and to compile a series of drawings (or photographs) of the nebula's shape. It's quite possible to see the strange object change considerably in short periods of time. It probably lies some 750 parsecs away.

Other bright open clusters inhabit Monoceros. Of these, **M-50** (NGC-2323) is the best. It contains eighty stars in an area spanning 16′, creating a total magnitude of 5.9 — just on the limit of naked-eye visibility from a perfectly dark site. The cluster is attractive in binoculars as it contains many arcs of stars and a solitary ruddy star about 7′ south of the group's center. **NGC-2301** is a fine object for small telescopes, glowing at magnitude 6.0 and measuring 12′ across. This highly compressed group contains eighty stars of magnitudes 8.0 and fainter and is attractive even in small refractors and binoculars. The cluster **NGC-2353** is a 20′-diameter collection of thirty stars. Their combined light is equivalent to that of a magnitude 7.1 star. A favorite for those observers who know of it, **NGC-2506** is one of the richest groups of stars in its part of the sky, holding 150 stars of magnitude 10.8 and fainter in an area only 7′ across.

One of the prime examples of a triple star visible in small telescopes, **Beta** (*β*) **Monocerotis** should not be overlooked by even casual double-star aficionados. Lying about 11° south of the Rosette Nebula, Beta Mon contains a magnitude 4.7 primary and a 5.2 secondary separated by 7.3″ in p.a. 132°, and a magnitude 6.1 companion lying 10.0″ away in p.a. 124°. The three stars make a lovely sight surrounded by a rich, bright starfield; the only thing lacking is exotic star colors, since all three suns are brilliant white.

Canis Minor's prime attraction is the dazzling star **Procyon** (Alpha [*α*] Canis Minoris), a magnitude 0.4 luminary. It is the eighth brightest star in the sky. It is also the fifth closest of the naked eye stars, at a distance of 3.5 parsecs. Procyon is an easy optical double, with a magnitude 11.6 star 2′ away in p.a. 13°, but an impossible binary star — a physically associated magnitude 12.9 white dwarf — lies only 5″ away and is overwhelmed by the brilliance of Procyon. It is visible only in the largest telescopes.

Object	M#	Type	R.A. (2000)	Dec.	Mag.	Size/Sep./Per.	N★
Beta (β)		★³	6h 28.8m	− 7°02′	4.7,5.2,6.1	7.3″,10.0″	
NGC-2237-9,46		□E	6h 32.3m	+ 5°03′	—	80′x60′	
NGC-2244		⊙	6h 32.4m	+ 4°03′	4.8	24′	100
NGC-2261		□R	6h 39.2m	+ 8°44′	—	2′x1′	
S Mon		IV	6h 41.0m	+ 9°54′	4.2↔4.6	irr.	
NGC-2264		⊙□E	6h 41.1m	+ 9°53′	3.9	60′x30′	
NGC-2301		⊙	6h 51.8m	+ 0°28′	6.0	12′	80
NGC-2323	M-50	⊙	7h 03.2m	− 8°20′	5.9	16′	30
NGC-2353		⊙	7h 14.6m	−10°18′	7.1	20′	30
Procyon (α CMI)		★²	7h 39.3m	+ 5°14′	0.4,12.9	5″	
NGC-2506		⊙	8h 00.2m	−10°47′	7.6	7′	150

★²	*Double Star*
★³	*Triple Star*
IV	*Irregular Variable*
⊙	*Open Star Cluster*
□E	*Emission Nebula*
⊙□E	*Emission Nebula with Star Cluster*
□R	*Reflection Nebula*

N★ = number of stars.

8h
7h
6h
+20°
+10°
0°
-10°
-20°
GEMINI
CANIS MINOR
ORION
MONOCEROS
CANIS MAJOR
NGC-2264
S
R
NGC-2261
NGC-2237-9
NGC-2244
NGC-2301
M-50
NGC-2506
NGC-2353
IC 2177
Galactic Equator
Sirius

Above: The Eagle Nebula (M-16) in Serpens Cauda is a rich complex of dust, glowing hydrogen gas, and newly formed stars. Though only a large telescope will show much of the gas, the cluster of new stars is far more prominent. Photo by Alfred Lilge. Left: This scanty sprinkling of stars is IC 4665, a loose open cluster in Ophiuchus — use low magnifications when scanning for this one! Photo by Jamey Jenkins. Bottom left: Although a globular cluster, M-12 has a loose appearance that makes its individual stars easy to resolve. Photo by Jay Anderson. Bottom center: NGC-6369 is a dim planetary nebula — a faint cousin of Lyra's famed Ring Nebula. Photo by Martin Germano. Bottom right: M-62 is a tighter and richer globular than M-12; look for a condensed core with a soft periphery. Photo by T.L. Dessert.

Ophiuchus

Oph
Ophiuchi

Serpens Cauda

Ser
Serpentis

Lying adjacent to the main line of the summer Milky Way, the large, box-shaped constellation **Ophiuchus**, the Serpent-Bearer, holds many globular star clusters, open clusters, planetary nebulae, and unusual stars — and even contains one galaxy! Sandwiched between Ophiuchus and the bright star clouds toward the galactic center is **Serpens Cauda**, the tail of the Serpent, home to a variety of objects.

The brightest globular in Ophiuchus is **M-62** (NGC-6266), a magnitude 6.5 cluster lying about 7° southeast of Antares, nearly on the border with Scorpius. It measures 14.1′ of arc across, but backyard telescopes show it to be considerably smaller. (Globular cluster diameters are based on long photographic exposures, which reveal many tiny stars far removed from the cluster's center.) M-62 is a moderately compressed and asymmetrical globular — its major axis points toward position angle 75° — and the average magnitude of its brightest 25 stars is 15.9. M-62 lies 6 kiloparsecs (Kpc) from the Sun and measures 25 parsecs (pc) across; it contains a remarkable 89 known variable stars, most of which are RR Lyrae-type stars.

About 4.5° north of M-62 is the similarly bright globular **M-19** (NGC-6273), a cluster shining at magnitude 6.8 and measuring 13.5′ across. M-19 is a "looser" globular than M-62 — it is easier to resolve into individual stars. A 6-inch scope on good nights will resolve M-19's outer edges, whereas an 8-incher is needed for even a hint of resolution with M-62. On good nights, assuming your telescope is properly collimated, you can use magnifications of 20x to 30x per inch of telescopic aperture. M-19 lies at a distance of 10.1 kpc and is a large globular, measuring 40 pc across. Visible in the eyepiece and in amateur photos of M-19 is its highly oblate shape: its long axis is oriented almost due north-south, at p.a. 15°.

BEST VISIBLE DURING
SUMMER

An unusual pair of objects lies northeast of M-19, near the bright stars Theta (θ) and Omicron (ο) Ophiuchi. **NGC-6369**, a ring-shaped planetary nebula, is a fine sight in small telescopes. It glows at magnitude 11, measures 28″ across, and has a fairly high surface brightness (it can stand lots of magnification). Its 16th magnitude central star will challenge the largest backyard telescopes, but medium-sized instruments show a perfectly formed ring. A mere 1.5° north-northeast of θ Oph is the elusive dark nebula **Barnard 72**, known as the "S" Nebula. Its obscuring dust, backlit by stars and nebulosity, spans 30′ in the shape of the letter S. On a dark night, with a wide-field eyepiece, even small telescopes faintly show this patch of dust grains. But if the Moon is out or the transparency poor, you won't find B72.

The area contains two large, scattered open clusters: one lies in Ophiuchus, the other in Serpens Cauda. **IC 4756**, in Serpens, covers an area of more than 70′ and contains 80 stars of seventh magnitude and fainter. It is an impressive sight in binoculars, but loses appeal when subjected to the cramped fields of telescopes. **IC 4665**, just northeast of Beta (β) Oph, is somewhat smaller and much sparser yet: it contains 20 stars of magnitude 7 and fainter in its 55′ diameter. Again, binoculars and finder scopes provide the best views.

Straddling IC 4665 on opposite sides are the out-of-place galaxy **NGC-6384**, an Sb-type spiral, and **Barnard's Star**, a star with one of the highest known proper motions. The galaxy measures 4.0′ x 3.0′ and glows softly at magnitude 12.3, making it visible as a smudge in a 6-inch scope and as an elongated blur with a bright middle in an 8-inch instrument. Barnard's Star, a faint red dwarf only 6 light-years away (this is the second closest star system to the Sun), shines at magnitude 9.5 and is recognizable by its ruddy hue. Its great proper motion — more than 10″ of arc per year — results in part from its close proximity; over 351 years, Barnard's Star traverses an entire degree of sky! Sproul Observatory astronomer Peter van de Kamp has reported a "wobble" in the star's motion and maintains that it is due to a large planetary companion. This has yet to be proven.

Ophiuchus contains many more globular clusters, including two of the finest anywhere in the sky — **M-10** (NGC-6254) and **M-12** (NGC-6218). The two are nearly identical and lie only a few degrees apart. M-10 shines at magnitude 6.6 and measures 15.1′ across; M-12 radiates at magnitude 6.9 and is 14.5′ in

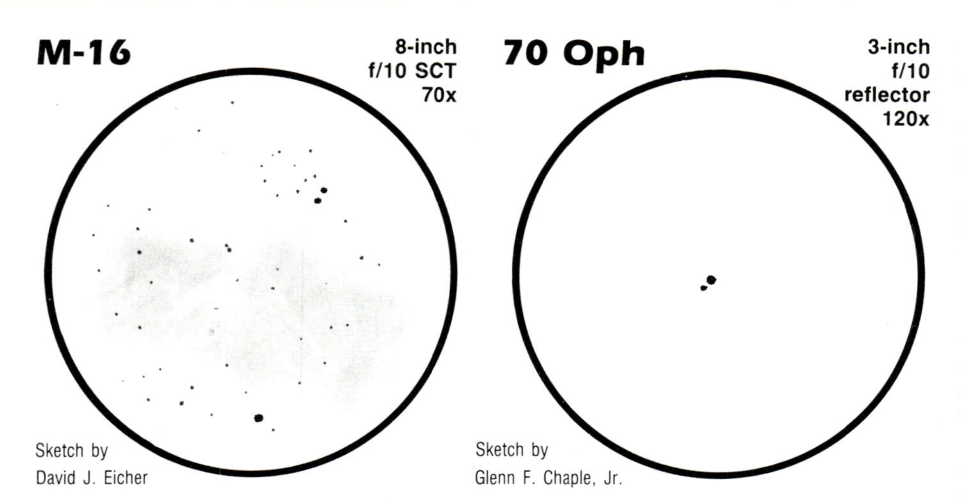

Sketch by David J. Eicher

Sketch by Glenn F. Chaple, Jr.

diameter. Both clusters make superb targets for all telescopes, and 6-inch telescopes at high power resolve their peripheries beautifully into gleaming pinpoints. (M-12 is slightly looser, but they are very nearly equal in resolvability.) On good nights, large amateur telescopes show stars across the entire faces of these clusters. M-10 lies 4.4 kpc away and measures 19 pc in physical diameter; M-12 is 5.9 kpc distant and fills a sphere measuring 25 pc.

Due east of the M-10/M-12 area is another bright globular, **M-14** (NGC-6402). This group of stars is a fainter, dimmer version of M-10 and M-12, but still is quite impressive in any telescope. At magnitude 7.5 and 11.7′ diameter, the group is a fuzzy spot in binoculars and shows up well in a 4-inch reflector. A 16-inch telescope at high power resolves M-14 across its face, leaving a misty grainy light suspended behind a curtain of glistening 15th magnitude stars. Although M-14 lies at the remote distance of 13.8 kpc, its large physical size of 47 pc causes it to appear as "one of the gang" among Ophiuchus globulars. It contains a high number of variable stars — 85, at last count.

Two more globulars, in the southern reaches of Ophiuchus, are easily visible in small telescopes. One is **M-9** (NGC-6333), a compact cluster that lies atop a prominent Milky Way dust cloud called Barnard 64. The cluster glows at magnitude 7.8, measures 9.3′ across, and can be partially resolved with amateur telescopes. It is 9.1 kpc away and measures 25 pc across. On dark nights, with a low-power eyepiece, the dark nebula appears as a black stain spilt onto the starry backdrop. The last globular is **M-107** (NGC-6171), an eighth-magnitude group spanning 10′ and lying at a distance of 6.7 kpc. It is a globular of average size, measuring 19 pc in extent.

Ophiuchus contains several double and variable stars; one is the famous double **70 Ophiuchi**. Discovered to be binary by William Herschel in 1779, 70 Ophiuchi's components are of magnitude 4.2 and 5.9, are separated by 1.8″ in p.a. 300° (1984), are yellowish and rose-colored, respectively, and are rapidly changing in separation and p.a. (The period is just over 88 years.) **X Ophiuchi** is a typical long-period variable whose magnitude range is 6.0 to 9.3 and whose period is 334 days — just short of one year. **RS Ophiuchi**, on the other hand, is one of seven so-called recurrent novae: in 1898, 1933, 1958, and 1967 this star underwent maxima in which it shone at fourth magnitude, becoming visible to the naked eye. Normally RS Oph glows dimly at magnitude 11.5. Systems like this one involve two stars, one of which may be a white dwarf. As hydrogen from the companion flows on to the white dwarf, the material heats up and explodes.

A fine planetary nebula in Ophiuchus is **NGC-6572**, a tiny, bright disk characterized by a strong blue-green hue. The nebula measures only 15″ x 12″ across and shines at 9th magnitude; a medium-sized scope on a dark night shows it easily and has a chance of revealing the nebula's central star, an O-type object of 12th magnitude.

Serpens Cauda contains one of the most unusual and complex regions of ionized hydrogen in our galactic neighborhood — **M-16** (NGC-6611), otherwise known as the Eagle Nebula. This large emission nebula is centered on a bright star cluster containing 60 stars of eighth magnitude and fainter in an area 25′ across. The star cluster makes for a fine binocular sight and, at sixth magnitude, is theoretically visible with the naked eye from a dark site. The challenging part of M-16 is spotting the glow from the gas that produced the cluster and now fluoresces under strong gusts from hot stellar winds. The nebulosity has a low surface brightness (and its visibility is further hindered by many bright stars in the field); in an 8-inch RFT at low power, it appears as a milky, greenish light. Larger scopes show more nebulosity and the dark globules that may be collapsing into protostars.

Object	M#	Type	R.A. (2000)	Dec.	Mag.	Size/Sep./Per.	N★	Mag.★
NGC-6171	M-107	●	16h 32.5m	−13°03′	8.2	10.0′		
NGC-6218	M-12	●	16h 47.2m	−01°57′	6.9	14.5′		
NGC-6254	M-10	●	16h 57.1m	−04°07′	6.6	15.1′		
NGC-6266	M-62	●	17h 02.2m	−30°07′	6.5	14.1′		
NGC-6333	M-9	●	17h 19.1m	−18°31′	7.8	9.3′		
Barnard 72		□D	17h 24.0m	−23°38′	—	30.0′		
NGC-6369		■	17h 29.3m	−23°46′	10.4	28″		16.0
NGC-6384		§	17h 32.3m	+07°04′	11.3	4.0′ x 3.0′		
NGC-6402	M-14	●	17h 37.6m	−03°15′	7.5	11.8′		
IC 4665		⊙	17h 46.2m	+05°43′	6.0	55.0′	20	7....
RS Oph		RN	17h 50.2m	−06°43′	4.0↔12.0	Irr.		
Barnard's Star		★	17h 57.9m	+04°24′	9.5	—		
70 Oph		★²	18h 05.4m	+02°32′	4.2, 5.9	1.9″ (1984)		
NGC-6572		■	18h 12.1m	+06°51′	9.0	14.8″		9.8
NGC-6611	M-16	⊙□E	18h 18.8m	−13°47′	6.5	25′	60	8....
X Oph		LPV	18h 38.4m	+08°50′	6.0↔9.3	334d		
IC 4756		⊙	18h 39.1m	+05°29′	6.0	70′	80	7....

★	*Star*
★²	*Double Star*
●	*Globular Cluster*
⊙	*Open Cluster*
LPV	*Long Period Variable*
RN	*Recurring Nova*
□E	*Emission Nebula*
■	*Planetary Nebula*
□D	*Dark Nebula*
§	*Spiral Galaxy*

N★ = number of stars, Mag.★ = magnitude range of cluster or magnitude of central star, (...) indicates many fainter.

18h
17h
NGC-6572
IC 4756
NGC-6384
IC 4665
Barnard's
Star
70
HERCULES
M-12
M-14
M-10
RS
OPHIUCHUS
Galactic Equator
SCUTUM
M-16
M-107
SERPENS CAUDA
M-17
M-18
M-9
M-23
M-25
M-24
M-21
Ecliptic
M-20
NGC-6369
B72
M-22
M-28
M-8
M-19
SCORPIUS
M-4
SAGITTARIUS
Galactic
Center
M-62
M-6
M-70
M-69
M-7

Left: Messier 42, the Orion Nebula, is one of the finest emission nebulae in the sky. Binoculars or finder telescopes show a one-degree patch of fuzzy light centered on the Hunter's Belt. Photo by David Ratledge. Below left: Messier 78 is one of the sky's brightest reflection nebulae; M-78 appears as a circular patch of light containing three bright stars. Photo by Lee C. Coombs. Below right: The Horsehead Nebula is a small dark nebula backlit by a dim strip of emission nebulosity, making it very difficult to see. Photo by Scott Rosen.

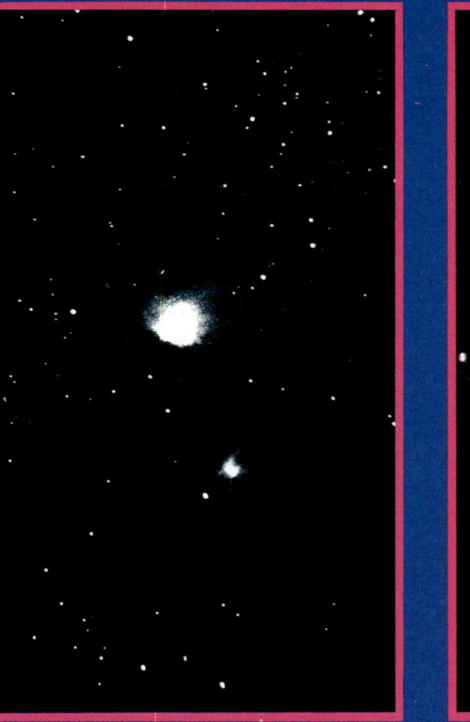

Orion

Ori
Orionis

Lepus

Lep
Leporis

Keith Ward

The constellation **Orion** is one of the most familiar sights in the entire sky and its most famous inhabitant — the Orion Nebula — is one of the most familiar deep-sky objects. Orion, the Hunter, contains many other attractions, including some fine double stars and star clusters, as well as multitudes of faint nebulae. Its neighbor to the south, Lepus the Hare, is a rather overlooked group of stars which also houses fine objects for backyard deep-sky observers.

Orion's prime attraction is **M-42** (NGC-1976), the Orion Nebula, which is just visible to the unaided eye as a fuzzy patch in the Hunter's sword. When viewed with a small telescope, M-42 transforms into a glowing, greenish-white nebulosity spanning over one degree in diameter. Distinctly shaped, it resembles a giant, luminous fan. A larger telescope's low-power field of view reveals subtle gradations of light and dark patches, and streamers of bright nebulosity so detailed that it is impossible to adequately describe or sketch.

The total magnitude of M-42 is about 3.5, and its dimensions are 85′ x 60′. If the calculated distance of 500 parsecs is correct, M-42 measures some 10 parsecs across and could contain more than 20,000 solar systems stacked end to end! Glowing ionized gases energized by hot young stars in its central part produce the Orion Nebula's light. M-42 is a typical emission nebula; these contain mostly hydrogen, some helium and small amounts of carbon, oxygen, and nitrogen.

Embedded within M-42 is the multiple star **Theta¹** (θ^1) **Orionis**, otherwise known as the Trapezium for its trapezoidal shape. The Trapezium is the most observed quadruple star in the sky. Its members have magnitudes of 5.4 (star C: spectral type O6), 6.3 (star D: B0), 6.7-7.7 (star A: A7, eclipsing binary), and 8.0-8.7 (star B: eclipsing binary of uncertain spectral type). The separations are as follows: A-B is 8.7″, B-D is 19.2″, D-C is 13.3″, and C-A is 12.9″. A small telescope easily resolves Theta¹, showing the four brilliant white components.

North of M-42 and visible within the same low-power field is the bright nebula **M-43** (NGC-1982), which looks like a bloated comma. An eighth-magnitude star within M-43 illuminates it, but the nebula's surface brightness makes it easily visible. M-43 doesn't show as much range in brightness or detail as M-42, but it has a bright ring around the star which in turn is surrounded by fainter nebulosity. M-43 is actually a piece of the Orion Nebula, seemingly detached because a dark nebula obscures their connection.

A Moon's diameter to the south of Zeta (ζ) Orionis, the easternmost star in the Hunter's Belt, is the famous dark nebula **Barnard 33**. This object is silhouetted against the bright nebula IC 434 and is best known as the Horsehead Nebula. Unlike M-42 and M-43, B33 is *very* difficult to see in backyard telescopes — its small size (5′) and low contrast against IC 434 conspire to make it a tricky target. Success in observing the Horsehead varies greatly and depends mainly on sky conditions and the skill of the observer. Some have seen it using a 5-inch wide field refractor, while others have failed to locate B33 even with a 40-inch telescope!

Moving north to Zeta, we find the bright nebulosity **NGC-2024** just east of the star in the same low-power field. NGC-2024 is a large, detailed nebula which shows up fairly well in telescopes of 6-inches or more aperture, despite considerable interference from second-magnitude Zeta. The nebula's most striking feature is a broad, dark band which conspicuously bisects the 20′ diameter circular disk. Using high power (which removes Zeta from the field), many observers report seeing subtle bright and dark patches. Under a dark sky, NGC-2024 is one of the finest emission nebulae in Orion — imagine how spectacular it would look if it wasn't located next to second-magnitude Zeta!

A curious planetary nebula in Orion is **NGC-2022** — a tiny disk of light measuring only 25″ across. The total magnitude of this object is only about 12, making it difficult to find in the starfield. To locate NGC-2022, wait for good conditions and use relatively high power to determine which "star" is actually a disk; under great skies with lots of magnification, you may even pick out the 14th magnitude central star.

One of the sky's easiest-to-observe reflection nebulae is bluish **M-78** (NGC-2068), which lies northeast of

BEST VISIBLE DURING
WINTER

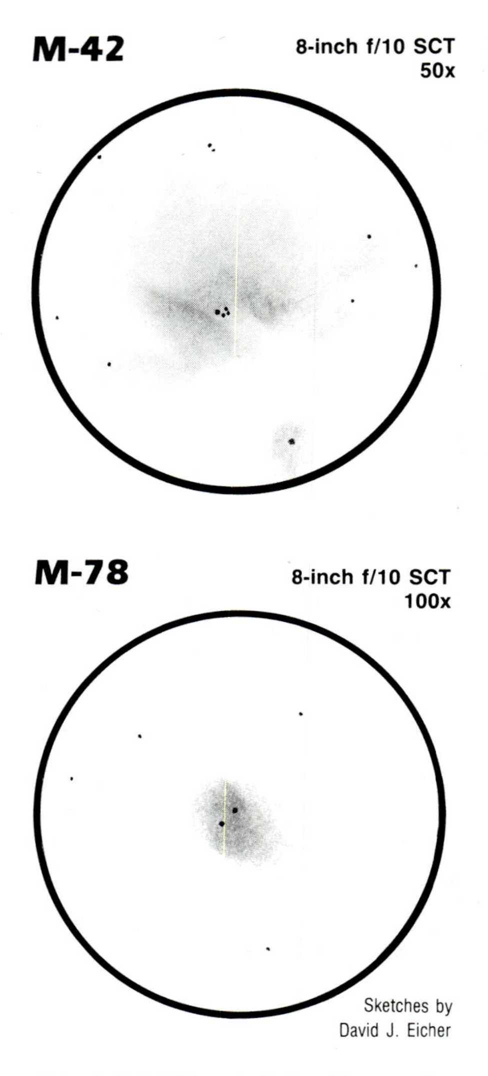

Sketches by
David J. Eicher

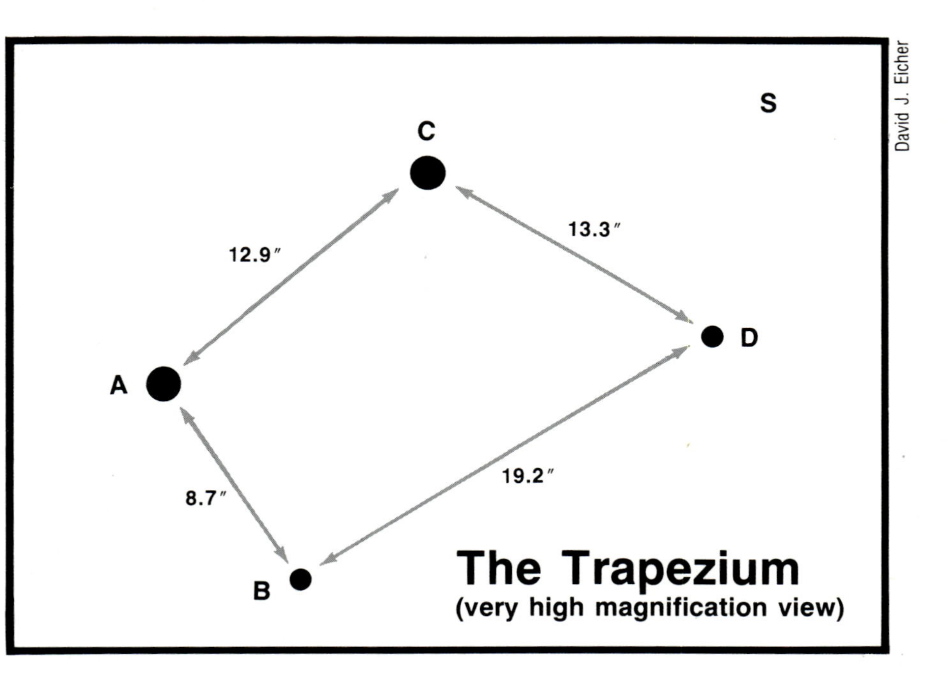

Orion's Belt. This nebulosity shines not by fluorescence, as do emission nebulae, but by starlight reflected off dark grains of material — that's why so few reflection nebulae are bright enough to observe with backyard equipment. At its distance of 500 parsecs, it's amazing we can see M-78 so well. The nebula measures 8′ x 6′ across — about the same size as the Crab Nebula in Taurus — and contains three bright stars, all of which contribute light to the nebula. A good 4-inch reflector at 100x shows a patch of light so subtle it appears like breath on a mirror; a 10-inch reflector reveals an elongated shape in M-78 and easily shows 5′ worth of nebulosity. North of M-78 in the same field is **NGC-2071**, a detached piece of nebulosity which shows up almost as well as M-78 itself.

High up in Orion's northernmost reaches, near the star cluster M-35 in Gemini, is an unusual pair of deep-sky objects: **NGC-2174** is a faint ruddy emission nebula surrounding the bright open cluster **NGC-2175**, which is easily visible in binoculars. Small telescopes show the latter as a condensed cluster of tiny stars that appear grainy. Large amateur telescopes operating under dark skies also show NGC-2174 as a fan-shaped nebulosity spanning some 25′.

The constellation Lepus contains two objects of interest to deep-sky observers. The first is **M-79** (NGC-1904) — one of the finest globular clusters visible during the winter. This fine object shines at magnitude 8.4 and measures 7.5′ across and is easily visible as a fuzzy spot in binoculars or finderscopes. Small telescopes show a hazy, unresolved ball of light with a golden-yellow color; backyard scopes in the 12-inch to 20-inch class resolve M-79's outer edges, confirming its stellar composition. Lepus also contains the bright galaxy **NGC-1964** — a magnitude 11.8 Sb-type spiral measuring 5.0′ x 1.6′ across. Its nearly edge-on angle allows us an oblique view of the galaxy's hub and core, but doesn't reveal much of NGC-1964's spiral arms. The very largest backyard telescopes, under near-perfect conditions, may show the faintest hint of detail within this distant spiral.

Object	M#	Type	R.A. (2000)	Dec.	Mag.	Size/Sep./Per.	N★	Mag.★
R Lep		LPV	4h 59.5m	−14°49′	5.9↔11.0	432d		
Rigel (β Ori)		★^{2}	5h 14.5m	−08°12′	0.1,6.7	9.4″		
NGC-1904	M-79	●	5h 24.3m	−24°31′	8.4	7.5′		
NGC-1964		§	5h 33.3m	−21°57′	11.8	5.0′x1.6′		
θ^1 Ori		★^{4}	5h 35.3m	−05°23′	5.4,6.3, 6.8,7.0	13.3″,19.2″, 8.7″,12.9″		
NGC-1976	M-42	□E	5h 35.4m	−05°23′	3.5	85.0′x60.0′		
NGC-1973/5/7		□R	5h 35.5m	−04°50′	—	45.0′x35.0′		
NGC-1981		⊙	5h 35.5m	−04°22′	8.0	18′	45	8....
NGC-1982	M-43	□E	5h 35.5m	−05°16′	9.0	10′x5′		
IC 434		□E	5h 41.1m	−02°25′	—	60′x10′		
Barnard 33		□D	5h 41.2m	−02°31′	—	5′		
NGC-2022		■	5h 42.0m	+09°04′	12.3p	28″x27″		
NGC-2023		□R	5h 41.7m	−02°14′	—	10′		
NGC-2024		□E	5h 41.9m	−01°51′	—	30.0′		
NGC-2068	M-78	□R	5h 46.7m	+00°03′	11.0	8.0′x6.0′		
NGC-2174/5		⊙□E	6h 09.7m	+20°30′	—	25′	—	8....

N★ = number of stars, Mag.★ = magnitude range of cluster or magnitude of central star, (...) indicates many fainter.

★^{2}	*Double Star*
★^{4}	*Quadruple Star*
LPV	*Long Period Variable*
⊙	*Open Star Cluster*
●	*Globular Star Cluster*
■	*Planetary Nebula*
□E	*Emission Nebula*
□R	*Reflection Nebula*
□D	*Dark Nebula*
§	*Spiral Galaxy*

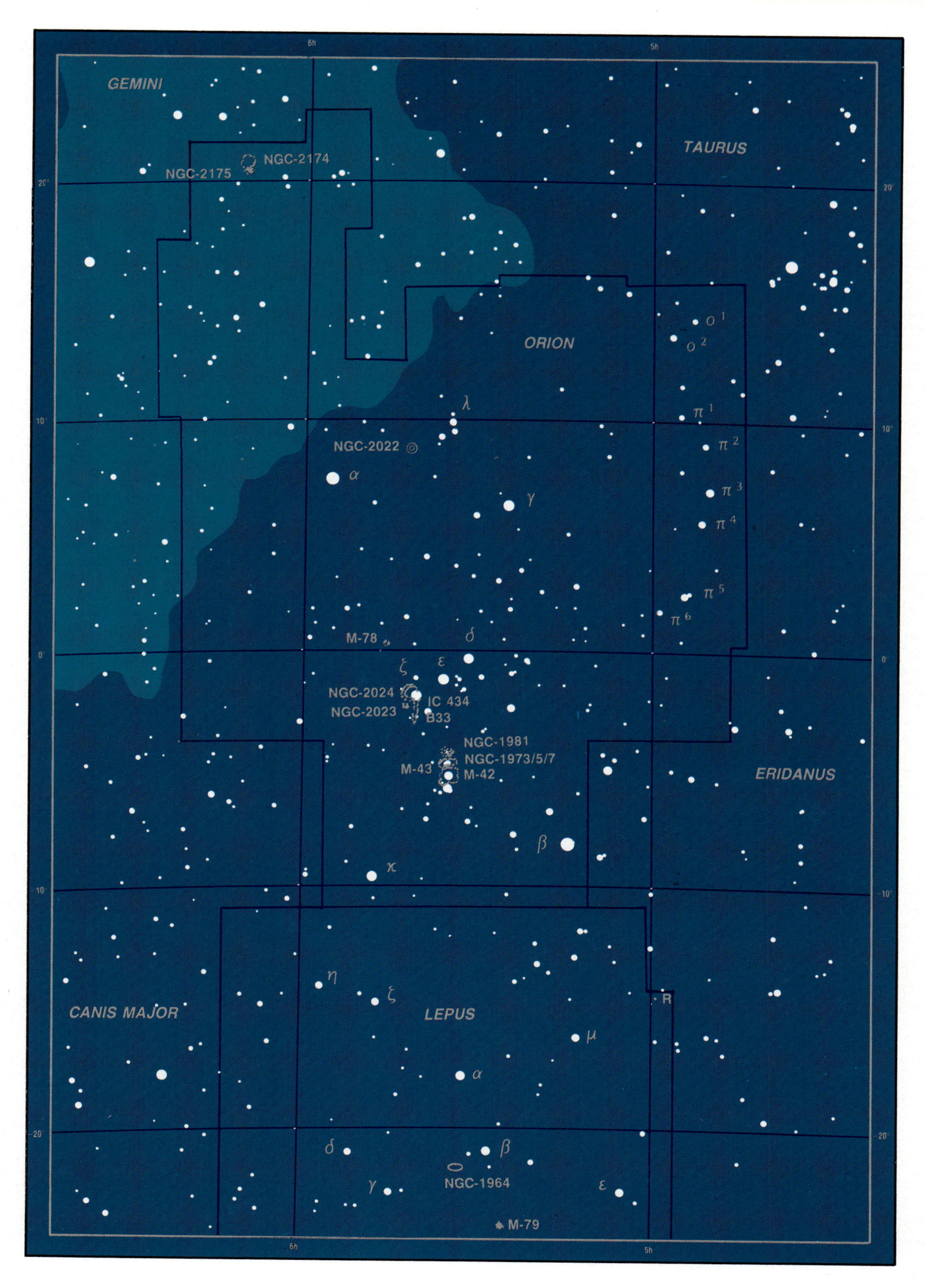
GEMINI
NGC-2174
NGC-2175
TAURUS
ORION
NGC-2022
M-78
NGC-2024
IC 434
NGC-2023
B33
NGC-1981
NGC-1973/5/7
M-43
M-42
ERIDANUS
CANIS MAJOR
LEPUS
R
NGC-1964
M-79

One of the finest barred spirals in the southern sky, NGC-6744 is visible as a 9th-magnitude nebulosity some 10′ in diameter. Photo courtesy Cerro Tololo Interamerican Observatory.

Pavo

Pav
Pavonis

Lying on the fringe of the southern Milky Way not far from the south celestial pole, **Pavo** the Peacock contains a hodgepodge of galactic and extragalactic deep-sky objects. The small constellation contains some galaxies like its neighbors Telescopium and Indus and some galactic objects like the bordering constellation Ara. Unfortunately, only two of Pavo's deep-sky objects appear impressive in small scopes, but the others warrant at least an occasional look.

One of the finest globular clusters in the sky is **NGC-6752**, a magnitude 5.4 collection of yellowish stars some 2° east of Omega (ω) Pavonis. This cluster is easily visible to the naked eye under a dark sky and appears bright in finder scopes. In a 3-inch telescope NGC-6752 appears as a knot of bright nebulosity some 3′ across surrounded by a faint haze. A 6-incher resolves the outer portions of this cluster and shows it much larger — spanning nearly 10′. NGC-6752 is a magnificent sight in large telescopes, showing up in a 10-inch scope as a blazing ball of light peppered with faint stars. A 16-inch scope shows countless stars extending nearly 20′ across that are so dense toward the center they merge into a yellowish nebulous haze.

NGC-6752 was probably first telescopically observed by J. Dunlop in 1828. One hundred and fifty years later, the expert observer E. J. Hartung characterized NGC-6752 as "a most lovely object; it is a moderately condensed type of globular cluster, the central region about 3′ wide." He also noted that many of the cluster's brightest stars are arranged in curved and looped arms and look distinctly reddish. An attractive foreground double star, which contains magnitude 7.7 and 9.3 stars separated by 3″, is visible on the cluster's edge. You might test the sky's steadiness by checking the lowest magnification that resolves this double star.

The other galactic objects in Pavo are two faint planetary nebulae coincidentally located very close to each other. If you observe with a 12-inch or larger telescope on a very dark night, you may spot **NGC-6630** and **IC 4723** together in the same low-power field of view. Lying 2° south of the 6th-magnitude star Nu (ν) Pavonis, this pair of planetaries is separated by less than 1° in a southeast-northwest orientation.

NGC-6630 and IC 4723 appear almost identical, each glowing dimly at 15th magnitude and measuring 19″ across its long axis. IC 4723 is very nearly circular, but NGC-6630 is slightly flattened — its dimensions are 19″ by 15″. Both of these planetaries are visible as faint, "out-of-focus stars" with high magnification. The surface brightnesses of these nebulae are much too low to show any color: they appear white or pale gray.

About 5° east of these planetaries is the multiple-arm barred spiral galaxy **NGC-6744**. One of the prettiest galaxies in the Southern Hemisphere, NGC-6744 shines at blue magnitude 9.0 and measures 15.5′ by 10.2′ across. This object is a treat for backyard observers: viewed with a 4-inch scope, NGC-6744 consists of a bright central oval of nebulosity about 2′ across surrounded by a very faint glow. Larger telescopes show more impressive views. For example, a 10-inch telescope reveals a large, oval halo of greenish nebulosity with the bright glow of the galaxy's hub at its center. In addition, this aperture on dark nights plainly shows spiral structure in the galaxy's halo along with several faint foreground stars. Larger scopes show striking detail in the galaxy's arms, including several bright knots of nebulosity and a spiral pattern obvious even to beginning observers. This galaxy should not be passed up.

The fine barred spiral **NGC-6684** lies several degrees east and slightly south of NGC-6744, just 6′ south of the 6th-magnitude star Theta (θ) Pavonis. This galaxy is a better-than-average target for small scopes because it shines at magnitude 10.5 and measures 3.7′ by 2.7′. NGC-6684 appears as a large oval halo of nebulosity with a bright, almost stellar nucleus. The star Theta Pavonis interferes considerably with observing this galaxy, but observing can be improved on nights of good seeing by using high powers and placing the star slightly outside the field of view.

Three degrees east of NGC-6744 is the elliptical galaxy **NGC-6776**. A 12th-magnitude disk measuring 1.9′ by 1.7′, NGC-6776 appears like a faint, unresolved globular cluster in most backyard telescopes. Its center is condensed and brighter than the rest of the galaxy, but overall NGC-6776 offers no detail for amateur observers.

Moving 4° north brings us back to the

BEST VISIBLE DURING
SUMMER

NGC-6752

12.5-inch
f/6 reflector
145x

NGC-6753

17.5-inch
f/4.5 reflector
269x

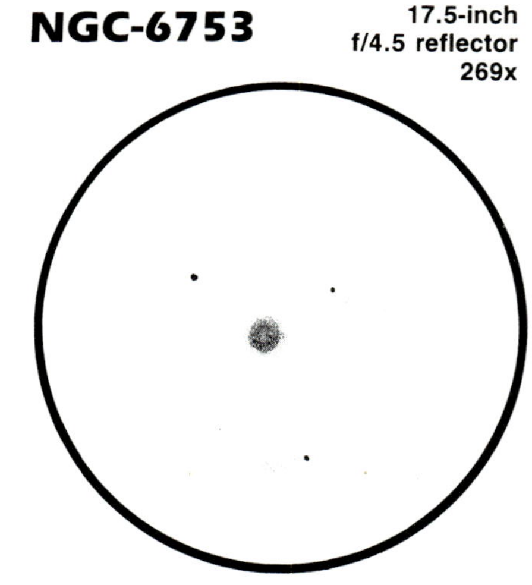

Sketches by Jeffrey Corder

NGC-6752, and you'll spot the spiral galaxy **NGC-6753**. Glowing at blue magnitude 11.9 and measuring 2.5′ by 2.2′, this little galaxy looks like an elongated smudge of gray light in small telescopes. An 8-inch instrument shows a pale oval nebulosity centered on a much brighter nuclear region.

Lying northwest of globular NGC-6752 region of globular NGC-6752. Look 2° southeast of the cluster and you'll see the faint barred spiral galaxy **NGC-6769**. This object shines at magnitude 11.8 and spans 2.5′ by 1.7′, which gives it the appearance of a pale, uniformly bright oval of nebulosity.

Now move your telescope 3.5° north of is the faint barred spiral **NGC-6699**. This galaxy has a blue magnitude of 12.7 and dimensions of 1.9′ by 1.8′, which make it appear like a faint, round ball of light. NGC-6699's appearance changes little when it is observed with different telescopes: the galaxy remains featureless.

On the other side of NGC-6752, some 4.5° distant, lies the spiral galaxy **NGC-6810**. With a blue-light magnitude of 12.3 and a size of 3.8′ by 1.1′, NGC-6810 shows no detail.

Four more noteworthy galaxies lie in the southeast corner of the constellation. Only 2° apart and bracketing the 6th-magnitude star Sigma (σ) Pavonis are the galaxies **NGC-6943** and **IC 5052**. On the northwest side NGC-6943 is a 12th-magnitude barred spiral some 4.1′ by 2.1′ across; it appears as a large, faint haze with a bright center. On the southeastern side IC 5052 is a 12th-magnitude edge-on barred spiral spanning 5.8′ by 0.9′. Although it is faint, this galaxy is a fine sight in most backyard scopes because it appears like a thin, silvery needle of light. The central bulge is not much wider than the spiral arms, and the galaxy's center is only slightly brighter than the arms. The field surrounding IC 5052 is very pretty and contains many interesting groups.

Southwest of the NGC-6943/IC 5052 pair is the very faint elliptical galaxy **NGC-6876**. This object glows dimly at magnitude 12.6 and measures 2.4′ by 1.8′; it appears as a small, round blob of light without any detail.

Eight degrees northeast of this little galaxy lies **NGC-7020**, a blue magnitude 12.4 lenticular galaxy measuring 4.3′ by 2.3′. Viewed with small scopes, this galaxy appears as an oval smear of gray light set amidst a pretty starfield.

Object	M#	Type	R.A. (2000)	Dec.	Mag.	Size/Sep./Per.	H
NGC-6630		■	18h 32.5m	−63°17′	15.0	19″x 15″	
IC 4723		■	18h 35.8m	−63°24′	15.0	19″	
NGC-6684		§B	18h 49.0m	−65°11′	10.5	3.7′x 2.7′	SB0
NGC-6699		§B	18h 52.1m	−57°19′	12.7_B	1.9′x 1.8′	S(B)b
NGC-6744		§B	19h 09.8m	−63°51′	9.0_B	15.5′x 10.2′	S(B)b+
NGC-6752		●	19h 10.9m	−59°59′	5.4	20.4′	
NGC-6753		§	19h 11.4m	−57°03′	11.9_B	2.5′x 2.2′	Sb
NGC-6769		§B	19h 18.4m	−60°31′	11.8	2.5′x 1.7′	S(B)b pc
NGC-6776		0	19h 25.4m	−63°52′	12.0	1.9′x 1.7′	E2
NGC-6810		§	19h 43.6m	−58°40′	12.3_B	3.8′x 1.1′	Sb−:
NGC-6876		0	20h 18.3m	−70°52′	12.6	2.4′x 1.8′	E3
NGC-6943		§B	20h 44.5m	−68°45′	12.0_B	4.1′x 2.1′	S(B)c:
IC 5052		§B	20h 52.1m	−69°12′	12.2_B	5.8′x 0.9′	SBd:
NGC-7020		§L	21h 11.4m	−64°03′	12.4_B	4.3′x 2.3′	S0

● *Globular Cluster*
■ *Planetary Nebula*
§ *Spiral Galaxy*
§B *Barred Spiral Galaxy*
§L *Lenticular Galaxy*
0 *Elliptical Galaxy*

H = Hubble type for galaxies
Subscript "P" denotes photographic magnitude; subscript "B" denotes blue magnitude.

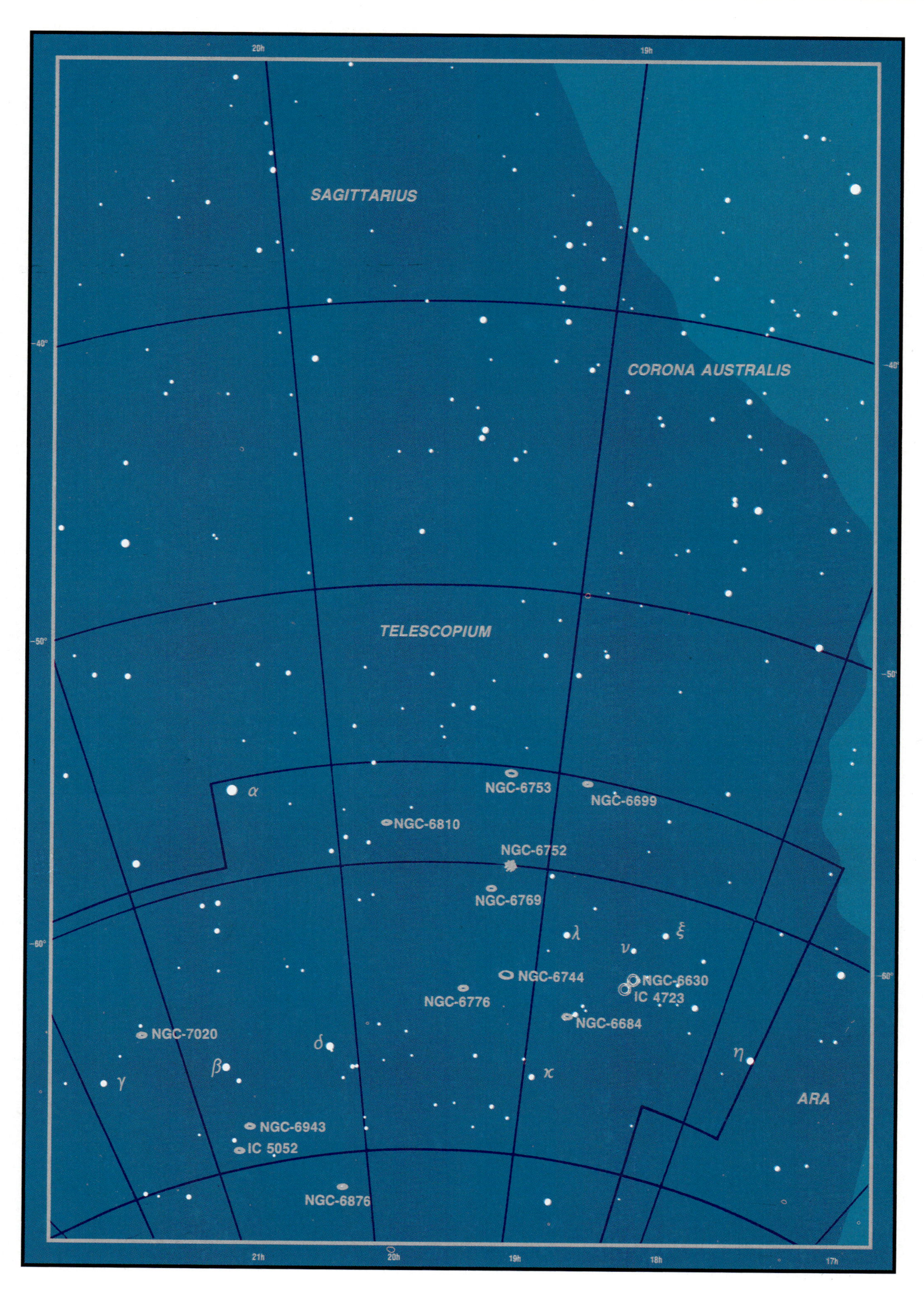
20h
19h
SAGITTARIUS
CORONA AUSTRALIS
-40°
TELESCOPIUM
-50°
α
NGC-6753
NGC-6699
NGC-6810
NGC-6752
NGC-6769
-60°
λ
ν
ξ
NGC-6744
NGC-6630
NGC-6776
IC 4723
NGC-6684
NGC-7020
δ
β
γ
κ
η
ARA
NGC-6943
IC 5052
NGC-6876
21h
20h
19h
18h
17h

Left: M-15 is one of the sky's brightest globular star clusters. Below left: NGC-7814, a 12th magnitude edge-on galaxy, shows its equatorial dust lane in 6-inch telescopes under a dark sky. Below: Stephan's Quintet, a group of interacting galaxies, appears as a misty patch of light in backyard telescopes. Below center: Although it has a low surface brightness, NGC-7479 is one of the sky's finest barred spiral galaxies. Bottom: NGC-7331 is a 10th magnitude galaxy resembling a fainter version of M-31. All photos by Bill Iburg.

Pegasus
Peg
Pegasi

Equuleus
Equ
Equulei

Keith Ward

Dominating the autumnal evening sky is the large asterism known as the Great Square. Composed of four bright stars — Markab, Scheat, Algenib, and Alpheratz — the Great Square is the centerpiece of **Pegasus** the Winged Horse, easily recognizable because it is surrounded by mostly blank sky. Pegasus contains more than the Great Square, however: it is far enough away from the Milky Way that it contains plentiful galaxies for backyard observing.

But the best deep-sky object in Pegasus is the bright globular cluster **Messier 15** (NGC-7078). It is quite easy to find, lying some 4° northwest of Enif (Epsilon [ε] Pegasi). Appearing as a magnitude 6.4 "fuzzy star" in finder telescopes or binoculars, M-15's appearance changes when viewed with a 4 or 6-inch telescope — at low power (40x) the globular's 12.3′ disk appears as a uniformly illuminated glow, while high power (100x) resolves its edges into dozens of tiny pinpoint stars. Larger telescopes do even better: a 12-incher at 175x resolves stars across M-15's disk to create a three-dimensional effect. A 17.5-inch telescope reveals hundreds of yellow stars — while small scopes tend to give a gray or greenish tint to the cluster.

M-15 is an unusual globular for several reasons. It is an intense X-ray source, leading astronomers to believe that it contains a central black hole "engine" that feeds off material from stars which stray too close to the cluster's core. (Even in a backyard telescope, you'll notice that M-15 has one of the most intensely brilliant cores of any globular.) It contains a huge number of variable stars — 110 in all; unfortunately, most are faint RR Lyrae types with short periods and amplitudes, making observing their brightness changes difficult. Curiously, M-15 is the only globular cluster containing a known planetary nebula, Pease 1. The nebula measures 3″ across, glows dimly at magnitude 14.9, and is surrounded by hundreds of stars; it is impossible to pick out with backyard telescopes.

Near M-15, about 3.3° northeast of Enif, is the erratic variable star **AG Pegasi**. This bizarre object is one of the brightest of the so-called symbiotic stars, systems showing bright hydrogen lines in their spectra. Astronomers believe that AG Peg contains a Wolf-Rayet star and a reddish M giant. The Wolf-Rayet object resembles the nucleus of a planetary nebula, and is surrounded by a dim blanket of nebulosity.

AG Peg has an unusual history of variability. Up until about 1850, it appeared as a steady 9th magnitude star. But between 1850 and 1870, it brightened to 6th magnitude. Then AG Peg remained relatively stable until 1920, when it began a rapid succession of spectral changes and gradual fading. Although it has a semi-regular period of 830 days, AG Peg now appears fairly stable and shines around 9th magnitude.

Pegasus contains a multitude of bright galaxies for backyard telescopes. The finest is **NGC-7331**, a 10th magnitude Sb-type spiral that looks like a miniature version of M-31, the Andromeda Galaxy. (Both M-31 and NGC-7331 are similar types of objects, but appear different because NGC-7331 is about 20 times more distant.) Lying 4.3° north and slightly west of the bright star Eta (η) Pegasi, NGC-7331 is visible in finder scopes under a dark, moonless sky. A 3-inch telescope shows this galaxy as a small oval of gray-green light spanning about 5′. A 6-incher at high power reveals a bright, roundish nucleus enveloped in the oval disk. Under good conditions a 10-inch scope shows a faint, uniformly-lit halo of nebulosity — NGC-7331's spiral arms, which measure 10.7′ x 4.0′.

Just ½° south-southwest of NGC-7331 is a faint group of galaxies collectively known as **Stephan's Quintet**, after M.E. Stephan, who discovered the group in 1877. The brightest member, NGC-7320, is a magnitude 12.7 spiral, the only object in the area visible with a 4-inch scope. The other members are 13th magnitude galaxies: NGC-7317, an elliptical; NGC-7318A, a peculiar elliptical; NGC-7318B, a peculiar barred spiral; and NGC-7319, another peculiar barred spiral. Although all five galaxies appear intertwined, redshift measurements show that the group lies about 90 megaparsecs' distant, while NGC-7320, the brightest member, is located a mere 13 megaparsecs away and simply lies along the same line of sight. Backyard telescopes generally show this group as a fuzzy patch of light some 10′ across, although 10-inch or larger scopes show NGC-7320 as a bright oval and resolve individual nebulous spots representing the light from

BEST VISIBLE DURING
AUTUMN

M-15

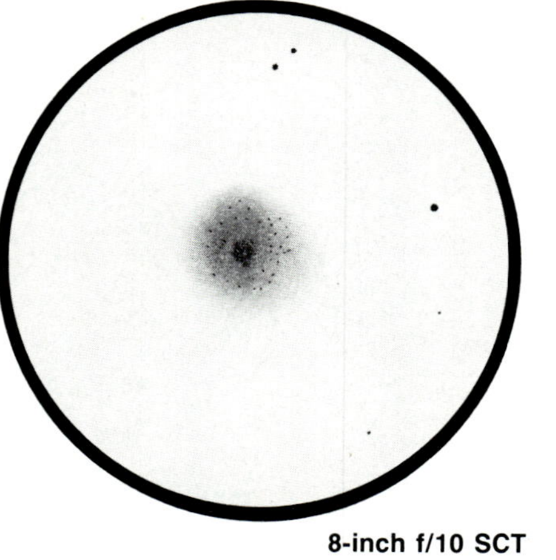

8-inch f/10 SCT
50X

NGC-7331

8-inch f/10 SCT
50X

Sketches by David J. Eicher

four distant galaxies.

About 2° south and slightly west of the bright double star Pi (π) Pegasi is the little galaxy **NGC-7217**. An Sb-type spiral with tightly-wound arms, NGC-7217 shines at magnitude 10.2 and appears nearly face-on, measuring 3.7′ x 3.2′ across. It has a relatively high surface brightness, and stands out as a fuzzy spot even in 2-inch scopes. Larger telescopes show a bright core surrounded by a circular gray haze of nebulosity.

Ten degrees due south of the NGC-7331/Stephan's Quintet area is another bright galaxy, **NGC-7332**. Shining at 11th magnitude and extending 4.2′ x 1.3′, NGC-7332 is a highly elliptical E7 galaxy that is visibly lens-shaped in backyard telescopes. It has a brilliant central condensation and tiny nucleus with a smooth, evenly-illuminated streak of nebulosity crossing through it. In the same field of view is the fainter galaxy NGC-7339, an edge-on barred spiral dimly glowing at 13th magnitude.

The area of sky below the Great Square also holds several bright galaxies. One is **NGC-7448**, a magnitude 11.7 Sc spiral lying 1½° northwest of Markab (Alpha [α] Pegasi), the star at the southwestern corner of the Square. NGC-7448 measures 2.7′ x 1.3′ across and exhibits a bright nucleus surrounded by an irregular faint haze. **NGC-7479**, found about 3° south of Markab, is a beautiful barred spiral. It shines at magnitude 11.0, measures 4.1′ x 3.2′ across, and has a bright core recognizable as a galaxy in scopes as small as 4-inches' aperture. An 8-incher shows NGC-7479's bright core encapsulated in a faint glow which seemingly sprouts two low surface brightness spiral arms winding their way into an oval from both sides of the nucleus. Because it is bright and more-or-less face-on to our line of sight, NGC-7479 is one of the best examples of a barred spiral galaxy. In this same area is the pair of galaxies **NGC-7619** and **NGC-7626**, the brightest members of the Pegasus I cluster, a small group of galaxies 50 megaparsecs away. NGC-7619 is an 11th magnitude elliptical some 2.9′ x 2.6′ across; NGC-7626 is a elliptical measuring 2.5′ x 2.0′ and shining at magnitude 11.2. Both galaxies lie in the same high-power field and appear as fuzzy splotches of light devoid of detail.

NGC-7741 is an 11.4 magnitude barred spiral lying within the Great Square. It spans 4.0′ x 2.8′ and is visible in large finder scopes as an elongated patch of misty light; small telescopes show a bright nucleus surrounded by a dim halo of nebulosity. Also inside the Square is a group of three diverse objects. **NGC-7814**, some 3° northwest of Algenib (Gamma [γ] Pegasi), is a bright edge-on Sb-type spiral. It glows at magnitude 10.5 and measures 6.3′ x 2.6′, making it easily visible as a spike of nebulosity in 3 or 4-inch scopes. A good 6-inch telescope at high power shows the galaxy's broad equatorial dust lane, which appears to split NGC-7814 in two. A degree west of this galaxy is the variable star **U Pegasi**, an eclipsing binary. This star varies between magnitudes 9.2 and 9.9 over a very short period — about 9 hours — so that observing it early and late during the same night may show significant changes. A degree west of U Peg is the small open cluster **NGC-7772**, a group of 7 stars of magnitude 11 and fainter.

Equuleus the Colt — a tiny constellation due west of Enif — doesn't cover much sky, but it does contain **Delta (δ) Equulei**, a close double star with a rapid orbit. Composed of twin 5th magnitude stars, Delta Equ completes one orbital revolution in just 5.7 years. Closest approach occurs again during 1987, when the two stars will be separated by only 0.35″. At present, this star is a difficult object for backyard scopes, but splitting it makes a fine test for large, high-resolution instruments.

Object	M#	Type	R.A. (2000)	Dec.	Mag.	Size/Sep./Per.	H
NGC-7814		§	0h 03.3m	+16°09′	10.5	6.3′x2.6′	Sb$^-$
Delta (δ) Equ		★^{2}	21h 14.5m	+10°00′	5.5	0.3″	
NGC-7078	M-15	●	21h 30.0m	+12°10′	6.4	12.3′	
AG Peg		IV	21h 51.1m	+12°22′	6.4↔8.2	Irr	
NGC-7217		§	22h 07.9m	+31°22′	10.2	3.7′x3.2′	Sb$^-$
NGC-7331		§	22h 37.1m	+34°25′	9.5	10.7′x4.0′	Sb
NGC-7332		()	22h 37.4m	+23°48′	11.0	4.3′x1.3′	E7
NGC-7448		§	23h 00.1m	+15°59′	11.7	2.7′x1.3′	Sc
NGC-7479		§B	23h 04.9m	+12°19′	11.0	4.1′x3.2′	SB(b)$^+$
NGC-7619		()	23h 20.2m	+ 8°12′	11.1	2.9′x2.6′	E1
NGC-7741		§B	23h 43.9m	+26°05′	11.4	4.0′x2.8′	SB(c)
NGC-7772		⊙	23h 51.0m	+20°09′	10	1.6′	
U Peg		EV	23h 55.4m	+15°40′	9.2↔9.9	0.37d	

H = Hubble classification type for galaxies

★^{2}	*Double Star*
EV	*Eclipsing Variable*
IV	*Irregular Variable*
⊙	*Open Star Cluster*
●	*Globular Star Cluster*
§	*Spiral Galaxy*
§B	*Barred Spiral Galaxy*
()	*Elliptical Galaxy*

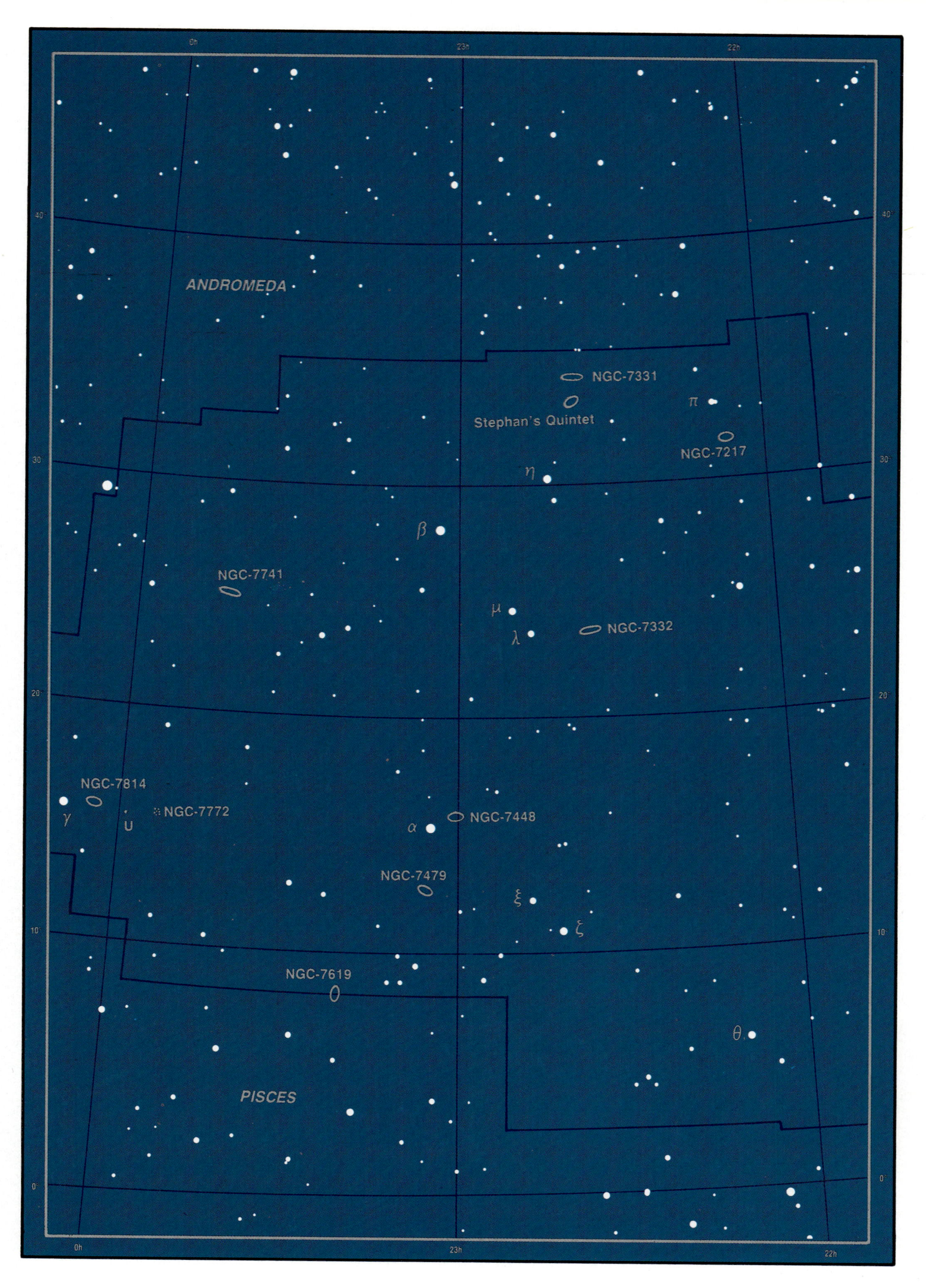
0h
23h
22h
40°
ANDROMEDA
NGC-7331
Stephan's Quintet
π
NGC-7217
30°
η
β
NGC-7741
μ
λ
NGC-7332
20°
NGC-7814
γ
U
NGC-7772
NGC-7448
α
NGC-7479
ξ
ζ
10°
NGC-7619
θ
PISCES
0°

Top left: M-76, the Little Dumbbell Nebula, is one of the faintest and most trying Messier objects to see. Photograph by Martin C. Germano. Top right: Hundreds of blue and white stars cover the field surrounding the Double Cluster; photo by Bill Iburg. Middle right: The 11th magnitude galaxy NGC-1023 shows its bright central hub to 6-inch telescopes on good nights. Jack Newton photograph. Below: One of the most difficult to find of the large emission nebulae, the California Nebula is visible only with superb skies. Photo by David Healy. Middle left: Lee Coombs recorded the stars of M-34, which fill a low-power eyepiece and even offer a good binocular view.

Perseus

Per
Persei

Keith Ward

Perseus the Hero is a rich area of sky representing in part our view toward the neighboring Perseus Arm of the Galaxy. It is chock full of bright stars and clusters; flanked by the sparkling Pleiades cluster (M-45) in Taurus and the elusive nebulae IC 1805 and IC 1848 in Cassiopeia; it even contains some rather exotic objects that will tax your telescope's optics and your observing experience.

As you look up toward Perseus, the combined soft glow of the **Double Cluster** (NGC-869 and NGC-884) is sure to catch your attention. So commandingly bright are these two star groups that they warranted the designations "h" (NGC-869) and "χ" (NGC-884) from early celestial cartographers. The two clusters lie side by side half a degree apart, so they require a wide field of view to be seen together — try 30x with a 6-inch or 40x with an 8-inch scope. Both clusters measure 35′ in diameter: NGC-869 is the brighter at magnitude 4.4, while NGC-884 glows at magnitude 4.7. With binoculars or a small telescope, the field of view appears peppered with two areas of bright blue and white stars, the brightest of which are 7th magnitude objects. With an 8-inch scope at 35x, a rich Milky Way field surrounds several hundred bright A- and B-type supergiants. A few dim red stars show up in and around the clusters — these are M-type supergiants.

NGC-884 is more distant than NGC-869, and much older. The former is 11.5 million years old and lies 2.5 kpc distant, while the latter is only 6.4 million years old and 2.15 kpc away. Each is about 70 light-years across and contains around 5,000 solar masses. Since each cluster contains only a few hundred stars, they are truly supermassive and superluminous.

BEST VISIBLE DURING
WINTER

Less than 2° northwest of the Double Cluster is **NGC-957**, a smaller but still impressive group. This cluster lies at a distance of 2.3 kpc and is apparently physically linked to the Double Cluster.

The largest conglomeration of stars in the constellation is not a true cluster at all, but rather a stellar association called **Perseus OB-3**. This group is roughly centered on Mirfak (Alpha [α] Persei), contains 70 stars between 5th and 10th magnitude, and covers several square degrees of sky. These stars probably constituted an open cluster some 4 million years ago, and have since expanded away from each other while moving through space. The group measures some 300 light-years across and lies 170 parsecs off. If you scan this area with a pair of binoculars, you'll be delighted at the rich groupings of bright blue-white stars.

In the midst of the Per OB-3 group is the bright open cluster **NGC-1245**, containing 40 stars fainter than 11th magnitude. Its total magnitude is around 7: In small telescopes it appears as a generally hazy circular group due to the faintness of its members. A 4-incher shows a patch of nebulosity mixed with a few stars; scopes in the 10-inch to 12-inch range reveal a multitude of tiny granular stars across its 10-arcsecond diameter like salt spilt on black velvet.

Another fine open cluster in Perseus is **NGC-1528**, an aggregation of 80 stars fainter than 7th magnitude about 25′ in diameter. This is a bright, obvious group in binoculars, which show it as a mottled grainy haze with several bright stars popping out. It is easy to observe at a total magnitude of 6, and a small telescope shows a fairly rich grouping, many of the fainter stars being arranged in pairs and triplets about the general hazy background.

The finest lone star cluster in this area is the large group **M-34** (NGC-1039). This cluster holds 80 stars fainter than 7th magnitude in an area spanning 20′ across — it is a grand binocular target appearing as a misty spot of stars and nebulosity. A small telescope such as a 4-inch, when used at low power, shows a large, scattered group of bright stars including the doubles OΣ 44 and h1123 near the group's center. With a 6-inch scope the group shows three "arms" of stars radiating outward from its center. This group is just visible with the naked eye under dark skies. It lies around 500 parsecs off, measures 18 light-years across, and, at 100 million years, is much older than either of the Double Cluster components.

Several unusual stars lie in this area, one of which is **Algol** (Beta [β] Persei), the prototype eclipsing variable star. Regularly as clockwork, Algol undergoes a dimming stage lasting 10 hours once every 2.86739 days. The reason: its whitish B8 primary star is eclipsed by the much fainter G or K-type companion. The system lies 30 parsecs away; the primary measures 2.6 million miles in diameter and the

California Nebula

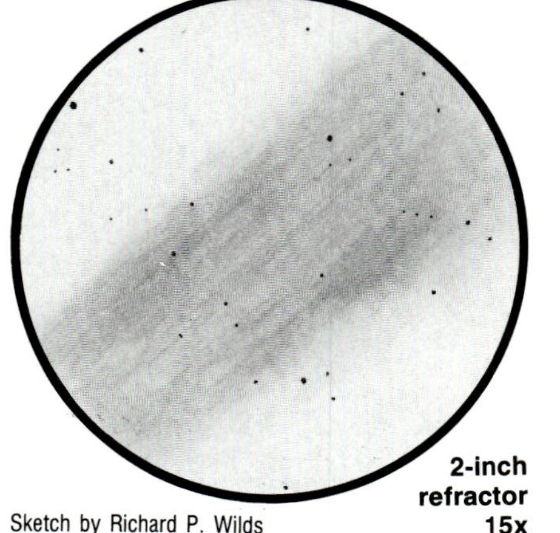

Sketch by Richard P. Wilds

NGC-1023

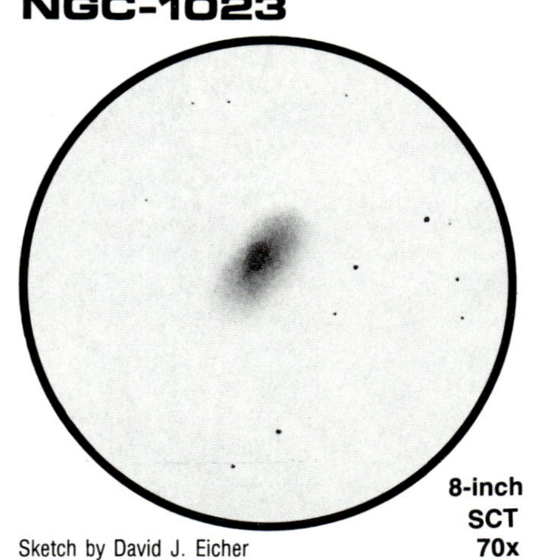

Sketch by David J. Eicher

companion some 3 million miles. Thus when the companion's orbit carries it into our line of sight to the primary, it blocks the brighter B8 star and plunges its brightness from magnitude 2.1 to 3.4.

To observe Algol's changes in brightness, go out with a pair of binoculars and make a sketch of the star and its surrounding field. After you do, estimate the brightness of the neighboring stars — especially those close to Algol's brightness — and record them. By comparing stars close in brightness with Algol over a period of several nights, you'll see that this "Demon Star" does indeed regularly fluctuate in brightness.

On February 21, 1901, a 13th magnitude anonymous star in Perseus violently exploded in the century's first nova outburst. The amateur astronomer T.D. Anderson of Edinburgh, Scotland found the 2nd magnitude out-of-place speck of light, and within 2 days **GK Persei** increased its light output sixfold to shine at magnitude 0.2 on February 23. Within three weeks it faded 4 magnitudes, oscillating in brightness in a strange fashion. After another four months the oscillations stopped and the star faded out of sight. Soon thereafter Dr. Max Wolf at Heidelberg discovered a nebula surrounding the star, growing at a fantastic rate of 2″ per day! Apparently, however, the star was merely illuminating a dark nebula which already existed. GK Persei lies at a distance of 450 parsecs; it still shows irregular variations in light, which brands it a recurrent nova. It hovers around 13th magnitude, but may rise to 11 or fall to 14 at irregular intervals. Look for the star — who knows *what* you may find.

A fine multiple star is **Zeta (ζ) Persei**, a magnitude 2.8 B1-type star which serves as the keystone to the Zeta Persei group, an association of young stars. This star lies 340 parsecs away and is associated with some nebulosity, which suggests a youthful age. It displays several companions in small telescopes: 12.9″ away at position angle 209° lies a 9th magnitude star that has been described as "ashy," the primary being greenish white in color. Some 33″ off in p.a. 287° is an 11th magnitude star, and 195″ away in p.a. 195° is a 9.5 magnitude companion.

Nearby Xi (ξ) Persei is responsible for illuminating the large elusive **California Nebula** (NGC-1499). At 145′x40′, it is one of the largest emission nebulae in the northern sky, but is not nearly as easily observed as the Orion, Lagoon, or even the Rosette Nebula. The brightest parts of this nebula may shine around 14th magnitude; sky darkness and transparency dictate success with the California. In most locations it is invisible in 6-inch, 10-inch, and even 18-inch RFTs, but at a superb site it is observable with binoculars or a 2-inch telescope. A nebula filter seems to help greatly with seeing this object. Be patient; if you do spot the California Nebula, you are among a select group of deep-sky observers.

Two fine planetary nebulae lie in Perseus: **M-76** (NGC-650/1), the Little Dumbbell Nebula, is a miniature version of M-27 in Vulpecula. Its twin lobes of nebulosity glow at 11th magnitude, making it one of the finest planetaries in the sky. Train a low-power eyepiece a degree north-northwest of Phi (ϕ) Persei and you'll see the double patch of fuzzy light. If conditions permit, "crank up" the magnification to something like 25x per inch of aperture and you'll get a fine sight. This expanding shell of gas lies about 550 parsecs away and measures, as most planetaries do, about a light-year across. Another planetary in the constellation, **IC 2003**, is a 12th magnitude object only 5″ across — compared to M-76's diameter of 140″ — and appears stellar under bad seeing. Wait for a good night and search for it with medium power eyepieces.

Two fine galaxies lie within Perseus' boundaries: **NGC-1023** and **NGC-1275**. The former is a bright E7-type lens-shaped galaxy shining at 11th magnitude, and is a fine sight in even 4-inch scopes. It lies in a rich field of stars and shows its bright elongated nucleus and faint outer envelope of nebulosity to any user of a good 6-inch scope. NGC-1275, also known as radio source Perseus A, is a tough target for amateur scopes. This unusual disturbed galaxy, which some astrophysicists speculate may house a supermassive black hole, measures only 42″ across and dimly glows at magnitude 13. An 8-inch scope on a good night shows a tiny grainy spot of weak light.

Object	M#	Type	R.A. (2000)	Dec.	Mag.	Size/Sep./Per.	N★	Mag.★
NGC-650/1	M-76	■	01h 41.9m	+51°34′	11.0	140″x70″		16.5
NGC-869		⊙	02h 19.0m	+57°09′	4.4	35′	340	7...
NGC-884		⊙	02h 22.5m	+57°07′	4.7	35′	300	7...
NGC-957		⊙	02h 32.5m	+57°31′	7.2	10′	40	11-15
NGC-1023		0	02h 40.3m	+39°05′	11.0	4.5′x1.3′		
NGC-1039	M-34	⊙	02h 42.0m	+42°47′	5.5	20′	80	8...
Algol (α)		EV	03h 08.2m	+40°57′	2.1↔3.4	2.867d		
NGC-1245		⊙	03h 14.6m	+47°14′	6.9	20′	40	11...
NGC-1275		#	03h 19.7m	+41°31′	13.0	0.7′x0.6′		
GK		N	03h 31.2m	+43°54′	14.0	Irr.		
Zeta (ζ)		★²	03h 54.2m	+31°53′	3.0,9.5	12.9″		
IC 2003		■	03h 56.4m	+33°53′	12.0	5″		18.0
NGC-1499		□E	04h 04.3m	+36°25′	14.0?	145′x40′		
NGC-1528		⊙	04h 15.2m	+51°15′	6.0	25′	80	8...

N★ = number of stars, Mag.★ = magnitude range of cluster or magnitude of central star, (...) indicates many fainter.

★²	*Double Star*
EV	*Eclipsing Variable*
N	*Nova*
⊙	*Open Cluster*
■	*Planetary Nebula*
□E	*Emission Nebula*
0	*Elliptical Galaxy*
#	*Peculiar Galaxy*

5h
4h
3h
2h
1h
+60°
+50°
+40°
+30°
CAMELOPARDALIS
CASSIOPEIA
IC 1805
IC 1848
NGC-957
NGC-869
NGC-884
NGC-1528
M-76
NGC-1245
ANDROMEDA
GK
M-34
NGC-1275
NGC-1023
NGC-1499
TRIANGULUM
IC 2003
TAURUS
Pleiades
ARIES
4h
3h
2h

Although it has a relatively low surface brightness, M-74 is one of the best examples of a face-on spiral in the sky. Under a dark sky, a 6-inch telescope reveals a subtle arm structure. Photo by Bill Iburg.

Pisces

Psc
Piscium

Keith Ward

Situated between Pegasus and Cetus in the barren star fields of autumn is the dim, winding constellation **Pisces** the Fish. Pisces lies far enough from the plane of the Milky Way that it contains no bright deep-sky objects that are part of the Galaxy, except for a few double and variable stars and one of the brightest white dwarfs. Instead, Pisces is inhabited by galaxies, quite a few of which are bright enough to be viewed with backyard telescopes.

The finest galaxy in Pisces is undoubtedly the face-on Sc-type spiral **M-74** (NGC-628). Lying 1.5° east-northeast of Eta (η) Piscium, M-74 is one of the best examples of a face-on spiral in the sky. M-74 is bright but its light is spread out over its large angular size such that it suffers from a low surface brightness. It shines at 9th magnitude and covers some 10.2′ by 9.5′. A 3-inch telescope at 50x shows a bright, round, featureless patch of nebulosity some 3′ across. An 8-inch scope shows a central hub surrounded by a faint 8′-diameter halo of nebulosity — the galaxy's spiral arms. On very dark nights, larger telescopes — especially those over 12-inches in aperture — show some subtle detail in the arms, strongly hinting at their spiral nature. Don't be discouraged if you have trouble locating or observing M-74 at first; it has a reputation, due to its low surface brightness, as being one of the most difficult Messier objects.

Discovered by Pierre Mechain in September 1780, the galaxy was sighted again by Messier a month later. Mechain's notes on M-74 described it as a nebula that "contains no star; it is fairly large, very obscure and extremely difficult to observe. . . . One can make it out with more certainty in fine, frosty conditions." M-74 has an unusual observational history behind it because of a mistake made by John Herschel in his *General Catalogue*, published in 1864. In it Herschel inexplicably classed M-74 as a globular cluster, a mixup carried over into the *New General Catalogue*. M-74's spiral structure had been observed as early as 1848 by Lord Rosse in Ireland.

BEST VISIBLE DURING
AUTUMN

Following an imaginary line drawn from M-74 through Eta and continuing another 4° southwest you will find another nearly face-on Sc spiral, **NGC-514**. This galaxy glows at magnitude 11.9 and spans only 3.5′ by 2.9′. It looks like a large planetary nebula when viewed with a small telescope. A 3-inch scope shows only a 1′-diameter, uniform disk of light dimly glowing with a pale gray hue. A 10-inch telescope reveals a moderately bright center surrounded by a 3′-diameter, slightly oval halo of very faint nebulosity. With this aperture there is still no detail to suggest spiral arms.

Some 8° south of the NGC-514 area lies a trio of bright galaxies, each of which is a worthwhile sight in backyard telescopes. **NGC-470**, lying just 1° southwest of the 5th-magnitude star 89 Piscium, is nearly identical in appearance to NGC-514. It glows at magnitude 11.9 and is slightly smaller at 3.0′ by 2.0′, but viewed with a small telescope it offers the same slightly oval, plain disk of gray light as that of NGC-514. Within the same low-power field of view is the diminutive galaxy NGC-474, a 13th-magnitude lenticular measuring only 0.4′ across.

About 1.5° northeast of the NGC-470/NGC-474 pair is the peculiar galaxy **NGC-520**. This is one of the strangest, most contorted objects in the sky. In photos made with large telescopes, NGC-520 appears twisted around itself, with streamers and wisps of material trailing outward in nearly every conceivable direction. Astronomers believe that an enormous gravitational influence has shredded the galaxy into the bizarre form we see.

NGC-520 glows at magnitude 11.2 and covers some 4.8′ by 2.1′ of sky. Small telescopes show it as an irregular patch of faint light some 3′ by 1′ in extent, looking like a normal edge-on spiral. Large backyard instruments, however, show NGC-520 as a different entity altogether. A 17.5-inch scope reveals a mottled, 4′-by-1.5′ nebulosity that is not uniformly illuminated and appears "twisted." This unusual mottled appearance is indicative of NGC-520's peculiar nature.

NGC-488 is the brightest and most easily observed of the Pisces trio. Measuring 5.2′ by 4.1′ and shining at magnitude 10.3, this galaxy is a face-on Sb-type spiral. NGC-488's spiral arms are rather faint and diffuse, but the galaxy's central hub has an exceptionally high surface brightness that is unmistakable in any telescope. A 3-inch scope at 50x shows a bright round spot of nebulosity some 2.5′ across; a 10-inch instrument shows the same bright

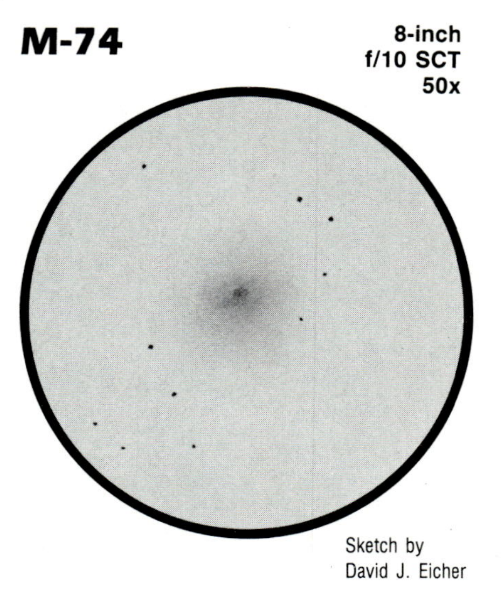

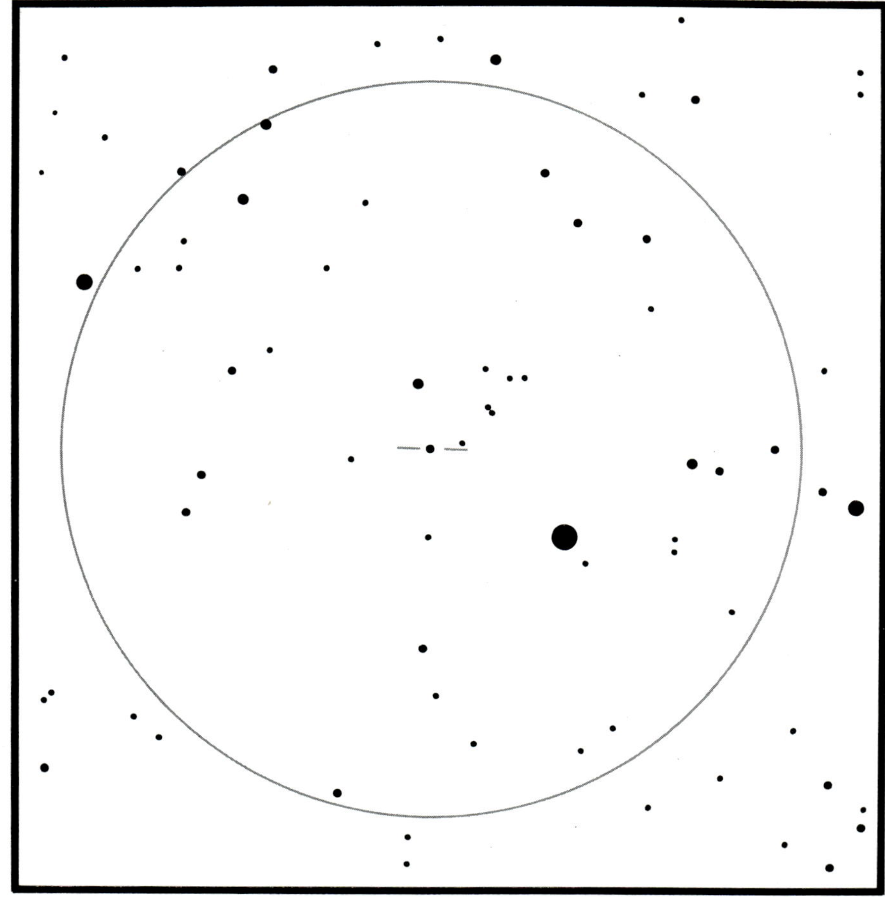

The white dwarf Wolf 28, also known as van Maanen's Star, lies in Pisces at 0h49.1m +5°25′ (2000). The star dimly glows at magnitude 12.4 and lies in a rather rich field, but this finder chart should enable you to spot it. This circle represents 1°, and north is at top.

disk of nebulosity centered inside a ring of faint light some 4′ across. NGC-488 is best observed in a telescope over 12-inches in aperture, when subtle hints of spiral structure appear.

One of the sky's brightest white dwarf stars lies 1.5° southeast of the 4th-magnitude star Delta (δ) Piscium. Catalogued by the famous German astronomer Max Wolf and known contemporarily as **Wolf 28**, this star was identified as a white dwarf by Adrian van Maanen in 1917 and is consequently also known as "van Maanen's Star." With a magnitude of 12.4, this little star — whose luminosity is 1/5,800 that of the Sun — is one of the few white dwarfs visible in backyard telescopes. The diameter of Wolf 28 is about 12,480 kilometers, close to the size of Earth. Its distance is 4.2 parsecs.

Two bright galaxies lie within the central part of Pisces. **NGC-95** is found about 45′ southeast of a 7th-magnitude double star; it is a magnitude 12.6, Sc-type spiral measuring 1.9′ by 1.5′. This galaxy is faint enough that it appears as a featureless smudge of wispy light in most backyard instruments. NGC-128 lies 2° northeast of 44 Piscium, just inside the Pisces-Cetus border. NGC-128 is a magnitude 11.6 lenticular galaxy whose size (3.4′ by 1.0′) gives it a rather low surface brightness.

One of several galaxies found in the western part of Pisces, **NGC-7782** is a faint Sb-type spiral located some 3° northwest of the 4th-magnitude star Omega (ω) Piscium. **NGC-7619** and **NGC-7626** are a pair of ellipticals that lie right on the border with Pegasus and form the core of a remote cluster of galaxies called the Pegasus I Cluster. **NGC-7541** is a 12th-magnitude Sc-type spiral measuring 3.5′ by 1.4′. It lies 2.5° northwest of the bright star Gamma (γ) Piscium and is easily observed at low power in a 6-inch scope.

Object	M#	Type	R.A. (2000)	Dec.	Mag.	Size/Sep./Per.	H
NGC-95		§	0h 22.2m	+10°30′	12.6	1.9′x 1.5′	Sc
NGC-128		§L	0h 22.2m	+ 2°52′	11.6	3.4′x1.0′	S0 pec
Wolf 28		★WD	0h 49.1m	+ 5°25′	12.4	—	
Zeta (ζ)		★^{2}	1h 13.7m	+ 7°35′	5.6,6.2	23.0″	
NGC-470		§	1h 19.7m	+ 3°25′	11.9	3.0′x2.0′	Sc
NGC-488		§	1h 21.8m	+ 5°15′	10.3	5.2′x 4.1′	Sb$^-$
NGC-514		§	1h 24.1m	+12°55′	11.9	3.5′x 2.9′	Sc
NGC-520		#	1h 24.6m	+ 4°44′	11.2	4.8′x 2.1′	pec
NGC-628	M-74	§	1h 36.7m	+15°47′	9.2	10.2′x 9.5′	Sc
Alpha (α)		★^{2}	2h 02.0m	+ 2°46′	4.2,5.1	1.5″	
NGC-7541		§	23h 14.7m	+ 4°32′	11.7	3.5′x 1.4′	Sc
NGC-7619		0	23h 20.2m	+ 8°12′	11.1	2.9′x 2.6′	E1
NGC-7626		0	23h 20.7m	+ 8°13′	11.2	2.5′x 2.0′	E2 pec
TX		IV	23h 46.6m	+ 3°29′	5.5-6.0	irr.	
NGC-7782		§	23h 53.9m	+ 7°58′	13.1_B	2.4′x 1.4′	Sb

★^{2}	*Double Star*
★WD	*White Dwarf*
IV	*Irregular Variable*
§	*Spiral Galaxy*
0	*Elliptical Galaxy*
§L	*Lenticular Galaxy*
#	*Peculiar Galaxy*

H = Hubble type for galaxies

Subscript "P" denotes photographic magnitude; subscript "B" denotes blue magnitude.

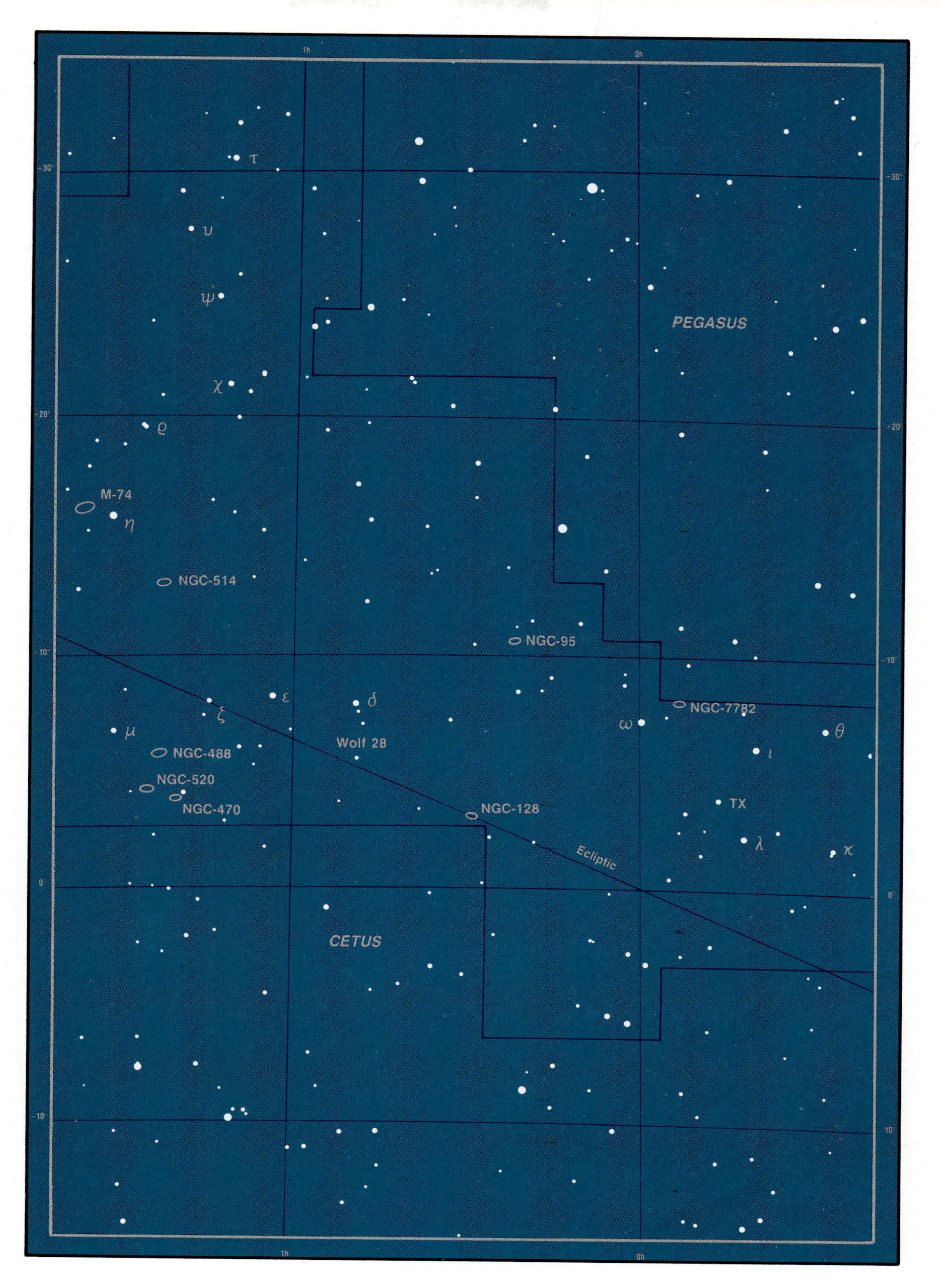
1h
0h
+30°
+20°
+10°
0°
-10°
τ
υ
ψ
χ
ϱ
PEGASUS
M-74
η
NGC-514
NGC-95
ε
δ
ζ
μ
NGC-488
NGC-520
NGC-470
Wolf 28
NGC-7782
ω
θ
ι
TX
NGC-128
λ
κ
Ecliptic
CETUS

Above: The Lagoon Nebula, at bottom, i one of the sky's brightest and mos distinctive emission nebulae; its bisectin dark lane and cluster of blue-white star show in any small telescope. The Trifi Nebula, at top, shows a somewhat dimme patch of emission nebulosity trisected b dust bands, and a bluish patch c reflection nebulosity. Photo by Bill Iburg Top left: M-24, the Small Sagittarius Sta Cloud, is a detached portion of Milky Wa surrounded by regions of dark nebulosity Photo by John Gleason. Middle left: Th globular cluster M-22 is one of the fines in the sky, rivaling M-13 in Hercules. Phot by Roger Sliva. Bottom left: The Omeg Nebula, M-17, has one of the highes surface brightnesses of all emissio nebulae. Photo by Jack Marling.

Sagittarius

Sgr

Sagittarii

The bright, southerly summer constellation **Sagittarius** marks the richest and most varied hunting ground for galactic deep-sky objects. This is because 0° galactic latitude and 0° galactic longitude — the direction of the center of our Galaxy — lies at a point roughly between the open cluster M-6 in Scorpius and the bright emission nebula M-8 in Sagittarius. Thus, when we look toward Sagittarius, we are looking toward the center of the Milky Way, the densest concentration of stars, gas, and dust in the sky.

M-8 (NGC-6523), the Lagoon Nebula, is one of the most spectacular of the emission nebulae; it glows at fifth magnitude and covers an area measuring 60′ x 35′. On a dark night the Lagoon Nebula is visible as a large, fuzzy naked-eye nebulosity, 4.7° west and slightly north of Lambda (λ) Sgr. Binoculars show **NGC-6530**, a bright open cluster inside M-8, and a large oval envelope of wispy greenish light. Small scopes show the nebulosity in two distinct sections, separated by the broad, dark, irregular dust band that gives the nebula its nickname.

Just a degree northwest of M-8 is **M-20** (NGC-6514), better known as the Trifid Nebula. M-20 is another emission nebula, but smaller and much dimmer than the sprawling Lagoon. The Trifid complex shines at eighth magnitude and measures 29′ x 27′ in extent; this excludes a faint patch of bluish reflection nebulosity on the northern tip of the emission nebulosity. Under dark skies the Trifid is a fine sight in binoculars, appearing as a round hazy spot slightly smaller than the Moon's diameter. With a 6-inch telescope, the Trifid becomes an exciting object: it shows a milky greenish irregular outline, trisecting dark lanes, and the just visible ghostly blue-green reflection gas.

BEST VISIBLE DURING SUMMER

Less than a degree northeast of the Trifid is the bright open cluster **M-21** (NGC-6531), a seventh-magnitude group of 50 stars compressed into an area spanning 12′. This group contains three dozen stars between magnitudes 8 and 12, making it a fine binocular sight. It is compact enough that large telescopes still show it well, providing you keep to your low-power eyepieces.

Moving back northward, past M-8 and M-20, we come to the brilliant open star cluster **M-23** (NGC-6494). This group holds 120 stars inside a Moon's diameter, and shines with a total magnitude of 7. Thus it is a fine binocular object — appearing as grainy, glistening sand on the black velvet of space — and a grand sight in any low-power telescopic field. The distance to M-23 is about 660 parsecs, its linear diameter some 5 parsecs.

Nearly due east of M-23 is one of the largest and brightest Messier objects — the starcloud **M-24**, also known as the Small Sagittarius Starcloud. It is a dense patch of Milky Way, detached from its surroundings by lanes of dark nebulae. The cloud shines at magnitude 4.5, and measures 120′ x 40′ across. Its entire area fits into a binocular field, making for a spectacular sight. Telescopes don't show the whole cloud, but several telescopic objects lie within and around the piece of Milky Way Galaxy.

The open cluster **NGC-6603**, which appears as a condensation in the rich background of starcloud M-24, measures 4′ across and contains 50 stars of 14th magnitude and fainter, giving it a total magnitude of 11.4. Telescopes operating at high power show this misty spot as being slightly nebulous, giving the impression of an unresolved globular. The object looks similar to NGC-2158 in Gemini, the little cluster sitting beside M-35. Also within the cloud is the bright, tiny planetary nebula **NGC-6567**, which glows at magnitude 11.5 and measures 11″ x 7″ in diameter. It is rather difficult to locate among the richness of the stellar background, but medium powers reveal the nebula's fuzziness. Seeing 6567's 15th magnitude central star is a difficult task even for large telescope owners: it is easily overpowered by the nebulosity. Another object immersed in M-24 is the dark nebula **Barnard 92**, which measures 15′ across and lies on the starcloud's northwest edge. On good dark nights it is visible as an obvious "hole" in the glittery backdrop of stars.

About 3.5° east of M-24 is the large and commanding open cluster **M-25** (IC 4725), a group which spans over a degree and contains 80 stars, including the bright Cepheid variable **U Sagittarii**. Binoculars or low-power telescopes must be used to view the entire group, whose combined magnitude is 6.2 (making it visible to the naked eye on dark nights). The Cepheid varies in brightness quite regularly from magnitude 6.3 to 7.1 over a period of 6.745 days. The cluster provides many bright comparison stars, so watching U Sgr and charting its fluctuations in brightness make for a good project.

Two bright Messier objects lie north of M-24. The first is the open cluster **M-18** (NGC-6613), a group of 18 stars loosely arranged in an area 7′ in diameter. The cluster's total magnitude is 8.0, making it bright in binoculars but not particularly impressive in larger telescopes because of its lack of concentration. The second is the high-surface-brightness nebula **M-17** (NGC-6618), most commonly known as the Omega Nebula. This is one of the sky's brightest and most easily observed emission nebulae; the familiar omega-shaped bar that gives M-17 its name is visible in binoculars, and wisps and streamers curving in all directions pop out in a large backyard telescope. M-17 is probably as distant as M-8 — about 1700 parsecs.

Sagittarius contains a large number of globular clusters, since it is in the direction of the galactic center, on which the halo of globulars is centered. The brightest and largest of these — one that rivals its big brother M-13 in Hercules — is **M-22** (NGC-6656), a magnitude 6.5 ball of stars measuring an enormous 17.0′ across and containing perhaps half a million stars. M-22 resolves fairly easily with high power, particularly with 6-inch or larger scopes. At 200x in an 8-inch scope, M-22 is a great big ball of sparkles, shimmering with feeble light from one edge of the field to the other. Located only a few degrees southwest, quite close to the star Lambda Sgr, is the smaller yet still impressive globular **M-28** (NGC-6626). Glowing at eighth magnitude, this cluster measures one-quarter degree in diameter. It is a fine sight in any telescope, and partially resolves at high power in a 5-inch telescope. Forming a triangle with globulars M-22 and M-28 is the 10th magnitude planetary NGC-6629, which measures 16″ x 14″ and lies in a rich center-of-the-Galaxy field, making it difficult to pick out of the background.

Four fine globulars lie inside the constellation's "Teapot": **M-54** (NGC-6715), **M-69** (NGC-6637), **M-70** (NGC-6681), and **NGC-6624**. All of these are roughly equivalent in size but differ somewhat in brightness. Because they are pretty small, they are difficult to resolve into stars. M-54 spans 6′ and shines at magnitude 9; M-69 is a 7.5 magnitude globular measuring 4′ across; M-70 is an 8th magnitude globular some 4′ across; and 6624 shines at magnitude 8.5 and measures 3′ from one end to the other.

Farther east are two more globular clusters. **M-55** (NGC-6809) is a large, bright, and exceptionally loose globular whose southerly declination hurts its status with Northern Hemisphere observers. Were it placed high in the sky, this gem of a globular — seventh magnitude and 15′ in diameter — would rate among the finest. A low-power eyepiece shows a large, rather low-surface-brightness ball of stars that easily separates into distinct objects with additional magnification. **M-75** (NGC-6864), the last of the bright globulars in Sagittarius, shines at eighth magnitude and measures only 3′ across. It is roughly comparable in size and brightness to the four globulars that are located within the Teapot.

The elusive galaxy **NGC-6822**, also called Barnard's Galaxy (E.E. Barnard discovered it visually with a 5-inch refractor), and the tiny blue-green planetary nebula **NGC-6818** form an unusual pair of deep-sky objects up in the northeastern reaches of Sagittarius. The galaxy's total magnitude is 9, but that light is spread over an area measuring 20′ x 11′, making its surface brightness dismally low. Dark skies and wide fields of view are the answer to viewing this fellow Local Group member, which probably lies at a distance of about 500 kiloparsecs. The planetary measures only 22″ x 15″, shines at 10th magnitude, and generally looks stellar save for its greenish disk. At medium power, this object's slight fuzziness and telltale color should give it away.

Object	M#	Type	R.A. (2000)	Dec.	Mag.	Size/Sep./Per.	N★	Mag.★
NGC-6494	M-23	⊙	17h 57.0m	−19°01′	7.0	25′	100	9....
NGC-6514	M-20	□ER	18h 01.9m	−23°02′	9.0	29′ x 27′		
Barnard 86		□D	18h 03.1m	−27°50′	—	4.5′ x 3′		
NGC-6520		⊙	18h 03.5m	−27°54′	9.0	5′	25	9...12
NGC-6523	M-8	□E	18h 04.7m	−24°20′	5.0	80′ x 35′		
NGC-6530		⊙	18h 04.7m	−24°20′	7.5	10′	25	7...
NGC-6531	M-21	⊙	18h 04.8m	−22°30′	7.0	10′	50	9...12
NGC-6544		●	18h 07.2m	−25°01′	9.0	1′		
NGC-6567		□	18h 13.7m	−19°04′	11.5	11″ x 7″		15.0
Barnard 92		□D	18h 15.6m	−18°19′	—	15′		
NGC-6603		⊙	18h 18.4m	−18°26′	11.4	4.5′	50	14...
	M-24	SC	18h 18.4m	−18°26′	4.5	120′ x 40′		
NGC-6613	M-18	⊙	18h 19.9m	−17°08′	8.0	7′	12	9...10
NGC-6618	M-17	□E	18h 20.9m	−16°11′	6.9	46′ x 37′		
NGC-6624		●	18h 23.7m	−30°21′	8.5	3′		
NGC-6626	M-28	●	18h 24.6m	−24°52′	8.0	6′		
NGC-6637	M-69	●	18h 31.4m	−31°21′	7.5	4′		
IC 4725	M-25	⊙	18h 31.6m	−19°15′	6.0	35′	50	6...10
U		CV	18h 31.8m	−19°08′	6.3↔7.1	6.745d		
NGC-6656	M-22	●	18h 33.3m	−23°58′	6.0	18′		
NGC-6681	M-70	●	18h 43.3m	−32°18′	8.0	4′		
NGC-6715	M-54	●	18h 55.2m	−30°28′	9.0	6′		
NGC-6809	M-55	●	19h 40.1m	−30°56′	7.0	15′		
NGC-6818		■	19h 43.9m	−14°10′	10.0	22″ x 15″		15.0
NGC-6802		§Irr	19h 44.9m	−14°46′	9.0	20′ x 10′		
NGC-6864	M-75	●	20h 06.1m	−21°55′	8.0	3′		

CV	*Cepheid Variable*
⊙	*Open Cluster*
●	*Globular Cluster*
SC	*Star Cloud*
□E	*Emission Nebula*
□R	*Reflection Nebula*
□D	*Dark Nebula*
■	*Planetary Nebula*
§Irr	*Irregular Galaxy*

N★ = number of stars, Mag.★ = magnitude range of cluster or magnitude of central star, (...) indicates many fainter.

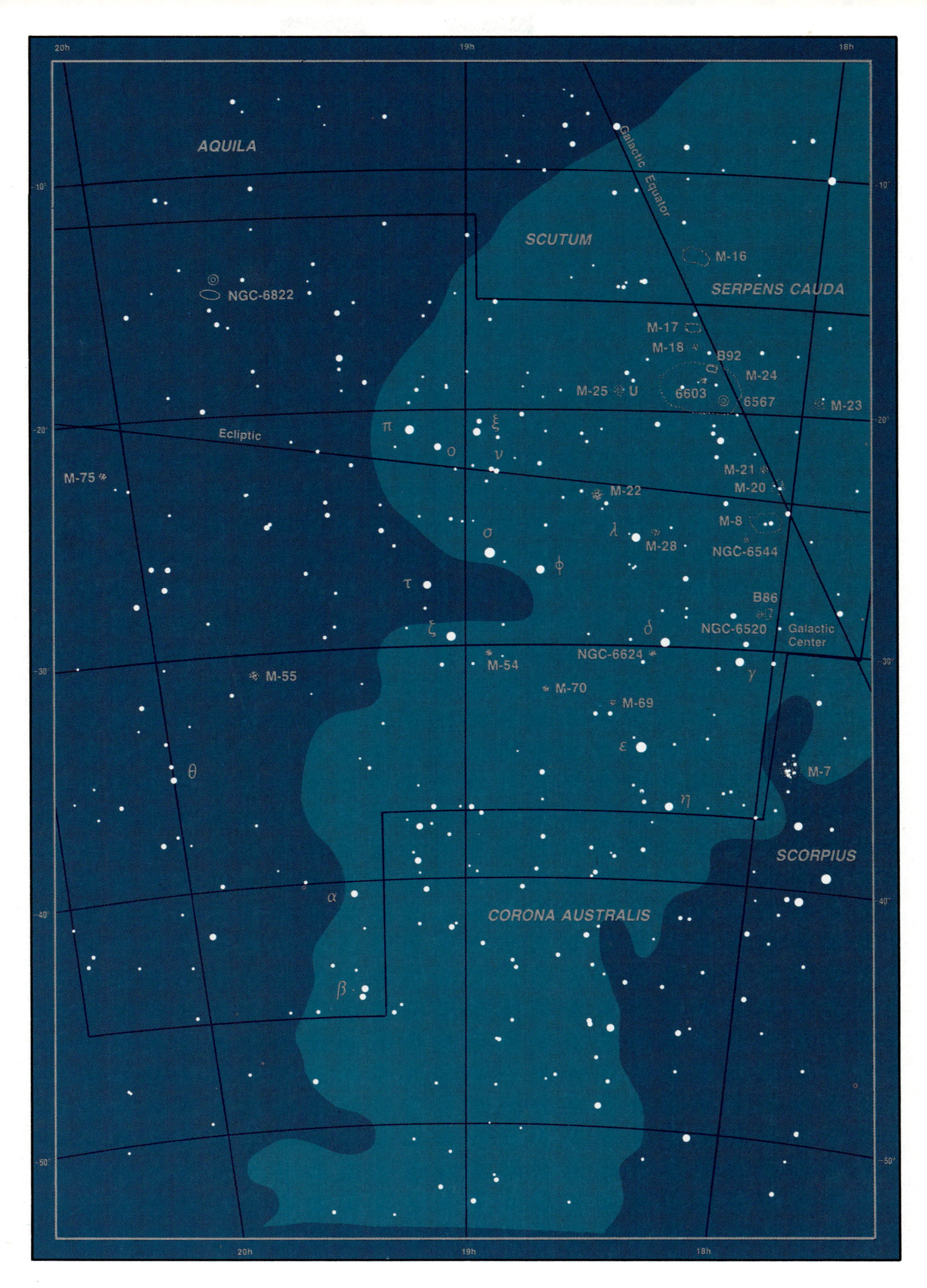

20h
19h
18h
AQUILA
SCUTUM
SERPENS CAUDA
Galactic Equator
M-16
NGC-6822
M-17
M-18
B92
M-24
M-25
U
6603
6567
M-23
Ecliptic
M-75
M-21
M-20
M-22
M-8
M-28
NGC-6544
B86
NGC-6520
Galactic Center
NGC-6624
M-54
M-55
M-70
M-69
M-7
SCORPIUS
CORONA AUSTRALIS
10°
20°
30°
40°
50°

Left: The region of Antares contains abundant star clusters and wispy nebulosity. In Bill Iburg's photo are Antares and globulars M-4 and NGC-6144 at bottom, globular M-80 at upper right, and a haze of reflection nebulosity covering the field. Top right: Jack Marling's photo of the large open star cluster NGC-6231 shows it as a clump of uniformly bright stars. Bottom left: NGC-6302, the Bug Nebula, is a curious bipolar object in central Scorpius. Photo by Martin C. Germano. Bottom right: David Healy's photo of the huge clusters M-6 and M-7 shows a section of sky strewn with stars.

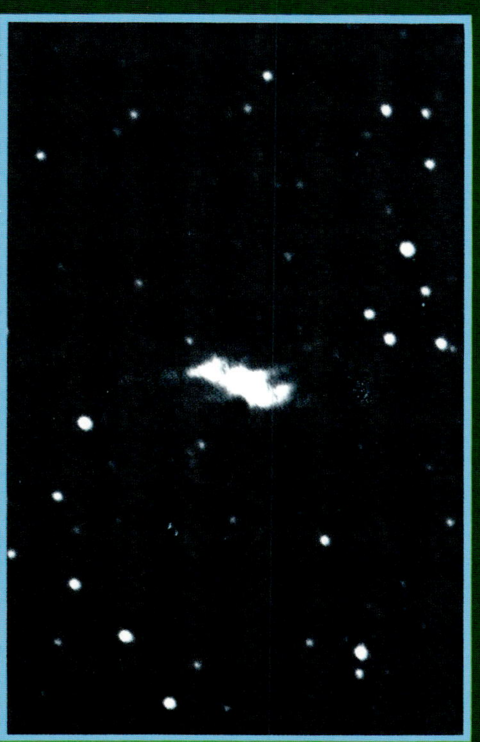

Scorpius

Sco

Scorpii

Keith Ward

Scorpius the Scorpion, one of the richest areas of the entire sky, holds a multitude of bright and impressive stars, star clusters, wispy nebulae and pulsating variables. This is not surprising, since looking toward Scorpius affords a view in the direction of our Galaxy's center — the nucleus of the Milky Way. Bounded by the bright stars of Sagittarius and Centaurus, Scorpius offers binocular fields peppered with tiny stars and telescopic views of most types of deep-sky objects.

The brightest, most obvious object in the constellation is the open cluster **M-7** (NGC-6475), a giant group of 130 stars, more than half of which are brighter than 10th magnitude. With a combined magnitude of about 3.3, the stars of M-7 form one of the brightest of all open star clusters, noticeable to the naked eye as a fuzzy patch even under suburban skies. It lies about 4° northeast of 1.6 magnitude Shaula (Lambda [λ] Sco). M-7 is mentioned in Ptolemy's star catalog and is cited as a nebulous patch in the 16th century Latin translation of the *Almagest*. It is so bright and large that the best views often come in finder telescopes with magnifications of around 10x. As M-7 covers an area 50′ across, any telescope must operate at low powers to contain all of it. The faint and distant globular cluster NGC-6453 lies in the same field as M-7 and appears like an 11th magnitude spot of grainy nebulosity only 1′ across.

The brightest individual member of M-7 shines at magnitude 5.6 and is of spectral type G8; most of the other bright members are type A and B stars. The cluster lies some 273 parsecs distant. Computations show it to be approximately 260 million years old — considerably older than young clusters like the Pleiades, but far younger than such aged galactic objects as the globular clusters.

BEST VISIBLE DURING
SPRING

Some 3.5° northwest of the great M-7 lies another of Scorpius' bright clusters, **M-6** (NGC-6405). Although it is a naked-eye object, the group's telescopic "discovery," made in 1746, is credited to Philippe Loys de Cheseaux. It is sometimes called the Butterfly Cluster because of its arrangement of winglike strings of stars of similar brightness. M-6 contains 132 stars in an area 26′ across, and glows at magnitude 4.6 — again, use low powers with telescopes and scrutinize the group with binoculars if you have them. The brightest star in M-6, **BM Scorpii**, is an unusual long-period variable displaying an orange tint. (BM Sco's spectral class varies between K0 and K3.) It fluctuates between magnitudes 6.0 and 8.1 in a decidedly irregular manner: as you observe M-6, look for the bright orange star in the northeast section of the cluster and, once every few weeks or months, compare its brightness to stars surrounding it. BM Sco's period is roughly 850 days; in time you'll create a record of the star's behavior and gain experience in estimating subtle differences in stellar brightness. Nearly 450 parsecs distant, M-6 lies nearly twice as far away as its brilliant neighbor M-7.

South of M-7, just 4.5′ east of the magnitude 3.5 star G Scorpii, lies the little globular cluster **NGC-6441**. This eighth-magnitude object measures just 3′ across and is composed of multitudes of 17th magnitude and fainter stars: it has a high surface brightness, but is partially obliterated by the bright star practically lying on top of it. After you locate the fuzzy patch of grey light, switch to moderately high powers to increase contrast between sky background and globular, and try to keep the star just outside the scope's field of view. Although it isn't a highly impressive object, 6441 is always easy to find.

A slightly bigger and brighter globular lies in a more barren region at the base of the Scorpion's stinger — **NGC-6388**, a seventh-magnitude group offering a fine sight for small telescopes. This cluster is marginally larger than 6441; it appears as a grainy mottled disk at high power in a 5-inch reflector. Large telescopes, in the 12-inch-and-up class, fail to resolve 6338, whose tiny 17th magnitude members remain hopelessly merged.

Moving northeastward up the Scorpion's body, we find the dazzling open star cluster **NGC-6231**. This assemblage lies half a degree north of the bright star Zeta (ζ) Scorpii, and includes some 120 stars in its 15′ diameter. NGC-6231 glows at sixth magnitude, making it within naked-eye range. It contains a remarkably high number of intensely luminous O- and B-type supergiants — at fifth and sixth magnitude, they appear as bluish-white beacons in small telescopes. On a dark clear night, NGC-6231's brilliance in low power fields will startle you: it looks like luminescent buckshot, sprayed onto the velvet black

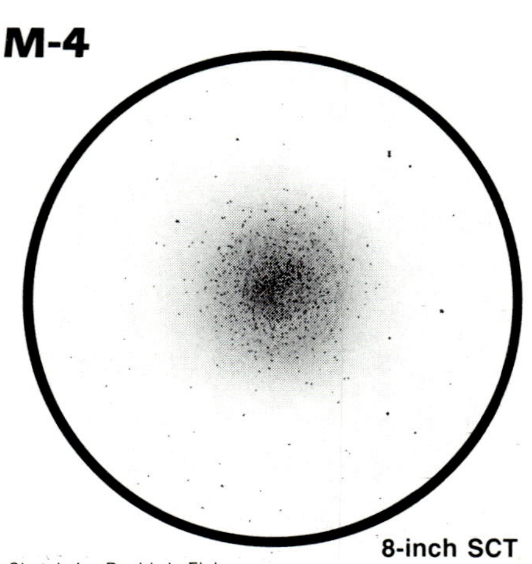

Sketch by David J. Eicher

background of interstellar space. Were it as close as the Pleiades, it would cover the same amount of sky but outshine the Pleiads by a factor of 50!

Just northeast of NGC-6231 is the looser conglomeration of stars known as **H 12**. A stellar association containing at least 200 stars, H 12 is involved with the faint nebulosity IC 4628, which is invisible in small telescopes. The group is designated the I Scorpii Association; it indicates the course of one of the Milky Way's spiral arms, about 1700 parsecs distant, in the direction of the galactic center.

An unusual isolated object is the bipolar nebula **NGC-6302** (otherwise known as the Bug Nebula, due to its funny crab-like appendages). A mere 2′ x 1′ in extent, this slender streak of fuzz glows rather dimly. Its central illuminating star shows a B6-type spectrum and shines at 10th magnitude. The exact nature of 6302 is questionable, but its bipolar structure suggests a young planetary nebula whose parent star "burped" a ring of ionized gas and later drove off its outer envelope with fierce stellar winds.

One of the most beautiful regions of the sky surrounds **Antares**, Alpha (α) Scorpii. This area contains several globulars, fine double stars, and extensive wispy, low-surface-brightness reflection nebulosity. Antares itself is a ruddy M1 supergiant, of magnitude 0.9, and one of the easiest in which to detect color. ("Antares" in Greek means "Rival of Mars.") This aged star shows slight, erratic variations in brightness, and is about 600 million miles across, or about 700 times the size of the Sun. Since its density is exceedingly low, however, Antares is not supermassive — only about 10 or 15 times the mass of Sol. The great red giant has a tiny greenish companion star, about 3″ away at position angle 275° (1959); the little star's magnitude is 6.5, making it extremely difficult to spot in Antares' overpowering glare. It can nevertheless be glimpsed in a 6-inch reflector under good seeing conditions.

Near Antares are three fine globulars, one of which is exceptional. **M-4** (NGC-6121) is one of the most easily resolved such clusters in small scopes because of its large diameter of 23′ and loose, unconcentrated structure. On nights of fine transparency, a good 4-inch telescope at high power can resolve the entire face of the group into pinpoint multitudes of stars. Its combined magnitude is about 7.4, and its brightest individual stars about magnitude 11. Some of these stars appear to form a central "bar" cutting across M-4's central region, a feature distinct in most telescopes. This globe contains 10,000 known stars down to 19th magnitude, and undoubtedly many fainter ones. At a distance of 1.8 kiloparsecs, it is one of the closest globulars to the solar system. Forming a triangle with Antares and M-4, 1.5° northwest of Antares, is the globular **NGC-6144**. This group shines at 10th magnitude and spans 3′; it is visible as a smallish grainy nebulosity in 5-inch telescopes.

The globular cluster **M-80** (NGC-6093) is an eighth-magnitude object measuring 7′ across; it lies north of the M-4/6144 region and is a downright impressive disk of stellar images. Its constituent stars glow softly at 14th magnitude and fainter; observers with instruments of medium or large aperture can expect to resolve its outer periphery with high powers on nights of steady transparency. It contains the nova T Scorpii, which suddenly erupted from anonymity to seventh magnitude in 1860.

The fine double and multiple stars **Akrab** (Beta [β] Scorpii) and **Nu** (ν) **Scorpii** offer bright, colorful sights for small-telescope users. One of the most enticing bright doubles in the sky, Akrab is plainly visible in a 2-inch refractor. The individual components shine at magnitudes 2.6 and 4.9, and are separated by 13.7″ in p.a. 23° (1958); both are early-type bluish-white stars. The pair lies at a distance of just under 150 parsecs, making its luminosity about 2700 times that of the Sun. The Beta Sco system contains a fainter, more distant star that is a spectroscopic double, making Beta a quadruple star system. Nu Scorpii, which lies 1.5° east and slightly north of Beta, is an observable quadruple system. It consists of magnitude 4.0 and 6.2 stars separated by 41.4″ in p.a. 336° (1955); each star in this double is itself a double, resulting in a grand total of four.

Perhaps the finest variable star in Scorpius is **RR Scorpii**, a long-period variable ranging from magnitude 5.1 (naked-eye object) to magnitude 12.3 (just visible in small scopes) in a period of some 279 days. As with BM Sco, make a sketch and magnitude estimate of this star and its surrounding field every few weeks on a regular basis. You'll record a little bit of the star's "life history," documented about 600 years after the fact.

Object	M#	Type	R.A. (2000)	Dec.	Mag.	Size/Sep./Per.	N★	Mag.★
Beta (β)		★^{2}	16h 05.5m	−19°48′	2.6,4.9	13.7″		
Nu (ν)		★^{4}	16h 12.9m	−19°29′	4.0,6.2	1.2″		
					7.0,7.3	2.3″		
NGC-6093	M-80	●	16h 18.0m	−22°59′	8.4	5.1″		14...
NGC-6121	M-4	●	16h 23.7m	−26°31′	7.4	22.8′		11...
NGC-6144		●	16h 27.3m	−26°03′	10.0	3.2′		14...
Antares (α)		★^{2}	16h 29.5m	−26°26′	0.9,6.5	3.0″		
NGC-6231		⊙	16h 54.2m	−41°48′	6.0	15.0′	120	6....
H 12		⊙	16h 56.2m	−40°43′	8.5	40.0′	200	
RR		LPV	16h 56.6m	−30°35′	5.1↔12.3	279d		
NGC-6302		□E	17h 13.9m	−37°06′	12.0	2′x1′		
NGC-6388		●	17h 36.3m	−44°45′	7.0	4.0′		17....
NGC-6405	M-6	⊙	17h 40.0m	−32°13′	4.6	26.0′	132	6....
BM		LPV	17h 40.9m	−32°13′	6.0↔8.1	850d		
NGC-6441		●	17h 50.2m	−37°03′	8.0	3.0′		17....
NGC-6475	M-7	⊙	17h 54.0m	−34°49′	3.3	50.0′	130	6....

N★ = number of stars, Mag.★ = magnitude range of cluster or magnitude of central star, (...) indicates many fainter.

★^{2} *Double Star*
★^{4} *Quadruple Star*
● *Globular Cluster*
⊙ *Open Cluster*
LPV *Long Period Variable*
□E *Emission Nebula*

18h
17h
16h
-10°
-20°
-30°
-40°
-50°
OPHIUCHUS
β
ω
Ecliptic
δ
M-21
M-20
M-8
M-80
M-19
NGC-6144
σ
M-4
α
π
SAGITTARIUS
τ
Galactic Center
M-62
ρ
RR
BM
M-6
ε
M-7
NGC-6441
λ
υ
NGC-6302
μ
κ
ι
H 12
NGC-6231
ζ
θ
η
NGC-6388
Galactic Equator
NORMA
CORONA AUSTRALIS
ARA

Top: NGC-253 is one of the brightest, largest, and most detailed galaxies in the sky. It appears as a large silvery needle of light in small scopes. Photo by Jack B. Marling.

Above: NGC-55 is an edge-on barred spiral galaxy that is as long as the Moon is wide. Easily visible in small scopes, it shows mottled patches of light and dark. Photo by Martin C. Germano.

Left: NGC-288 is a bright globular cluster visible in small scopes as a speckled haze some 10′ across. High power with large backyard scopes completely resolves this fine group of suns. Photo by Martin C. Germano.

Sculptor

Scl
Sculptoris

Phoenix

Phe
Phoenicis

Keith Ward

Between the bright stars Fomalhaut in Piscis Austrinus and Achernar in Eridanus are the faint constellations **Sculptor** and **Phoenix**. Sculptor represents a sculptor's tools and Phoenix is the mythical bird that rises from the ashes of a funeral pyre. Since they lie away from the plane of the Galaxy in a sparse region of stars, both are unimpressive to the naked eye. Sculptor's brightest star is only a magnitude-4.4 object. Phoenix's brightest star is an orange-colored sun shining at magnitude 2.4, but the remaining naked-eye stars in that group are 3rd- and 4th-magnitude objects

Hidden in the apparently empty region, however, are a number of galaxies that show detail when viewed with medium and large backyard telescopes. Among them are NGC-253, the fourth best galaxy for small telescopes; NGC-55, an edge-on barred spiral; and the Sculptor system, a dwarf elliptical that is a member of the Local Group of galaxies.

Lying 7.5° south of Beta (β) Ceti and 3.5° northeast of the South Galactic Pole, **NGC-253** is easy to find by star-hopping across several 6th- and 7th-magnitude stars and the fuzzy, 8th-magnitude globular cluster, NGC-288. Shining at magnitude 7.1 and spanning 25.1′ by 7.4′ (the galaxy's long axis nearly covers the width of the Moon), NGC-253 is bettered only by the Magellanic clouds and M-31 as the best galaxy for small telescopes.

Although NGC-253 is large, it shines so brightly that its surface brightness is high. This affords backyard viewers with even 3- or 4-inch telescopes a good view of an object that obviously looks like a galaxy, despite the fact that its declination is −25° and it never rises more than a few degrees above the treetops for many Northern Hemisphere viewers.

To get the best view of NGC-253, employ a low-power eyepiece. First inspect the overall shape of this galaxy and use averted vision to see obvious details within the core and spiral arms. Then switch to moderately high power (about 15x per inch of telescopic aperture) and look for mottled patches (dust in the spiral arms), the tiny, stellar nucleus, and subtle spiral structure.

NGC-253 is the brightest member of a group of galaxies called the Sculptor group, which includes NGC-55, NGC-300, and NGC-7793. These galaxies lie approximately 3.5 megaparsecs away and, together with NGC-45 and NGC-247 in Cetus, comprise one of the closest galaxy groups outside our own Local Group.

Some 2° southeast of NGC-253 lies the impressive globular **NGC-288**. This object covers 13.8′ of sky and shines at magnitude 8.1, making it readily visible in finder scopes as a hazy patch of light. Small scopes show a large disk of gray light peppered by occasional resolved stars across its face.

NGC-288 is very popular with observers since it is a "loose" cluster and resolves nicely into myriad stars when viewed at high power with 6- to 12-inch telescopes. At moderate powers the cluster takes on a speckled appearance as its brightest stars "pop" out of the background glow. Does your telescope show you a mottled, partially resolved NGC-288?

In south-central Sculptor, about 8.5° south-southeast of NGC-253, lies one of the oddest members of our Local Group of galaxies. The dwarf elliptical known as the **Sculptor system** is an extreme example of its class; it is only about 50 times larger than a globular cluster and it contains a very low density of stars.

Only 90 kiloparsecs distant — therefore close as galaxies go — the Sculptor system is notoriously difficult to view with any telescope. Its brightest stars glow dimly at magnitude 18, and despite the fact that it has a total magnitude of 10.5, that light is spread out over a large area. It may be visible on very dark, clear nights as a faint haze roughly 1° in diameter in the largest backyard telescopes.

In northeastern Sculptor, near Tau (τ) Sculptoris, lies **NGC-613**, a bright barred spiral. Measuring 5.8′ by 4.6′, 10th-magnitude NGC-613 is easily visible as a nebulous, slightly oval patch of light in a 3-inch telescope. In a 6-inch scope at high power the galaxy appears as a bright, condensed core surrounded by a fainter, unevenly lit halo of nebulosity some 4′ across. Larger scopes show mottling in the nebulous halo, suggesting NGC-613's barred spiral arms.

Near the star Eta (η) Sculptoris is a barred spiral designated **NGC-134**. Compare this object carefully with NGC-613. They have the same revised Hubble classification, shine at virtually the same brightness, and measure about the same angular size (NGC-134 is a little larger). Which do you find *you* favor?

BEST VISIBLE DURING
AUTUMN

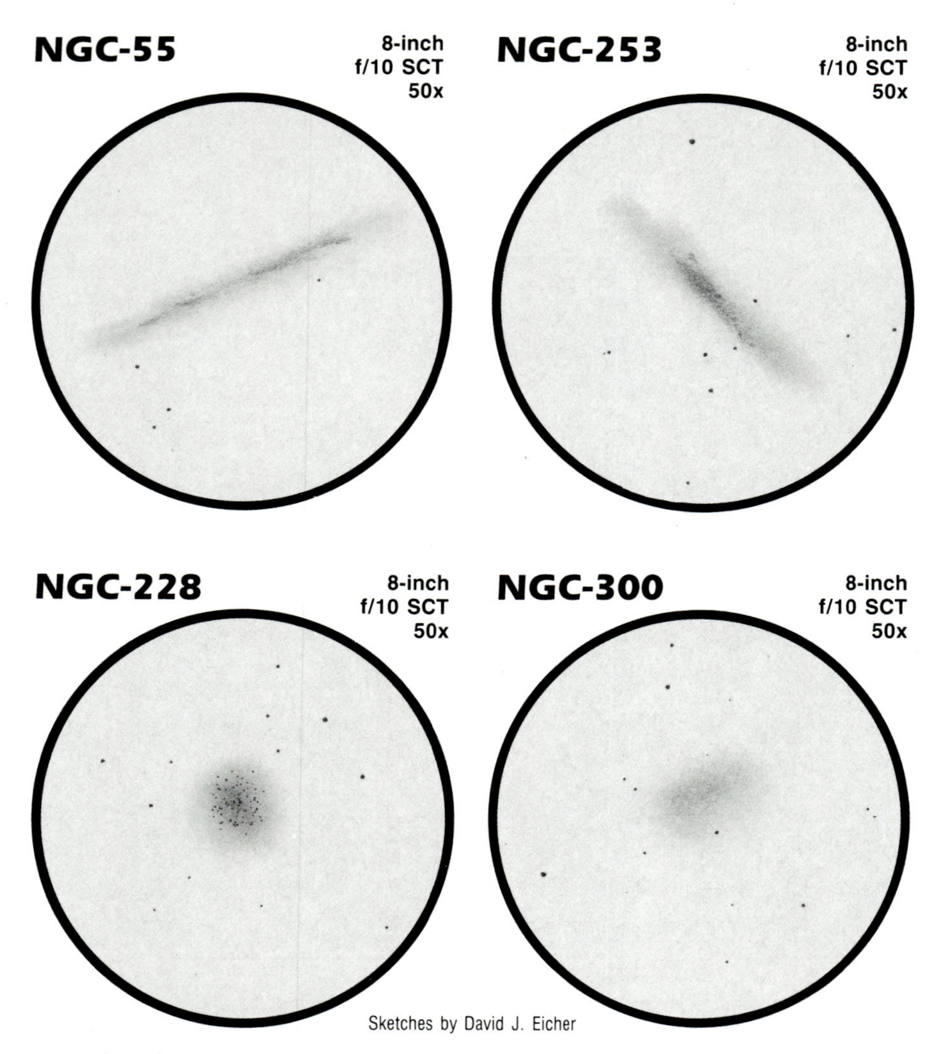

Sketches by David J. Eicher

South and a little west of the Sculptor system, near the border with Phoenix, is a large galaxy with a low surface brightness called **NGC-300**. This object spans some 20.0′ by 14.8′ and shines with a total magnitude of 8.7 in blue light and roughly one magnitude brighter in the yellow-green part of the spectrum that the eye is most sensitive to. Known for its relatively bright and recognizable core, NGC-300's spiral arms are extremely faint. Under ordinary conditions they may require a 10- or 12-inch telescope to make out as a faint haze.

The largest galaxy in Sculptor is **NGC-55**, a huge barred spiral oriented edge-on to our line of sight. To find NGC-55, look about 3.5° north-northwest of Alpha (α) Phoenicis. The galaxy spans 32.4′ by 6.5′ across and shines at magnitude 8.2 in blue light, making it visible with any optical aid as a long, thin sliver of silvery light.

A 6-inch telescope reveals that NGC-55 is mottled by dust lanes and that the nucleus is encapsulated within a small, bright oval core. Larger scopes show progressively more detail in the arms and around the nucleus, and nearly 30′ worth of galaxy — a distance equivalent to the width of the Moon!

Five more impressive galaxies lie in Sculptor. **NGC-24** is a magnitude-11.5 Sb-type spiral measuring 5.5′ by 1.6′ across. It lies near the northern border with Cetus and is visible in large finder scopes as a misty patch of weak light. **IC 5332**, an Sd-type spiral, lies in the southwestern part of the constellation near a 7th-magnitude double star. This galaxy appears like a small, unresolved globular cluster since it shines at 10th magnitude and spans 6.6′ by 5.1′.

One degree east of the bright star Beta (β) Sculptoris is the barred spiral **NGC-7713**, a blue magnitude 11.6 object some 4.3′ by 2.0′ across. **NGC-7755**, another barred spiral, lies in the center of a triangle formed by Mu (μ), Zeta (ζ), and Delta (δ) Sculptoris. It glows at blue magnitude 11.8 and covers 3.7′ by 3.0′. Nearby is **NGC-7793**, a large, close Sd-type spiral glowing at 9th magnitude.

Phoenix contains two bright objects for small telescopes. **NGC-625** is a bright little barred spiral lying 2° northeast of Gamma (γ) Phoenicis. Small scopes show it as a condensed, bright nuclear region 30″ across surrounded by a 3.0′ by 1.3′ diameter haze of greenish light. Its overall magnitude is 12.3 in blue light, making its V magnitude close to 11.

The unusual Cepheid variable star **SX Phoenicis**, some 8° west of Alpha Phe, can be interesting to observe with a small telescope. Its magnitude range is 6.8 to 7.5, so that you must carefully compare its brightness with surrounding stars to accurately determine the star's brightness. The period for SX Phe is 0.05 days, or just 1.2 hours, so you can see it change noticeably during a single observing session.

Object	M#	Type	R.A. (2000)	Dec.	Mag.	Size/Sep./Per.	H
NGC-24		§	0h 09.9m	−24°58′	11.5	5.5′x 1.6′	Sb
NGC-55		§B	0h 14.9m	−39°11′	8.2	32.4′x6.5′	SBm:
NGC-134		§B	0h 30.4m	−33°15′	10.1	8.1′x 2.6′	S(B)b+
NGC-253		§	0h 47.6m	−25°17′	7.1	25.1′x 7.4′	Sc pec
NGC-288		•	0h 52.8m	−26°35′	8.1	13.8′	
NGC-300		§	0h 54.9m	−37°41′	8.7_B	20.0′x 14.8′	Sd
Sculptor System		0	0h 59.8m	−33°42′	10.5_B	—	dE3
NGC-613		§B	1h 34.3m	−29°25′	10.0	5.8′x 4.6′	S(B)b+
NGC-625		§B	1h 35.1m	−41°35′	12.3_B	3.0′x 1.3′	SBm:
IC 5332		§	23h 34.5m	−36°06′	10.6	6.6′x 5.1′	Sd
NGC-7713		§B	23h 36.5m	−37°56′	11.6_B	4.3′x 2.0′	SBd
SX Phe		CV	23h 46.5m	−41°35′	6.8↔7.5	0.05d	
NGC-7755		§B	23h 47.9m	−30°31′	11.8_B	3.7′x 3.0′	S(B)b
NGC-7793		§	23h 57.8m	−32°35′	9.1	9.1′x 6.6′	Sdm

CV	*Cepheid Variable*
•	*Globular Cluster*
§	*Spiral Galaxy*
§B	*Barred Spiral Galaxy*
0	*Elliptical Galaxy*

H = Hubble type for galaxies

Subscript "P" denotes photographic magnitude; subscript "B" denotes blue magnitude.

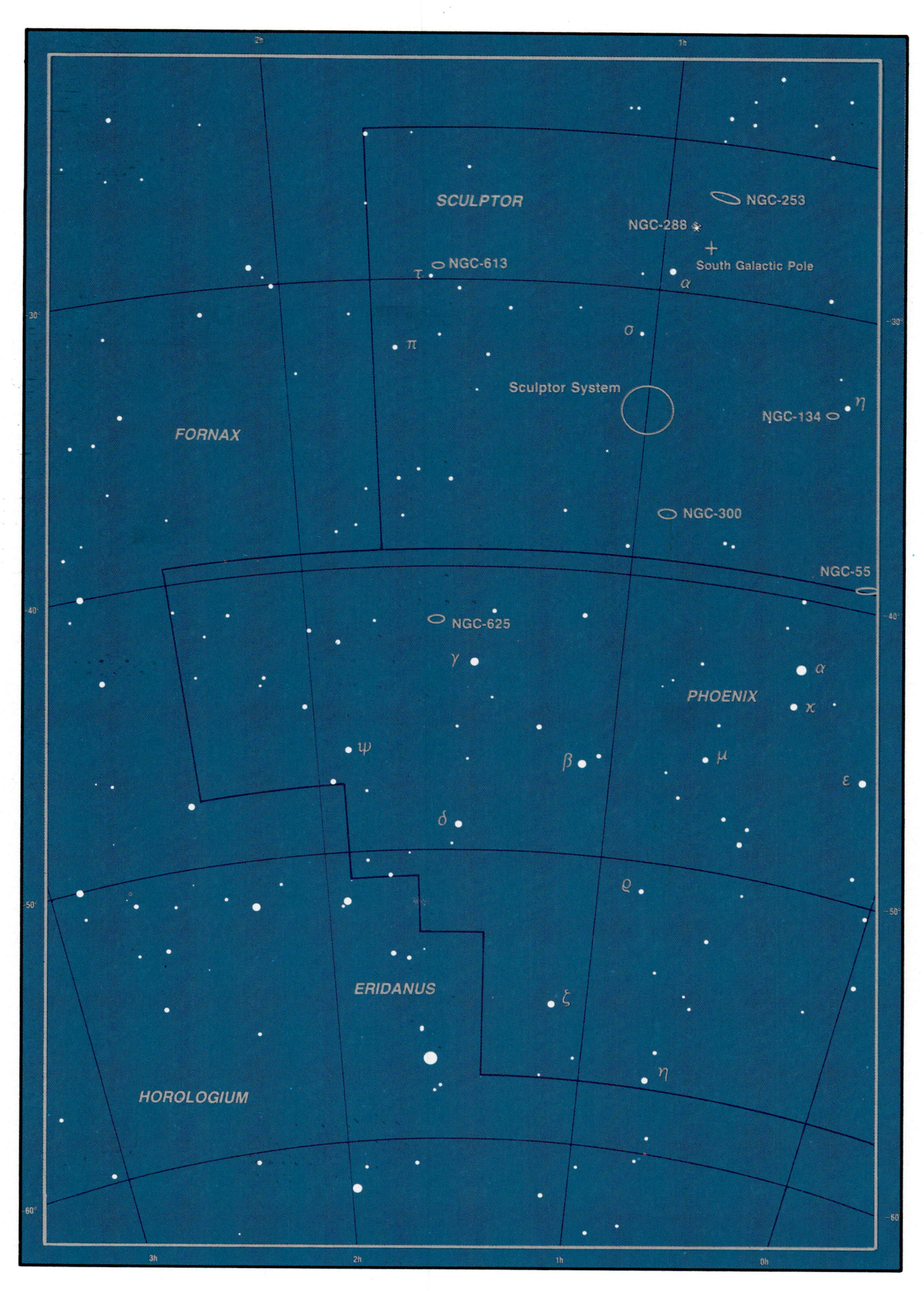
SCULPTOR
NGC-253
NGC-288
South Galactic Pole
NGC-613
Sculptor System
NGC-134
FORNAX
NGC-300
NGC-55
NGC-625
PHOENIX
ERIDANUS
HOROLOGIUM
3h
2h
1h
0h
-30°
-40°
-50°
-60°

Above: The Pleiades cluster contains dozens of bright blue-white stars visible in binoculars. Large backyard telescopes show its faint reflection nebulosity, principally surrounding the star Merope. Photo by Mace Hooley. Left center: The Crab Nebula, M-1, is the sky's best example of a supernova remnant. Photo by Jack B. Newton. Bottom left: NGC-1647 is a large and scattered cluster of 200 stars. Photo by Lee C. Coombs. Bottom right: NGC-1746, another large cluster in Taurus, contains only 20 stars in an area spanning 42'. Photo by Lee C. Coombs.

Taurus

Tau
Tauri

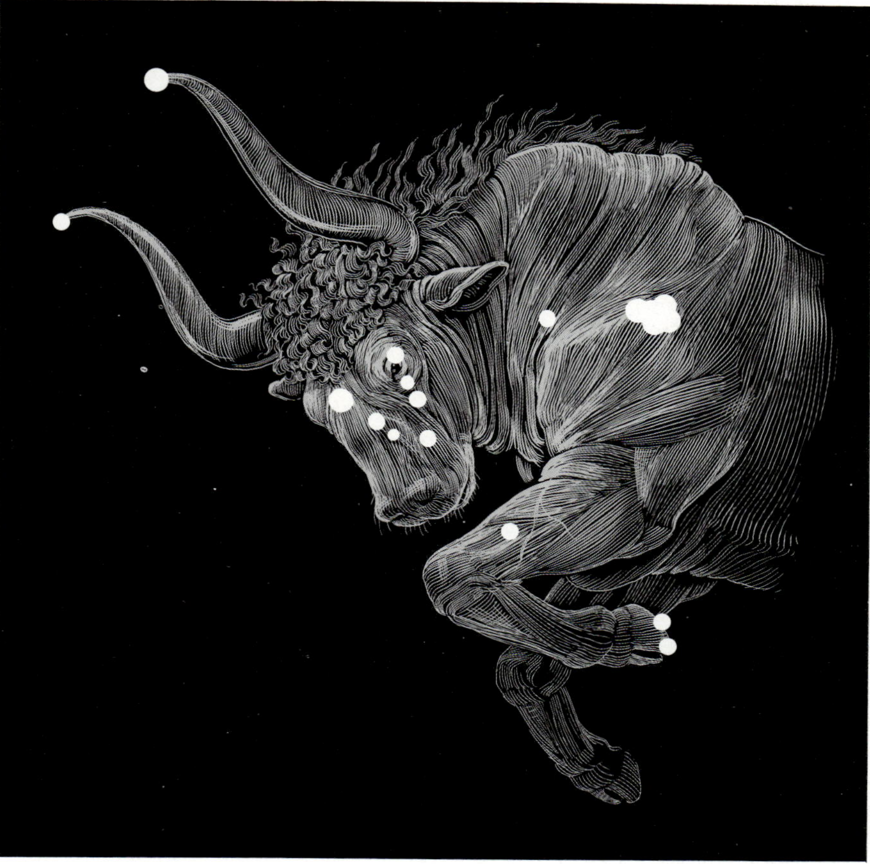

Keith Ward

Lodged between Orion and Perseus in the bright Milky Way is **Taurus** the Bull, a constellation containing two of the sky's brightest star clusters. Along with these, Taurus also contains many unusual variable stars of all types.

Lying at a distance of about 40 parsecs, **the Hyades** is the second closest cluster to us, the nearest being the Ursa Major Moving Group. The Hyades contains several hundred stars, 132 of which are brighter than ninth magnitude, in an area spanning 5.5° across. The clusters' stars come in many colors and there are several fine double stars within the group, making the Hyades a wonderful area of sky to scan with binoculars.

The bright star in this region, **Aldebaran**, is not a member of the cluster but rather a foreground object. It lies at the paltry distance of only 21 parsecs — about halfway to the center of the Hyades. Aldebaran is not only the 13th brightest star in the sky, but it's also a double star. The primary, a K5 subgiant, shines at magnitude 0.9; 30.4″ away in position angle 112° lies its companion — a dwarf M2 star feebly shining at magnitude 13.4. Because of the great difference in brightness between the two stars, spotting the red dwarf is not always easy.

The most brilliant star actually inside the Hyades cluster is Theta² (θ^2) Tauri, a magnitude 3.3 A7 subgiant. Other bright Hyades stars that form the cluster's distinctive "V" shape include Aldeberan, Gamma (γ) Tauri (magnitude 3.7, type G9 subgiant), Delta (δ) Tauri (magnitude 3.8, K0 subgiant), and Epsilon (ε) Tauri (magnitude 3.5, K0 subgiant).

Because the Hyades cluster is so close to us and contains such a broad range of stars, it is a veritable gold mine in the sky for astrophysicists: using proper motion studies, spectroscopy, photometry, and photography, they have gleaned huge amounts of invaluable data from the Hyades cluster.

About 12° northwest of the Hyades is the brilliant **Pleiades** star cluster (Messier 45). Although not as bright or large as the Hyades — magnitude 1.2 *vs.* 0.5 and just under 2° *vs.* 5.5° in size — the Pleiades also contains hundreds of stars and seems brighter because it is more condensed. The brightest stars in the Pleiades are the so-called Seven Sisters from Greek mythology: Alcyone (Eta [η] Tauri), Atlas (27 Tauri), Electra (17 Tauri), Maia (20 Tauri), Merope (23 Tauri), Taygeta (19 Tauri), and Celaeno (16 Tauri).

The Pleiades is a young cluster containing bluish-white stars; the brightest objects in the group are all B-type subgiants or main sequence stars. Several are Be stars showing peculiar emission lines in their spectra. At a distance of about 120 parsecs, or roughly three times that of the Hyades, the stars are not as easily studied as the Hyades, but they provide a wealth of information for astronomers nonetheless.

Since the Pleiades cluster covers nearly 2° of sky, the best way to observe it is with a pair of binoculars, a rich-field telescope, or even a finderscope. These instruments allow you to see the entire cluster at once, affording a good impression of the dipper shape formed by the brightest members. A large telescope will reveal dozens of additional stars, but you'll be limited to eyeing part of the cluster and scanning around to see the whole thing. On dark nights, a large scope may also show faint reflection nebulosity surrounding several of the stars, particularly Merope. This is extremely faint dust caused to shine by reflected starlight from the Pleiades' members. Some observers describe its delicacy as resembling "breath on a mirror."

Taurus abounds with other, albeit smaller, star clusters. One of these is **NGC-1647** a group measuring 45′ across and shining with the light of a magnitude 6.4 star. NGC-1647 would normally appear far more prominent, but its proximity to the bright Hyades and Pleiades clusters causes it to be overlooked by many observers. Containing 200 stars, NGC-1647 is a rich cluster that is easily observable in any telescope. Its brightest stars, which shine at eighth magnitude and are bluish-white in color, appear as fainter versions of the Pleiades stars. Accordingly, NGC-1647 lies at a distance of about 500 parsecs — some 4.5 times farther away than the Pleiades.

Another bright cluster is **NGC-1746**, a group similar in size and brightness to NGC-1647. Covering 42′ of sky and shining at magnitude 6.1, this cluster appears as a "washed-out" version of NGC-1647 in the sense that it has only one-tenth as many stars. Yet the 20 stars

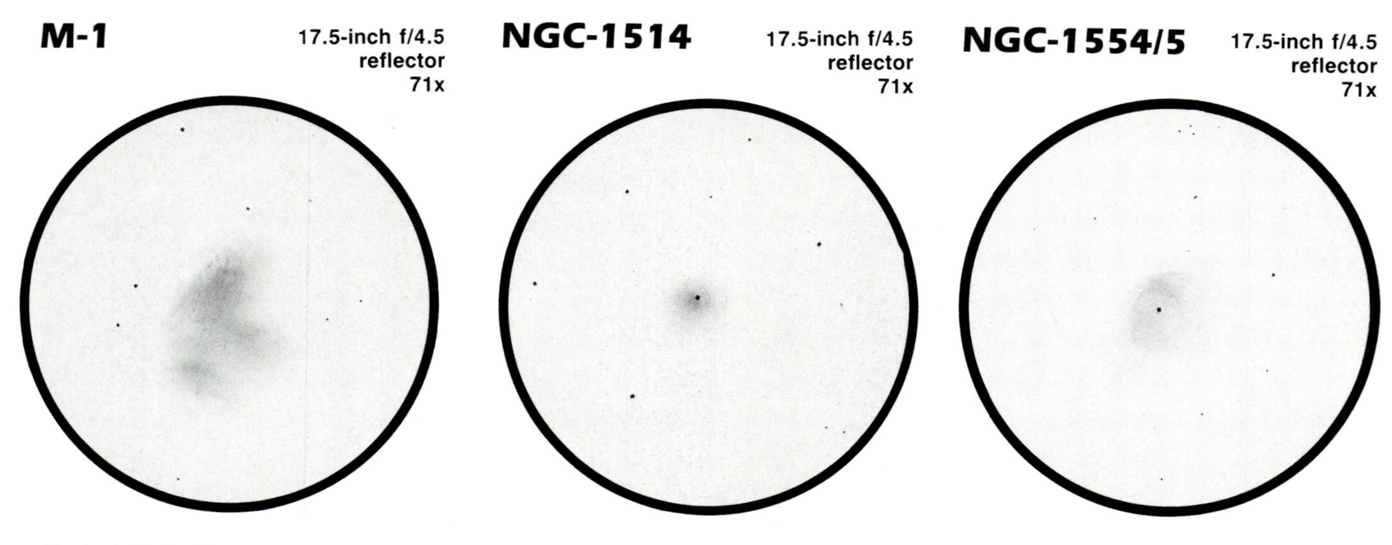

Sketches by David J. Eicher

in NGC-1746 are bright enough to equal the combined magnitude of the 200 stars found in NGC-1647. NGC-1746 lies at a distance of 420 parsecs.

Two smaller clusters found only a few degrees apart are **NGC-1807** and **NGC-1817**. NGC-1807, a seventh-magnitude object, spans 17′ of sky. It contains 20 stars that are eighth magnitude and fainter and presents a fine sight in small telescopes operating at low power. NGC-1817 is slightly fainter and smaller but is also richer, containing three times as many stars. The brightest members of NGC-1817 are 11th-magnitude type A stars. The distance to NGC-1817 is 1.75 kiloparsecs; the distance to NGC-1807 is not well-established.

One of the oddest deep-sky objects in Taurus is the faint, variable reflection nebula **NGC-1554/5**. This low surface brightness object is normally bright enough to spot with a telescope 12 inches or larger. Occasionally this nebula becomes brighter, but normally it appears much like the Pleiades reflection nebulosity — so subtle that the slightest amount of atmospheric haze renders it invisible. NGC-1554/5's variability stems from its illuminating star, **T Tauri** — a young nebular variable enshrouded in a cocoon of warm gas and dust. As the star irregularly varies between magnitudes 9.4 and 13, the nebula gains and loses surface brightness.

Taurus contains several other unusual variable stars. A bright eclipsing variable is **Lambda (λ) Tauri**. This star varies regularly from magnitude 3.4 to 4.1 every 3.95 days when one of its components passes in front of the other. **RV Tauri**, an odd irregular star, fluctuates between magnitudes 9.5 and 13 over a period of some 79 days. **RR Tauri**, another irregular variable, moves between magnitudes 9.9 and 13 in a completely chaotic fashion. **Zeta (ζ) Tauri** varies quite irregularly from magnitude 2.9 to 3.0 because it is a so-called shell star, a hot young star encapsulated by a warm cloud of gaseous nebulosity.

The lone bright planetary nebula in Taurus is **NGC-1514**, an object that shines at photographic magnitude 10 and covers a semicircular arc some 1.9′ in diameter. NGC-1514's surface brightness is rather low and the nebula surrounds a ninth-magnitude central star. These factors conspire to make observing NGC-1514 difficult with very small instruments.

Taurus contains the sky's best example of a supernova remnant — **M-1**, the Crab Nebula (NGC-1952). The Crab, so named by Lord Rosse for its wispy filamentary structure, is the visible remains of a 1054 A.D. supernova documented by Chinese and Amerindian astronomers. Small telescopes show the Crab Nebula as an 8′ x 6′ oval glow without structure; large backyard instruments show an irregular shape and changing surface brightness over the nebula's face. The Crab's magnitude is about 9, making it visible as a fuzzy spot even in finderscopes under a dark, moonless sky.

Object	M#	Type	R.A. (2000)	Dec.	Mag.	Size/Sep./Per.
Pleiades	M-45	⊙	3h 47.0m	+24°07′	1.2	110′
Lambda (λ)		EV	4h 00.7m	+12°29′	3.4↔4.1	3.95d
NGC-1514		■	4h 09.2m	+30°47′	10p	1.9′
NGC-1554/5		□R	4h 21.8m	+19°32′	—	var.
T		IV	4h 22.0m	+19°32′	9.4↔13	irr.
Hyades		⊙	4h 27.0m	+16°00′	0.5	330′
Alpha (α)		★²	4h 35.9m	+16°31′	0.9, 13.4	30.4″
NGC-1647		⊙	4h 46.0m	+19°04′	6.4	45′
RV		IV	4h 47.1m	+26°11′	9.5↔13	79d
NGC-1746		⊙	5h 03.6m	+23°49′	6.1p	42′
NGC-1807		⊙	5h 10.7m	+16°32′	7.0	17′
NGC-1817		⊙	5h 12.1m	+16°42′	7.7	16′
NGC-1952	M-1	SNR	5h34.5m	+22°01′	8.5	8′x6′
Zeta (ζ)		★ˢ	5h 37.6m	+21°09′	2.9↔3.0	irr.
RR		IV	5h 39.5m	+26°23′	9.9↔13	irr.

⊙	*Open Star Cluster*
■	*Planetary Nebula*
□R	*Reflection Nebula*
SNR	*Supernova Remnant*
★²	*Double Star*
★ˢ	*Shell Star*
EV	*Eclipsing Variable*
IV	*Irregular Variable*

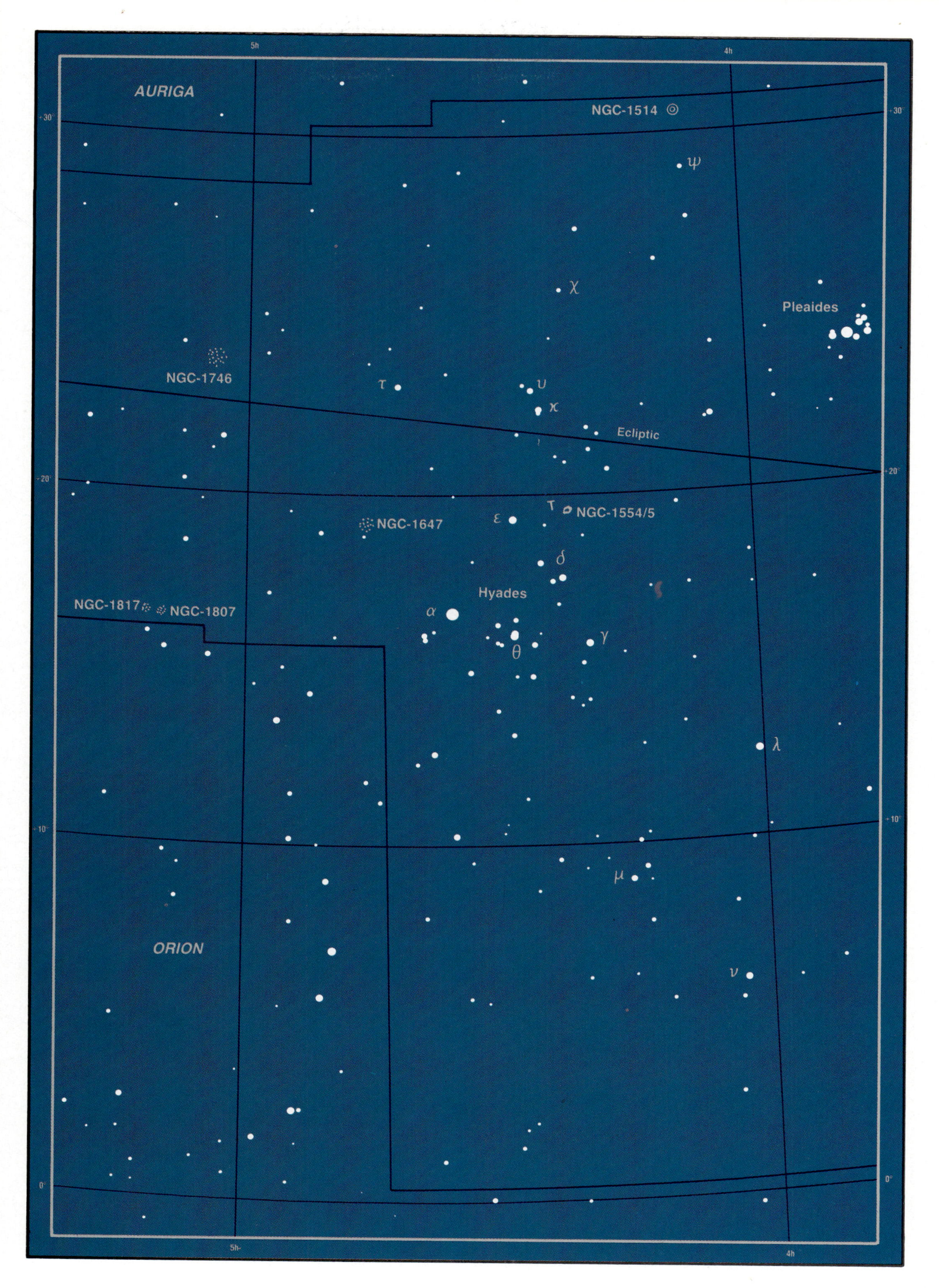

5h
4h
AURIGA
NGC-1514
30°
ψ
χ
Pleaides
NGC-1746
τ
υ
κ
Ecliptic
20°
ε
NGC-1647
NGC-1554/5
δ
Hyades
NGC-1817
NGC-1807
α
γ
θ
λ
10°
μ
ORION
ν
0°
5h
4h

Above: NGC-2841 is an Sb-type spiral galaxy that has had three supernovae recorded in it — in 1912, 1957, and 1972. Photo by Martin Germano. Above right: M-81 and M-82 form a pretty pair in a low-power field of view. M-81 is a classic Sb-type spiral, while M-82 is an intriguing irregular. Photo by Jack Newton. Below: Another low-power pair is the Owl Nebula (M-97), a planetary nebula, and its edge-on spiral galaxy neighbor, M-108. Photo by Ronald Royer. Below right: M-101 is a large, open spiral galaxy that has also had several supernovae, in 1909, 1951, and 1970. A dim blur in small scopes, it displays bright knots in an amorphous field with instruments 12 inches or more in aperture. Photo by K.A. Brownlee.

Ursa Major

Ursae Majoris
UMa

The large spring constellation **Ursa Major**, the Great Bear, is best known for the bright pattern of seven stars that forms the Big Dipper asterism. These stars do not appear together in the sky by chance; they are members of a related group of stars called the **Ursa Major Moving Cluster**, the nearest open star cluster. Consisting chiefly of main sequence stars of type A0 through K3, the cluster occupies an ellipsoidal volume of space about 10 parsecs long and 6 parsecs wide. Some 17 stars — including Alpha (α) Coronae Borealis — are probable members. The group lies about 25 parsecs distant.

The most observed of these stars is **Mizar** (Zeta [ζ] Ursae Majoris), perhaps the most popular double star in the entire sky. Mizar is the middle star in the Dipper's handle and with its little neighbor Alcor — some 11.8′ away in position angle 72° — has served as a test of naked-eye vision for centuries. Should you split the two stars unaided by optical instruments, you can rest assured your eyesight is in good shape. Alcor is not physically associated with Mizar; the stars constitute a mere optical double. A telescope, however, shows that Mizar itself is composed of twin white stars. The two stars, which are gravitationally bound, shine at magnitudes 2.5 and 4.0; they are separated by 14.4″ in p.a. 151°, and each has a spectral type of A2.

Another double star in Ursa Major has the distinction of being the only non-"nebular" Messier object. This is **M-40**, one of Messier's biggest mistakes: the astronomer Johannes Hevelius had recorded a nebula at the position in 1660; Messier observed only two faint stars close together but nevertheless included the entry in his catalog of nebulous objects. The double star, which more recently acquired the name Winnecke 4, has magnitudes of 9.0 and 9.3, a separation of 50.1″ in p.a. 83°, and a primary star spectrum of G0. This double lies just 70′ northeast of 70 Ursae Majoris, and the faint galaxy NGC-4290, a barred spiral of 13th magnitude, lies nearby. This galaxy could not have been recorded as Hevelius' nebulosity in 1660, however.

The star **R Ursae Majoris** is a fine example of a long-period variable; it fluctuates in brightness slowly and methodically between magnitudes 6.7 and 13.4 over a 302-day period. In 1853 the English astronomer N. Pogson found that this star varies in luminosity. It is a pulsating red giant about 450 parsecs distant and is receding at a rate of 13 kilometers per second. By observing R UMa's brightness with respect to nearby stars every couple of weeks, after a year's observing, you'll be able to draw a graph of the star's light curve.

Ursa Major is not well-known for its stars, but rather for the hordes of galaxies that lie within its boundaries. Looking at the Big Dipper, we are peering out through a thin part of the Milky Way's disk — toward the north galactic pole — which minimizes the Galaxy's dusty interference and reveals bright spirals and barred spirals.

A group of fine galaxies surrounds the bright Sb spiral **M-81** (NGC-3031), one of the best handful of galaxies in the sky. M-81 covers a sky area of 18.0′ x 10.0′ and shines at magnitude 7.9, making it visible in finderscopes. It is the primary member of the M-81 group, a cluster of about a dozen galaxies, which at a distance of 2.3 megaparsecs (Mpc), may be the nearest group of galaxies to the Local Group. **M-82** (NGC-3034), an edge-on irregular galaxy of high surface brightness, and **NGC-2976**, a small, bright Sc/Sd galaxy with an irregular spiral pattern, are also members. M-82 has an integrated magnitude of 9.3 and measures 8.0′ x 3.0′ across, whereas 2976 glows at magnitude 10.9 and spans only 3.4′ x 1.3′. In 20x80 binoculars, M-81 and M-82 appear in a single field of view, separated by 38′. In a 4-inch telescope at low power, M-81 appears as a bright oval haze without detail and M-82 shows a slim grey needle of uniform light. An 8-inch scope with high power reveals a huge low-surface-brightness halo of nebulosity around M-81 and dusty patches crossing M-82's sharp surface. M-82 shows a highly condensed nucleus at high power. The little galaxy NGC-2976 lies 84′ south-southwest of M-81; it displays a bright oval without detail in an 8-inch scope.

Another Messier galaxy in Ursa Major, about 5.5° east of Mizar, is **M-101** (NGC-5457), a giant Sc-type spiral measuring 22.0′ x 20.0′ across and radiating at magnitude 8.2. Since it is face-on to our line of sight, M-101 has a low surface brightness, which makes small parts of it appear dim. On a dark, crisp night, an 8-inch telescope shows a bright

BEST VISIBLE DURING
SPRING

M-97

M-81 and M-82

8-inch f/10 SCT

170x

35x

Sketches by David J. Eicher

condensed core surrounded by a halo of hazy nebulosity, which fills an entire low-power field and includes several knotty patches. Under ideal skies, small low focal ratio telescopes — coupled with wide field oculars — can reveal a weak spiral shape in the elusive halo of nebulosity. At a distance of 5 Mpc, M-101 is one of the closest spiral galaxies to our own spiral Milky Way.

About 2.4° southeast of Merak (Beta [β] Ursae Majoris) is the fine planetary nebula **M-97** (NGC-3587), commonly known as the Owl Nebula. This is one of only four Messier planetaries, and one of the faintest objects in the Messier catalog; it is, however, visible as a dim circular disk in 3-inch telescopes. The name Owl Nebula originated with Lord Rosse, who, over a century ago, first detected two dark patches superimposed on the face of the nebula. On photographs and with large telescopes, these patches resemble eyes staring from the inky blackness of space. Under good conditions an 8-inch telescope at high power can show the eyes, but a 12-inch telescope is normally required to see them normally. M-97's central star is a 14th magnitude object, making it elusive in anything less than 12-inch scopes. The Owl Nebula measures 203″ in diameter and glows at 12th magnitude. Just 48′ northwest of the Owl Nebula lies the edge-on galaxy **M-108** (NGC-3556). Both objects fit neatly into low-power fields of view, the galaxy appearing as an 8′ long streak of grey-green nebulosity. Its total magnitude is 10.7, giving it a fairly high surface brightness; thus it can withstand high magnification, where it sometimes shows tiny dark structures resembling those in M-82.

The bright, theta-shaped barred spiral **M-109** (NGC-3992) lies a mere 40′ southeast of Phecda (Gamma [γ] Ursae Majoris), and is visible as a hazy nebulosity in the same low-power field. To best observe the galaxy, however, switch to higher magnifications and move the bright star well out of the field; this will eliminate the star's overpowering glare. With a 6-inch telescope you'll see a small sharp nucleus surrounded by a 6.4′ x 3.5′ area of mottled nebulosity. The galaxy's total magnitude is 10.6; it is bright enough that large telescopes begin to show hints of structure on nights of good transparency. Southwest of M-109 is another bright spiral, **NGC-3953**. This object measures 6.0′ x 2.8′ and shines at magnitude 10.8, making it easily visible as a small patch of grey light in a 2-inch refractor. Larger telescopes 6 and 8 inches in aperture reveal subtle low-surface-brightness nebulosity within the spiral, suggesting tightly wound spiral arms. Far south of the M-109/NGC-3953 area is the nice Sb/Sc-type spiral **NGC-4051**, renowned for its thick spiral arms, which are invisible to backyard telescopes. The galaxy, which shines at magnitude 10.9 and measures 4.2′ x 3.0′ in diameter, does show a condensed nucleus and faint outer haze under sufficiently dark skies.

Three fine galaxies inhabit the far western reaches of Ursa Major: **NGC-2841**, **NGC-3079**, and **NGC-3184**. NGC-2841 is large and bright, an easy target for binoculars and finderscopes. Its magnitude of 10.2 and dimensions of 6.2′ x 2.0′ make its surface brightness fairly high, which in turn makes it an impressive galaxy in small and large telescopes. It is a type Sb system with tightly wrapped multiple arms, which appear in backyard scopes as a mottled oval haze encapsulating a large bright central hub. Telescopes 6 inches in aperture show finely detailed dust patches on the outer parts of the haze under very good skies — these patches are routinely visible with 16-inch telescopes.The barred spiral NGC-3079 is a fine example of an edge-on galaxy, glowing at magnitude 11.2 and measuring a thin 8.0′ x 1.0′ across. Observationally it fits into the same class as M-82, but shows less detail. NGC-3184 is a type-Sc spiral measuring 5.5′ x 5.5′ and shining at magnitude 10.4; it lies in the field of Mu (μ) Ursae Majoris and shows as a low-surface-brightness haze surrounding a relatively bright core.

Object	M#	Type	R.A. (2000)	Dec.	Mag.	Size/Sep./Per.	H
NGC-2841		∮	9h 22m	+50°59′	10.2	6.2′x2.0′	Sb
NGC-2976		∮	9h 47m	+67°54′	10.9	3.4′x1.3′	Sc/Sd
NGC-3031	M-81	∮	09h 56m	+69°04′	7.9	18.0′x10.0′	Sa/Sb
NGC-3034	M-82	#	09h 56m	+69°42′	9.3	8.0′x3.0′	Ipec
NGC-3079		∮B	10h 02m	+55°43′	11.2	8.0′x1.0′	SBb
NGC-3184		∮	10h 18m	+41°25′	10.4	5.5′x5.5′	Sc
R UMa		LPV	10h 45m	+68°46′	6.7↔13.4	302d	
NGC-3556	M-108	∮	11h 12m	+55°41′	10.7	7.8′x1.4′	Sc
NGC-3587	M-97	■	11h 15m	+55°02′	12.0	180″	
NGC-3953		∮B	11h 54m	+52°20′	10.8	6.0′x2.8′	SBb
NGC-3992	M-109	∮B	11h 58m	+53°22′	10.6	6.4′x3.5′	SBb
NGC-4051		∮	12h 03m	+44°31′	10.9	4.2′x3.0′	Sb/Sc
Winnecke 4	M-40	★²	12h 22m	+58°05′	9.0,9.3	50.1″	
Mizar (ζ)		★²	13h 24m	+54°55′	2.5,4.0	14.4″	
NGC-5457	M-101	∮	14h 03m	+54°21′	8.2	12.0′x20.0′	Sc

H = Hubble classification type for galaxies

★²	*Double Star*
LPV	*Long Period Variable*
■	*Planetary Nebula*
∮B	*Barred Spiral Galaxy*
∮	*Spiral Galaxy*
#	*Peculiar Galaxy*

14h
13h
12h
11h
10h
M-82
M-81
NGC-2976
DRACO
+60°
α
R
M-40
Alcor
Mizar
ε
δ
β
M-108
M-97
γ
M-109
NGC-3953
+50°
χ
NGC-4051
ψ
CANES VENATICI
+40°
12h
11h

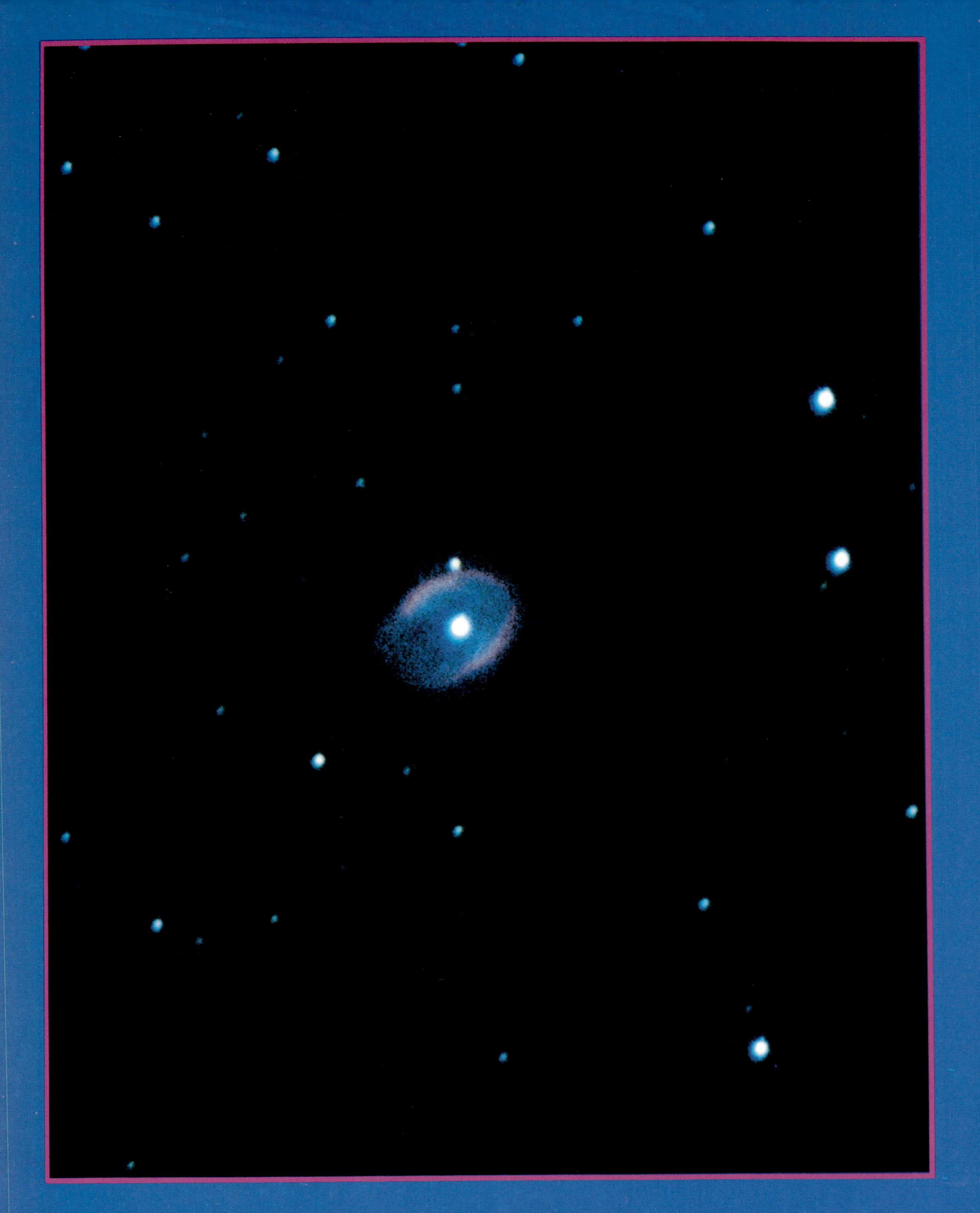

NGC-3132 is known as the "Eight-Burst Nebula" because of its complex multi-ring structure. In backyard telescopes it appears as a bright disk of nebulosity surrounding a 10th-magnitude central star. Photo by Jack B. Marling.

Vela

Vel
Velorum

Pyxis

Pyx
Pyxidis

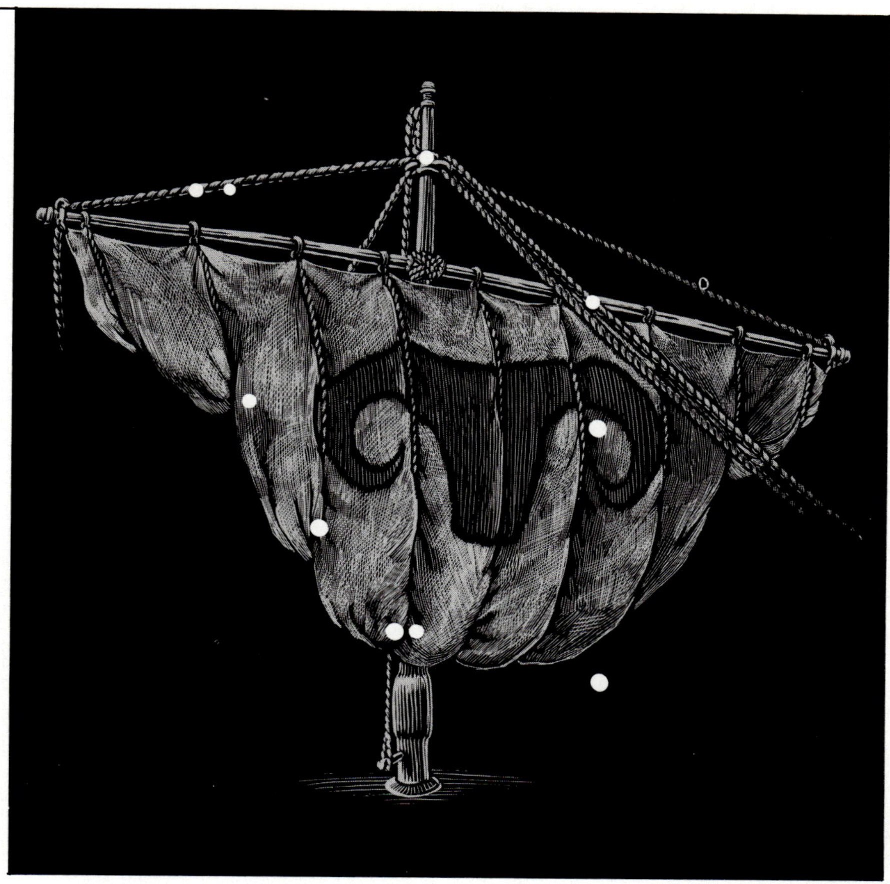
Keith Ward

Vela the Sails, a brilliant constellation containing more than four dozen naked-eye stars, lies square on the galactic equator in the thick of the southern Milky Way. To the south of Vela is the bright constellation Carina — the giant Eta Carinae Nebula lies just 3° south of Vela's southern boundary — and north of Vela lies the star-poor region dubbed **Pyxis** the Ship's Compass. All of these constellations were members of the antiquated star group Argo Navis, which was divided into separate constellations in the mid-eighteenth century.

One of Vela's most appealing objects is the bright planetary nebula **NGC-3132**. This object shines at photographic magnitude 8.2, and its central star glows at magnitude 10.1, which make this object one of the brightest planetary nebulae in the sky. To find NGC-3132, look 2.5° northwest of the 4th-magnitude star q Velorum. The nebula is located within several arcminutes of the Vela-Antlia border.

NGC-3132's appearance changes dramatically from telescope to telescope, though it is similar in many ways to the Ring Nebula in Lyra. In a finder scope only the central star is visible, while a longer-focus instrument like a 2-inch refractor shows a small, fuzzy disk of gray light surrounding the pinpoint star. A 6-inch scope at high power shows a broad, bright ring of nebulosity around the star, and some sections of the ring appear brighter than others. Astronomers have nicknamed NGC-3132 the "Eight-Burst Nebula" because its complex structure suggests a group of oval rings stacked on top of one another at differing tilt angles.

Curiously, the 10th-magnitude central star that lies in the ring's center isn't a true central star: a study conducted in 1977 showed that this is not the star illuminating the nebulosity. Instead, a 16th-magnitude dwarf star, some 1.7″ away from the 10th-magnitude sun, is thought to excite the gas to a state of fluorescence.

BEST VISIBLE DURING SPRING

About 7° south and 1.5° east of the Eight-Burst Nebula is a little-known globular cluster designated **NGC-3201**. This object is large and bright, spanning 18.2′ and glowing with the light of a magnitude 6.8 star. Unfortunately, it has received little attention from many Northern Hemisphere observers because of its southern declination. NGC-3201 is an extremely loose-structured globular that appears in a 6-inch telescope as a 5′-diameter clump of faint stars devoid of the rich central condensation characteristic of a typical globular cluster. The noted Southern Hemisphere observer E. J. Hartung described NGC-3201's stars as lying in "short curved rays like jets from a fountain." How many tiny stars can your telescope resolve in this unusual cluster during a night of particularly steady seeing and superb transparency?

Continuing another 6° south we come to the large open star cluster **NGC-3228**, a sparse group of fifteen stars splashed across a brilliant stellar background some 18′ across. Just visible to the naked eye as a 6th-magnitude patch of light, NGC-3228 is a fine object for observers with binoculars. Its brightest star shines at magnitude 7.9, while the rest dimly glow at 10th- and 11th-magnitude and fainter.

Another fine open cluster, **IC 2488**, lies on the southern border of Vela 1° west of N Velorum. Much richer than NGC-3228, this group of seventy faint stars has a combined photographic magnitude of 7.4 over the cluster's 15′ diameter. IC 2488's brightest stars are 10th-magnitude objects, so small telescopes show this cluster as a large haze peppered with a few resolved points of light. Ten-inch and larger scopes resolve several dozen stars and give a brighter overall image of the cluster, though they still show a milky nebulous veil strewn across the field of view.

The second brightest star in Vela (only Gamma [γ] Velorum is brighter) is the double star **Delta (δ) Velorum**, which is a fine target for small telescopes. Located in the southwestern corner of Vela practically on the constellation's border with Carina, the star forms a triangle with bright stars Kappa (κ) and N Velorum. The primary star in this binary system is a magnitude 2.1 yellow luminary and the secondary is a yellowish magnitude 5.1 star. The two suns are separated by 2.6″ in position angle 153° and are moving in a slow retrograde direction.

Just 2.5° northwest of Delta Velorum is the biggest, brightest open cluster in the constellation, **IC 2391**. This brilliant group of stars includes the magnitude 3.6 star Omicron (ο) Velorum and thirty others measuring nearly 1° across. The combined magnitude of all this light is 2.5, which means that IC 2391 is an easy naked-eye object even in relatively light-polluted

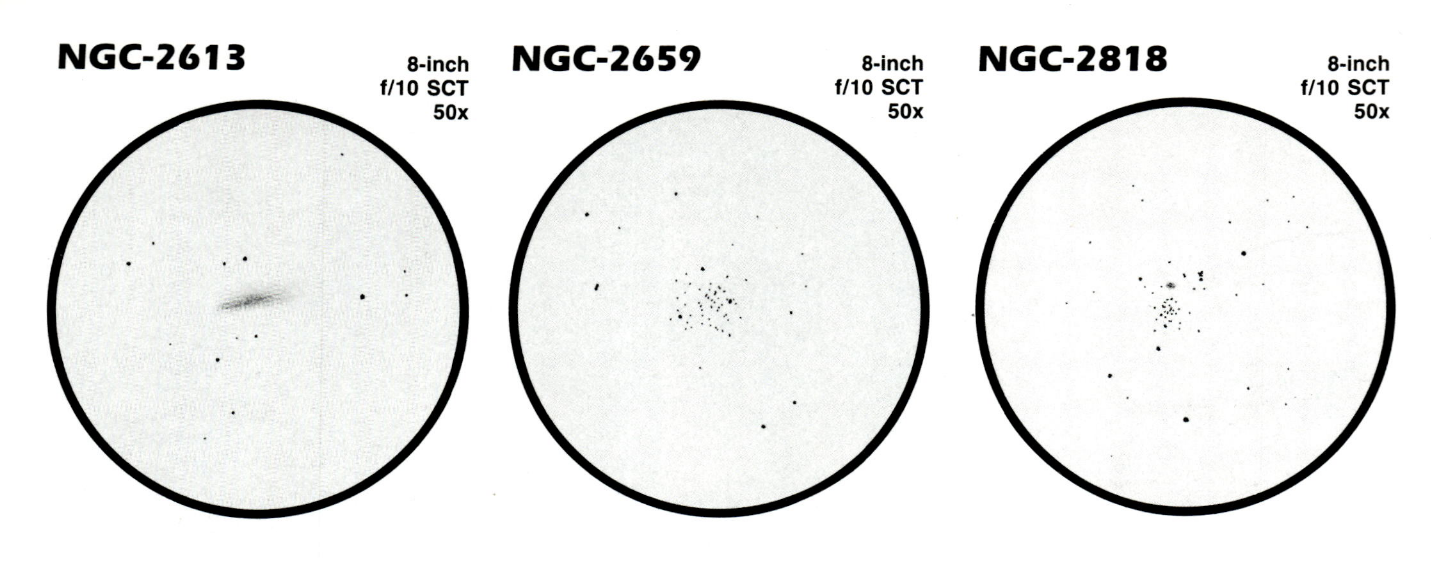

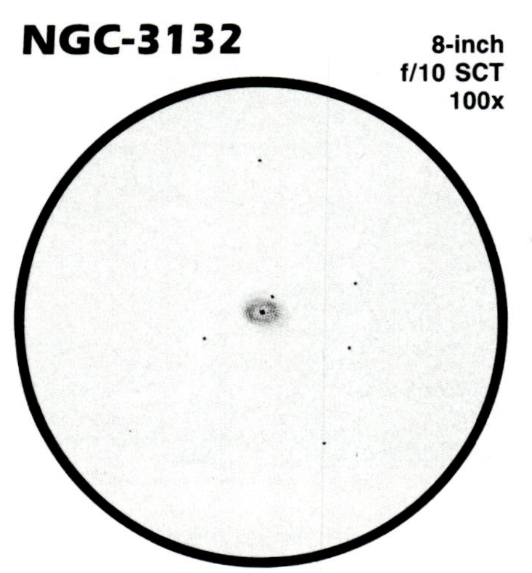

Sketches by David J. Eicher

areas. Any optical aid shows the smattering of stars as a big, coarse group of bright white stars and numerous faint stars scattered over a rich Milky Way field that is somewhat reminiscent of the Pleiades cluster in Taurus. The "Southern Pleiades" is one of the finest fields for binocular and small telescope viewers — especially star cluster devotees — and should be observed by everyone capable of seeing as far south as declination −53°.

Ten degrees north of the IC 2391 area is a smaller open cluster designated **IC 2395**. This is a rich cluster that contains forty stars in an area only 8′ across, which yield an integrated magnitude of 4.6. Easily a naked-eye object under a dark sky, IC 2395 appears as a slightly fuzzy "star" due to its small size. In finder scopes and binoculars its nature is more apparent, and in a 6-inch scope at high power it shows as a patch of nebulosity intermixed with faint stars that gives the impression of spilled powdered sugar on black velvet. This object is even more impressive when viewed with large backyard telescopes, which show it as a rich, resolved collection of three dozen faint stars floating in Vela's sparkling Milky Way background.

The final open cluster in Vela is the pretty group **NGC-2547**, an assemblage of eighty stars inside a diameter measuring 20′ across. A naked-eye object at magnitude 4.7, this star cluster offers a fine assortment of rings, chains, and arcs of stars.

Though the little constellation Pyxis offers less than its grandiose neighbor to the south, it does contain the open cluster **NGC-2818**. This object shines at magnitude 8.2 and spans some 9′; it contains forty faint stars in its gravitational grasp. A 13th-magnitude planetary nebula 40″ in diameter lies on the western edge of this cluster and is visible as a smudge of gray light at high power in backyard telescopes over 12 inches in aperture.

The fine galaxy **NGC-2613**, located in northern Pyxis some 3° southwest of a 6th-magnitude double star, concludes our survey of Vela and Pyxis. This spiral shines at magnitude 10.4 and covers 7.2′ by 2.1′, which make it a bright, elongated splash of nebulosity in small scopes. It is heavily obscured by dust in the Milky Way and would doubtlessly be more impressive if it lay well away from the plane of our galaxy.

Object	Type	R.A. (2000)	Dec.	Mag.	Size/Sep./Per.	N★	Mag.★
NGC-2547	⊙	8h 10.7m	−49°16′	4.7	20′	80	6.5
AI Vel	CV	8h 14.1m	−44°34′	6.4↔7.1	0.11d		
NGC-2613	§	8h 33.4m	−22°58′	10.4	7.2′x 2.1′		
NGC-2626	□ER	8h 35.6m	−40°40′	—	5′		
IC 2391	⊙	8h 40.2m	−53°04′	2.5	50′	30	3.6
IC 2395	⊙	8h 41.1m	−48°12′	4.6	8′	40	5.5
NGC-2659	⊙	8h 43.4m	−32°39′	9.2_p	12′	80	12_p
Delta (δ) Vel	★^{2}	8h 44.7m	−23°47′	2.1,5.1	2.6′		
T Pyx	IV	9h 04.7m	−32°23′	6.3↔14.0	7000:d		
NGC-2818	⊙	9h 16.0m	−36°37′	8.2	9′	40	11.3
IC 2488	⊙	9h 27.6m	−56°59′	7.4_p	15′	70	10_p
NGC-3132	■	10h 07.7m	−40°26′	8.2_p	47″		10.1
NGC-3201	●	10h 17.6m	−46°25′	6.8	18.2′		
NGC-3228	⊙	10h 21.8m	−51°43′	6.0	18′	15	7.9

★^{2}	*Double Star*
IV	*Irregular Variable Star*
CV	*Cepheid Variable Star*
⊙	*Open Cluster*
●	*Globular Cluster*
■	*Planetary Nebula*
□ER	*Emission and Reflection Nebula*
§	*Spiral Galaxy*

N★ = number of stars, Mag.★ = magnitude range of cluster or magnitude of central star

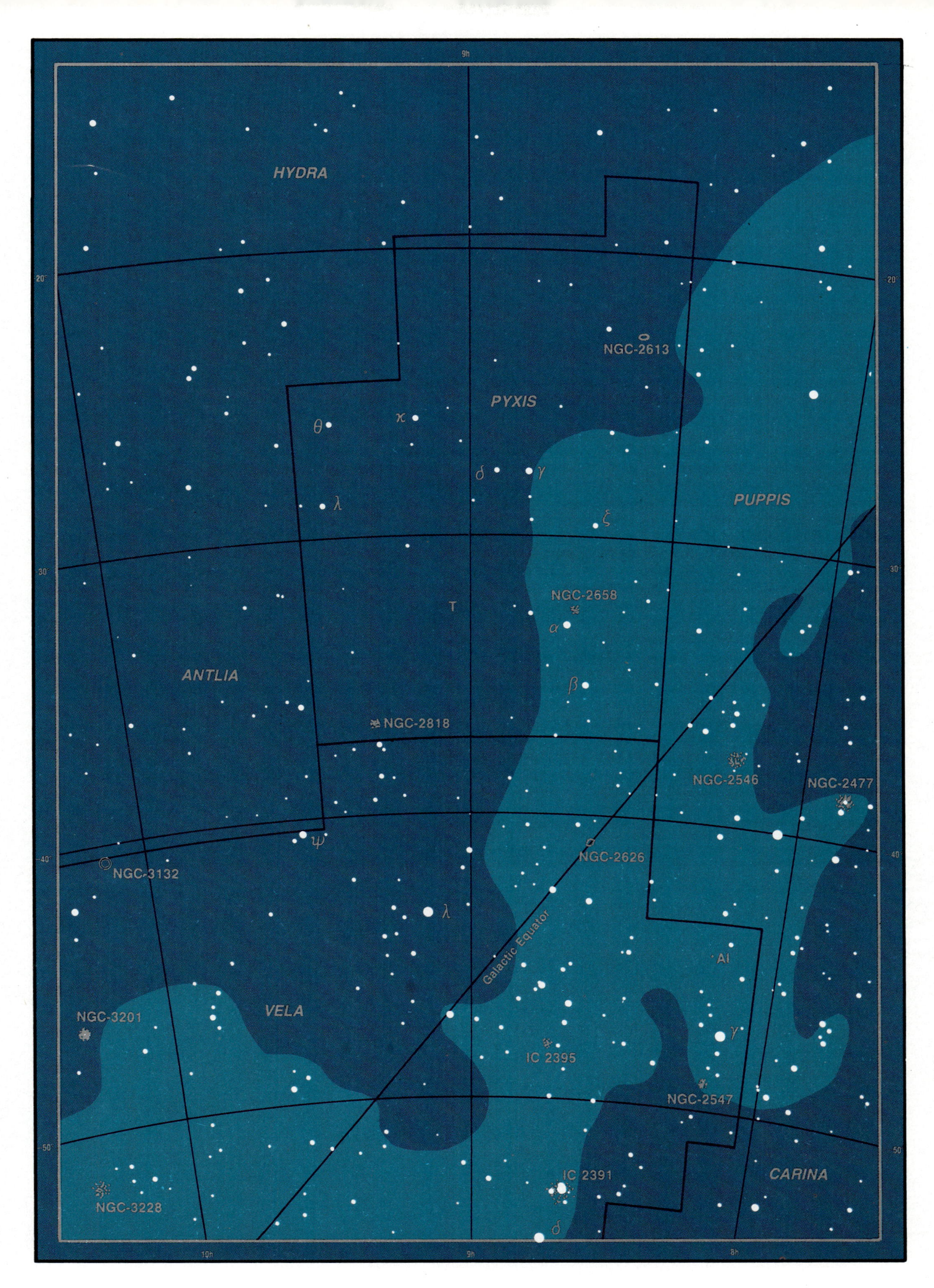
HYDRA
PYXIS
PUPPIS
ANTLIA
VELA
CARINA
NGC-2613
NGC-2658
NGC-2818
NGC-2546
NGC-2477
NGC-2626
NGC-3132
NGC-3201
IC 2395
NGC-2547
IC 2391
NGC-3228
Galactic Equator
9h
10h
8h
20°
30°
40°
50°

Top left: The Sombrero Galaxy, M-104, is a bright edge-on spiral presenting a broad dark lane to telescopes of 3-inches' or more aperture. Bottom left: Virgo's grand giant elliptical galaxy is M-87, a circular fuzzball in small scopes. Top: You'll need at least an 8-inch scope to detect M-61's subtle face-on spiral structure. Above: NGC-4535 is a face-on barred spiral shining at 10th magnitude. Photos by Jack Newton.

Virgo

Vir
Virginis

Virgo the Virgin is a galaxy-hunter's paradise; this constellation contains many bright galaxies visible in backyard telescopes. Virgo contains spirals, barred spirals, lenticulars, ellipticals, peculiars — you name it, it's there.

Why is Virgo filled with so many galaxies? This area of sky contains the so-called Virgo cluster of galaxies, the densest concentration of matter in our corner of the universe. This cluster, of which the Milky Way is a remote member, is in turn part of a gigantic group of galaxies called the Local Supercluster, a vast collection of relatively nearby groups of galaxies. Suffice it to say that Virgo's many galaxies will undoubtedly be some of the best visible in your backyard telescope.

Although many of the galaxies in Virgo are large and bright, the center of the Virgo cluster is actually quite distant as groups of bright galaxies go. Redshift measurements vary from galaxy to galaxy in the Virgo cluster, but it seems that the group's center lies at a distance of about 15 megaparsecs.

The best place to start observing is right in the heart of the Virgo cluster, practically on Virgo's northern border with Coma Berenices. Here lie **M-84** (NGC-4374) and **M-86** (NGC-4406), two bright elliptical galaxies that are visible together in a low-power field of view. Separated by 17′, these galaxies appear very similar in most telescopes although M-86 is slightly bigger and brighter than M-84, shining at magnitude 9.2 and measuring 7.4′ x 5.5′ in angular extent. Small scopes show both galaxies as ovals of grayish light with bright centers surrounded by fainter wispy halos of low surface brightness nebulosity. If you observe with a medium- or large-size telescope you'll see several more streaks of light in this same field.

BEST VISIBLE DURING
SPRING

These are the elliptical galaxy NGC-4435 and the spiral galaxies NGC-4388, NGC-4402, and NGC-4438.

About 1.5° southeast of the M-84/M-86 pair is the giant elliptical galaxy **M-87** (NGC-4486), an object of great interest to astrophysicists. As one of the largest and most luminous galaxies known, M-87 is also a strong source of radio and x-ray emission. This intense outpouring of energy corresponds to a luminous jet of material spiraling out from the galaxy's center. This jet was discovered photographically by Heber D. Curtis at Lick Observatory in 1918. Astronomers believe that the intensely energetic material shooting out from M-87's nucleus, and from the nuclei of many other "active" galaxies, results from matter being thrust into a supermassive black hole in the galaxy's core.

Observationally, M-87 looks like a brighter version of M-86; it shines at magnitude 8.6 and measures 7.2′ x 6.8′ across. The jet is far too faint to observe with backyard telescopes, but it can be photographed with large amateur scopes employing large image scales and long exposure times. Within the same low-power field as M-87 are two fainter elliptical galaxies, NGC-4476 and NGC-4378.

Some 1.5° to the northwest of the M-84/M-86 pair is the bright, highly-inclined spiral **NGC-4216**, which lies virtually on top of the Coma Berenices border. At magnitude 10, NGC-4216 is bright enough to spot as a streak of fuzzy light in large finder telescopes on dark nights; it is also easily observable in 2-inch and 3-inch refractors as a smudge of grainy nebulosity. Larger telescopes reveal a bright oval center surrounded by an 8.3′ x 2.2′ haze of very faint light. Inside the bright oval nuclear bulge is a condensed spot of light marking the nucleus. A bright star lies less than 1′ due east of the galaxy's nucleus.

Several interesting galaxies lie east of M-87. **M-90** (NGC-4569) is a bright spiral visible as a featureless, oval patch in small scopes. This galaxy measures 9.5′ x 4.7′ across and shines brightly at magnitude 9.5. A 6-inch telescope at high power shows M-90 as an intensely bright oval of milky nebulosity encased in a huge halo of fainter light. M-90's nuclear bulge is far brighter than its spiral arms. On steady nights, large backyard telescopes show a hint of detail in this galaxy revealing that its arms are mottled with dark nebulosity.

Forty-five arcminutes southwest of M-90 is the diminutive elliptical galaxy **M-89** (NGC-4552). This magnitude 9.8 galaxy is a round object measuring 4.2′ across; it is featureless and rather unexciting in most backyard telescopes. But nearly 1° southeast of M-89 is a much finer galaxy for backyard observers — **M-58** (NGC-4579). This object is a large and bright Sb-type spiral visible as a near-circular haze in small scopes. A 10-inch or 12-inch scope shows M-58 as a 10th-magnitude blob some 5.4′ x 4.4′ in extent with a bright nucleus wrapped in a featureless haze. Although M-58 isn't terrifically detailed, it's large and bright enough to be impressive through most backyard scopes.

Some 30′ southwest of **M-58** is a curious pair of interacting galaxies known as **NGC-4567** and **NGC-4568**, popularly called the Siamese Twins. Each galaxy shines at about magnitude 11 and measures some 3′ x 2′ across, making the pair easy to spot with 4-inch or larger telescopes. Although the two may appear as a single, very oddly shaped galaxy, this is in fact one of the sky's few bright pairs of physically connected galaxies.

About 2° east of the Siamese Twins are two close neighbors, **M-59** (NGC-4621) and **M-60** (NGC-4649). Both are large and bright ellipticals that appear rather similar in the eyepiece. M-60 is brighter by a magnitude and covers a bit more sky. It also has a small elliptical companion — NGC-4647 — just 4′ northwest in the same field of view. M-59 shines at 10th magnitude and measures 5.1′ x 3.4′ across. It appears as a nebulous ball with a bright, condensed nucleus.

Due south of the Virgo cluster's center is a group of three bright galaxies. **M-49** (NGC-4472), a fine example of an elliptical galaxy, is brighter and slightly larger than the giant elliptical M-87. **NGC-4526** is a 10th-magnitude, lens-shaped elliptical seen nearly edge-on; its dimensions are 7.2′ x 2.3′, causing it to appear as a misty brush stroke of nebulosity. **NGC-4535** is a beautiful example of a face-on barred spiral. Its 10th-magnitude light, however, is spread out into a low surface brightness that makes detail rather elusive.

Further south are three unusual galaxies. **M-61** (NGC-4303) is a 10th-magnitude face-on spiral measuring 6.0′ x 5.5′ across. As viewed in large backyard scopes, M-61 is one of the finest small spiral galaxies in the sky. A dark night will permit you to view traces of its delicate spiral arms and pinpoint nucleus. **NGC-4536**, a 10th-magnitude Sc-type spiral, appears as a bright, oval nuclear bulge surrounded by unevenly illuminated nebulosity. About 1.5° west of NGC-4536 is **3C 273**, the sky's brightest quasi-stellar object, or quasar. 3C 273 appears as a 13th-magnitude dot of light when at maximum (its light output is variable) and is difficult to find. But once you spot the correct bluish dot of light, you're seeing an object that lies over one billion parsecs away!

M-104 **17.5-inch f/4.5 reflector 270x**

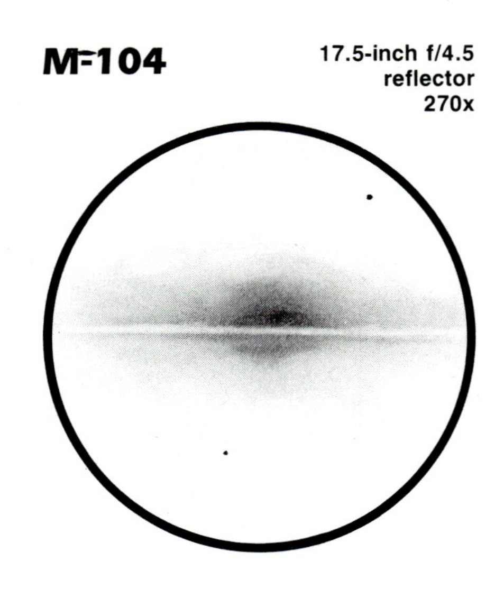

Sketch by David J. Eicher

Far south of this chain of galaxies, extending down from the cluster's center on Virgo's southern border with Corvus, is the finest galaxy in the constellation — **M-104** (NGC-4594). Known as the Sombrero Galaxy because of a prominent dark lane cutting across it, M-104 appears edge-on to our line of sight. The galaxy glows brightly at 8th magnitude and measures 8.9′ x 4.5′ across. M-104 is easy to observe with most any telescope and appears as a flattened saucer bisected by the dust lane. Large scopes reveal a bright disk in the galaxy's center separated by a huge semicircle of nebulosity extending outward on each side. The Sombrero Galaxy, on a good night, is quite a sight.

Object	M#	Type	R.A. (2000)	Dec.	Mag.	Size/Sep./Per.	H
NGC-4216		§	12h 15.9m	+13°09′	10.0	8.3′x2.2′	Sb
NGC-4303	M-61	§	12h 21.9m	+ 4°28′	9.7	6.0′x5.5′	Sc
NGC-4374	M-84	()	12h 25.1m	+12°53′	9.3	5.0′x4.4′	E1
NGC-4406	M-86	()	12h 26.2m	+12°57′	9.2	7.4′x5.5′	E3
3C 273		QSO	12h 29.1m	+ 2°03′	12.9v	stellar	
NGC-4472	M-49	()	12h 29.8m	+ 8°00′	8.4	8.9′x7.4′	E4
NGC-4486	M-87	()	12h 30.8m	+12°24′	8.6	7.2′x6.8′	E1
NGC-4526		()	12h 34.0m	+ 7°42′	9.6	7.2′x2.3′	E7
NGC-4535		§B	12h 34.3m	+ 8°12′	9.8	6.8′x5.0′	S(B)c
NGC-4536		§	12h 34.5m	+ 2°11′	10.4	7.4′x3.5′	Sc
NGC-4552	M-89	()	12h 35.7m	+12°33′	9.8	4.2′x4.2′	E0
NGC-4567		§	12h 36.5m	+11°15′	11.3	3.0′x2.1′	Sc
NGC-4569	M-90	§	12h 36.8m	+13°10′	9.5	9.5′x4.7′	Sb^+
NGC-4579	M-58	§	12h 37.7m	+11°49′	9.8	5.4′x4.4′	Sb
NGC-4594	M-104	§	12h 40.0m	−11°37′	8.3	8.9′x4.1′	Sb^-
NGC-4621	M-59	()	12h 42.0m	+11°39′	9.8	5.1′x3.4′	E3
NGC-4649	M-60	()	12h 43.7m	+11°33′	8.8	7.2′x6.2′	E1
NGC-5170		§	13h 29.8m	−17°58′	11.8_B	8.1′x1.3′	Sb
NGC-5364		§	13h 56.2m	+ 5°01′	10.4	7.1′x5.0′	Sb^+
NGC-5746		§	14h 44.9m	+ 1°57′	10.6	7.9′x1.7′	Sb

H = Hubble classification type for galaxies

§	*Sprial Galaxy*
§B	*Barred Spiral Galaxy*
()	*Elliptical Galaxy*
QSO	*Quasar*

14h
13h
12h
COMA BERENICES
M-90
M-86
M-84
NGC-4216
M-89
M-87
M-59
M-58
M-60
NGC-4567/8
ε
NGC-4535
M-49
NGC-4526
σ
NGC-5364
M-61
δ
NGC-4536
3C 273
ζ
η
γ
θ
Ecliptic
α
M-104
CORVUS
NGC-5170

Bibliography

The following lists books and periodicals containing useful information that may interest backyard astronomers. Most books listed here specifically emphasize observational astronomy, although some are general works including sections on observing the sky.

Ashbrook, Joseph. *The Astronomical Scrapbook.* 468 pp., hardcover. Cambridge University Press and Sky Publishing Corp., Cambridge, 1984. A fascinating collection of historical tales in astronomy, many of which relate to deep-sky pioneers or early deep-sky observing.

ASTRONOMY. Kalmbach Publishing Co., 1027 North 7th St., Milwaukee, Wisconsin 53233. Founded in 1973, ASTRONOMY is the largest-circulation English-language astronomy periodical. Its monthly columns contain much of interest to the deep-sky observer.

Berendzen, Richard, Richard Hart, and Daniel Seeley. *Man Discovers the Galaxies.* 228 pp., paper. Columbia University Press, 1984. A historical detective story of how astronomers unraveled the basic structure of the Milky Way Galaxy, including biographical information on Shapley, Hubble, Einstein, Jeans, Barnard, and many others.

Berry, Richard. *Discover the Stars.* 119 pp., paper. Harmony Books, New York, 1987. This book by the editor-in-chief of ASTRONOMY is an excellent introduction to observing the sky with the naked eye, binoculars, and small telescopes. Included are twelve all-sky maps and twenty-three close-up maps showing groups of interesting objects.

Bok, Bart J., and Priscilla F. Bok. *The Milky Way.* Fifth ed., 356 pp., hardcover. Harvard University Press, Cambridge, Massachusetts, 1981. Generally regarded as the standard semitechnical introduction to our Galaxy, *The Milky Way* is an extremely valuable sourcebook for those interested in the nature of galaxies.

Burnham, Robert Jr. *Burnham's Celestial Handbook.* Three vols., 2,138 pp., paper. Dover Publications, New York, 1978. An enormous constellation-by-constellation listing of stars and deep-sky objects that contains large amounts of information and many photographs and charts.

Couteau, Paul. *Observing Visual Double Stars.* 257 pp., paper. The Massachusetts Institute of Technology Press, Cambridge, Massachusetts, 1982. An excellent introduction to observing double stars.

Covington, Michael A. *Astrophotography for the Amateur.* 168 pp., hardcover. Cambridge University Press, Cambridge, 1985. A fine introduction to astrophotographic techniques for backyard telescopes.

Deep Sky. Kalmbach Publishing Co., 1027 North 7th St., Milwaukee, Wisconsin 53233. Founded in 1977 as *Deep Sky Monthly,* this magazine is now a quarterly devoted entirely to deep-sky observing. *Deep Sky*'s 192 annual pages feature articles, maps, charts, photographs, eyepiece drawings, and observing hints for clusters, nebulae, and galaxies. Departments cover double stars, variable stars, small telescope observing, and the latest research news in astronomy.

Eicher, David J., and the editors of *Deep Sky. Deep Sky Observing with Small Telescopes.* 316 pp., hardcover. Enslow Publishers, Hillside, New Jersey, 1988. An extensive single-volume manual to the historical, astrophysical, and observational aspects of deep-sky objects visible in 2-inch to 6-inch telescopes.

Ferris, Timothy. *Galaxies.* 191 pp., hardcover. Stewart, Tabori & Chang, New York, 1980. *Galaxies* is a folio-sized colorful picture book of galaxies that contains an unusually lucid and informative text on the astrophysical nature of these objects.

Freeman, Lenore. *A Star-Hopper's Guide to Messier Objects.* 23 pp., paper. Everything in the Universe, Oakland, California, 1983. Simple star-hopping charts for constellations containing Messier objects that shows how to find these objects easily.

Hartung, E. J. *Astronomical Objects for Southern Telescopes.* 238 pp., hardcover. Cambridge University Press, Cambridge, 1968. A compendium of high-quality observing notes on objects in the Southern Hemisphere; many are visible from the Northern Hemisphere as well.

Hirshfeld, Alan, and Roger W. Sinnott, eds. *Sky Catalogue 2000.0.* Cambridge University Press and Sky Publishing Corp., Cambridge, 1982-1985. Volume two (356 pp., hardcover) lists fundamental data on thousands of double and variable stars; 750 open clusters; 150 globular clusters; 283 bright nebulae; 150 dark nebulae; 564 planetary nebulae; 3,116 galaxies; and 297 quasars.

Hodge, Paul. *Galaxies.* 174 pp., hardcover. Harvard University Press, Cambridge, Massachusetts, 1986. An excellent introduction to the field of galaxy research and the properties of galaxies, this is a revision of a classic work originally written by Harlow Shapley.

Jones, Kenneth Glyn. *Messier's Nebulae and Star Clusters.* 480 pp., hardcover. Faber and Faber Ltd., London, 1968. A uniformly excellent set of descriptive notes, finder maps, and eyepiece sketches for each Messier object.

Jones, Kenneth Glyn. *The Search for the Nebulae.* 84 pp., hardcover. Alpha Academic, Giles, England, 1975. This book provides an account of observational notes and discussions of deep-sky objects made by the discovers and early observers of these objects.

Jones, Kenneth Glyn, ed. *The Webb Society Deep Sky Observer's Handbook.* Seven vols., 1,334 pp., paper. Enslow Publishers, Hillside, New Jersey, 1979-1987. A compilation of observing notes and sketches on double stars (vol. 1), planetary and gaseous nebulae (vol. 2), open and globular clusters (vol. 3), galaxies (vol. 4), clusters of galaxies (vol. 5), anonymous galaxies (vol. 6), and Southern Hemisphere objects (vol. 7).

Klein, Fred. *The Visibility of Deep-Sky Objects.* Five vols., 283 pp., paper. Klein Publications, Los Altos, California, 1984. A series of booklets offering a "visibility index" for 2,400 deep-sky objects.

Malin, David, and Paul Murdin. *Colours of the Stars.* 198 pp., hardcover. Cambridge University Press, Cambridge, 1984. The authors, professional astronomers at the Anglo-Australian Observatory and the Royal Greenwich Observatory, discuss colors of astronomical objects and the role these colors play in astronomical research.

Mallas, John H., and Evered Kreimer. *The Messier Album.* 216 pp., hardcover. Sky Publishing Corp., Cambridge, Massachusetts, 1978. This book provides a short description, photograph, and sketch for each Messier object.

Mayer, Ben. *Starwatch.* 144 pp., paper. Perigee Books, New York, 1984. A constellation-by-constellation introduction to stars and bright deep-sky objects using photographic maps.

Menzel, Donald, and Jay M. Pasachoff. *A Field Guide to the Stars and Planets.* 473 pp., paper. Houghton Mifflin Co., Boston, 1983. A general introduction containing small maps of the sky and a brief discussion of deep-sky objects.

Mitton, Simon. *Exploring the Galaxies.* 206 pp., paper. Charles Scribner's Sons, New York, 1976. A semitechnical introduction to the field of galaxies and galaxy research.

Mitton, Simon, ed. *The Cambridge Encyclopaedia of Astronomy.* 456 pp., hardcover. Crown Publishers, New York, 1977. Authoritative, precise encyclopedia covering all aspects of modern astronomy.

Murdin, Paul, and David Allen. *Catalogue of the Universe.* 256 pp., hardcover. Crown Publishers, Inc., New York, 1979. Photographically illustrated listing of bright and unusual objects in the sky.

Newton, Jack. *Deep Sky Objects: A Guide for the Amateur Astronomer.* 160 pp., hardcover. Gall Publications, Toronto, 1977. This reference book contains a photograph and map for each Messier object and a handful of bright NGC objects.

Newton, Jack, and Phillip Teece. *The Cambridge Deep-Sky Album.* 126 pp., hardcover. Cambridge University Press and AstroMedia Corp., Cambridge, 1984. A colorful photo album of Messier and bright NGC objects produced by two Canadian amateur astronomers.

Parker, Sybil P., ed. *The McGraw-Hill Encyclopedia of Astronomy.* 450 pp., hardcover. McGraw-Hill, New York, 1983. This is an extremely useful reference with lengthy entries that help with understanding complex subjects in astronomy.

Peltier, Leslie. *Leslie Peltier's Guide to the Stars.* 185 pp., paper. AstroMedia Corp. and Cambridge University Press, Milwaukee, 1986. A fine introduction to observing the sky with binoculars.

Peltier, Leslie. *Starlight Nights.* 236 pp., paper. Sky Publishing Corp., Cambridge, Massachusetts, 1965. *Starlight Nights* is the story of Leslie C. Peltier, one of the greatest amateur astronomers of all time. It contains no "how-to" observing information but is an interesting read.

Sky & Telescope. Sky Publishing Corp., 49 Bay State Road, Cambridge, Massachusetts 02138. *Sky & Telescope* is the oldest astronomy magazine in America and contains a "Deep Sky Wonders" column each month.

Sulentic, Jack W., and William G. Tifft. *The Revised New General Catalogue of Nonstellar Astronomical Objects.* 383 pp., hardcover. The University of Arizona Press, Tucson, 1973. A listing of the *Revised New General Catalogue,* the standard list of bright deep-sky objects originated by William and John Herschel, with positional, magnitude, and type descriptions.

Tirion, Wil. *Sky Atlas 2000.0.* Twenty-six fold-out folio charts. Cambridge University Press and Sky Publishing Corp., Cambridge, 1981. An excellent large-scale atlas showing 43,000 stars down to magnitude 8 and 2,500 deep-sky objects.

Tirion, Wil, Barry Rappaport, and George Lovi. *Uranometria 2000.0,* vol. 1. 259 pp., hardcover. Willmann-Bell, Inc., Richmond, Virginia, 1987. This first volume of a two-volume set is a sky atlas covering the northern hemisphere down to −6° declination. The work is more thorough than Tirion's *Sky Atlas 2000.0,* showing stars down to magnitude 9.5 and many more deep-sky objects.

Vehrenberg, Hans. *Atlas of Deep-Sky Splendors.* Third ed., 246 pp., hardcover. Treugesell-Verlag and Sky Publishing Corp., Dusseldorf, 1978. A marvelous photographic album depicting hundreds of deep-sky objects all recorded at the same scale for easy comparison.

Webb, Thomas W. *Celestial Objects for Common Telescopes.* Vol. 2, 351 pp., paper. Dover Publications, New York, 1962. First published in 1859, this work is historically interesting as a source of observations of double stars and nebulae.

Index